Transition Metal Catalysis in Macromolecular Design

ACS SYMPOSIUM SERIES **760**

Transition Metal Catalysis in Macromolecular Design

Lisa Saunders Boffa, EDITOR
Exxon Research & Engineering Company

Bruce M. Novak, EDITOR
North Carolina State University

American Chemical Society, Washington, DC

Library of Congress Cataloging-in-Publication Data

Transition metal catalysis in macromolecular design / Lisa Saunders Boffa, editor, Bruce M. Novak, editor.

    p. cm.—(ACS symposium series, ISSN 0097-6156 ; 760)

Includes bibliographical references and index.

ISBN 0–8412–3673–9

    1. Polymerization—Congresses. 2. Transition metal catalysts—Congresses.

    I. Boffa, Lisa Saunders, 1969– . II. Novak, Bruce M., 1955– . III. Series.

QD281..P6 T727   2000
547′.28—dc21   00–22803

The paper used in this publication meets the minimum requirements of American National Standard for Information Sciences—Permanence of Paper for Printed Library Materials, ANSI Z39.48–1984.

Copyright © 2000 American Chemical Society

Distributed by Oxford University Press

All Rights Reserved. Reprographic copying beyond that permitted by Sections 107 or 108 of the U.S. Copyright Act is allowed for internal use only, provided that a per-chapter fee of $20.00 plus $0.50 per page is paid to the Copyright Clearance Center, Inc., 222 Rosewood Drive, Danvers, MA 01923, USA. Republication or reproduction for sale of pages in this book is permitted only under license from ACS. Direct these and other permission requests to ACS Copyright Office, Publications Division, 1155 16th St., N.W., Washington, DC 20036.

The citation of trade names and/or names of manufacturers in this publication is not to be construed as an endorsement or as approval by ACS of the commercial products or services referenced herein; nor should the mere reference herein to any drawing, specification, chemical process, or other data be regarded as a license or as a conveyance of any right or permission to the holder, reader, or any other person or corporation, to manufacture, reproduce, use, or sell any patented invention or copyrighted work that may in any way be related thereto. Registered names, trademarks, etc., used in this publication, even without specific indication thereof, are not to be considered unprotected by law.

PRINTED IN THE UNITED STATES OF AMERICA

Foreword

THE ACS SYMPOSIUM SERIES was first published in 1974 to provide a mechanism for publishing symposia quickly in book form. The purpose of the series is to publish timely, comprehensive books developed from ACS sponsored symposia based on current scientific research. Occasionally, books are developed from symposia sponsored by other organizations when the topic is of keen interest to the chemistry audience.

Before agreeing to publish a book, the proposed table of contents is reviewed for appropriate and comprehensive coverage and for interest to the audience. Some papers may be excluded in order to better focus the book; others may be added to provide comprehensiveness. When appropriate, overview or introductory chapters are added. Drafts of chapters are peer-reviewed prior to final acceptance or rejection, and manuscripts are prepared in camera-ready format.

As a rule, only original research papers and original review papers are included in the volumes. Verbatim reproductions of previously published papers are not accepted.

ACS BOOKS DEPARTMENT

Contents

Preface ix

1. The Organometallic Polymerization of (Meth)acrylates: An Overview 1
   Lisa Saunders Boffa

NEW CHAIN-BUILDING MECHANISMS

2. New Applications of "Carbonylbis(triphenylphosphine)ruthenium"
   Catalysis in Polymer Synthesis 24
   William P. Weber, Jyri K. Paulasaari, Diyun Huang, Shashi Gupta,
   Timothy M. Londergan, Jonathan R. Sargent, and Joseph M. Mabry

3. Transition Metal-Catalyzed Alkyne Cycloaddition Polymerization 38
   Tetsuo Tsuda

4. Synthetic Control over Substituent Location on Carbon-Chain Polymers
   Using Ring-Opening Polymerization of Small Cycloalkanes 59
   J. Penelle

5. Nucleophilic Polymerization of Methyl Methacrylate Activated
   by a Brønsted Acid 77
   Masatoshi Miyamoto, Hirotomo Katano, Chika Kanda, and Yoshiharu Kimura

POLYMER MODIFICATION USING TRANSITION METALS

6. Modification of Polybutadiene by Transition Metal Catalysts:
   Hydroacylation of Polybutadiene 94
   Chul-Ho Jun, Hyuk Lee, Jun-Bae Hong, and Dae-Yon Lee

7. Metallocene Catalysts Used for the Hydrogenation of Polystyrene-b-
   Polybutadiene-b-Polystyrene Block Copolymers 108
   Raymond Chien-chao Tsiang, Wen-shen Yang, and Ming-der Tsai

8. E-Glass Fiber Supported Hydrosilation Catalysts 127
   L. G. Britcher and J. G. Matisons

OLEFIN/ALKYNE ADDITION POLYMERIZATION

9. New Catalysts for Polymerizations of Substituted Acetylenes 146
   Ben Zhong Tang, Kaitian Xu, Qunhui Sun, Priscilla P. S. Lee, Han Peng,
   Fouad Salhi, and Yuping Dong

10. Addition Polymerization of Norbornene: Catalysis of Monocyclopentadienyltitanium Compounds Activated with Methylaluminoxane 165
   Qing Wu, Yingying Lu, and Zejian Lu

11. Transfer and Isomerization Reactions in Propylene Polymerization with the Isospecific, Highly Regiospecific rac-Me$_2$C(3-t-Bu-1-Ind)$_2$ZrCl$_2$/MAO Catalyst 174
   Isabella Camurati, Anna Fait, Fabrizio Piemontesi, Luigi Resconi, and Stefano Tartarini

CONTROLLED RADICAL POLYMERIZATION

12. Living Radical Polymerization of Acrylates with Rhenium(V)-Based Initiating Systems: ReO$_2$I(PPh$_3$)$_2$/Alkyl Iodide 196
    Hiroko Uegaki, Yuzo Kotani, Masami Kamigaito, and Mitsuo Sawamoto

13. The Effect of Ligands on Copper-Mediated Atom Transfer Radical Polymerization 207
    Jianhui Xia, Xuan Zhang, and Krzysztof Matyjaszewski

14. Structural Chemistry and Polymerization Activity of Copper(I) Atom Transfer Radical Polymerization Catalysts 224
    Amy T. Levy and Timothy E. Patten

15. Atom Transfer Polymerization Mediated by Solid-Supported Catalysts 236
    David M. Haddleton, Arnaud Radigue, Dax Kukulj, and David J. Duncalf

16. Mechanistic Aspects of Catalytic Chain Transfer Polymerization 254
    Johan P. A. Heuts, Darren J. Forster, and Thomas P. Davis

INDEXES

Author Index 275

Subject Index 276

Preface

The use of organometallic catalysis as a tool for producing well-defined macromolecules has advanced rapidly since the development of metallocene and "living" transition metal-mediated polymerizations. In addition to the continued understanding and development of controlled addition and condensation polymerizations, recent years have seen the discovery of new catalytic chain-building mechanisms and the achievement of greater levels of control of polymer architecture. Catalytic reactions involving polymers as substrates have produced many new macromolecules, and advances in catalyst design have allowed controlled polymerization techniques to be applied under a wider range of conditions.

This book examines several themes in the area of transition metal-mediated polymer synthesis, with the special intent to demonstrate how this field has moved beyond the use of single-site catalysts to new applications and advanced techniques for architectural control. It is based on a symposium that took place at the 217th National American Chemical Society (ACS) Meeting in Anaheim, California, April 1999, entitled "Advanced Catalysis: New Polymer Syntheses and Modifications". We hope this book will serve as a useful introduction and tutorial to readers interested in polymer design through catalysis.

The book contains 16 chapters, beginning with an overview chapter designed to illustrate how organometallic techniques have advanced the design of one particular type of polymer (polyacrylates and -methacrylates). The remainder of the book is divided into four sections representing current areas of relevant research: New Chain-Building Mechanisms, Polymer Modification Using Transition Metals, Olefin–Alkyne Addition Polymerization, and Controlled Radical Polymerization. Each section begins with a brief description of the chapters to follow. We have attempted to include a balance of review, synthetic, mechanistic, and applied papers to fully represent the progress being made in organometallic polymer synthesis. We have also included two papers featuring unusual non-transition-metal-based polymerizations, which are in keeping with the spirit of this book.

Acknowledgments

We thank all of the chapter authors, primarily for their excellent contributions, but also for their cooperation and timeliness regarding manuscript submission. We are also grateful for the excellent support of Teresa Henline (North Carolina State University) and Pat Kocian (Exxon) with chapter reviews and submissions. Anne Wilson and Kelly Dennis at ACS were critical in helping this book take shape. Finally, we appreciate the patience of our colleagues, particularly Joe Sissano and Don Schulz at Exxon, during the preparation of the Advanced Catalysis symposium and this volume. Financial support for the Advanced

Catalysis symposium was provided by the ACS Division of Polymeric Materials: Science and Engineering, Inc., Exxon Chemical Company, Exxon Research & Engineering Company, The Goodyear Tire & Rubber Company, and Rohm & Haas. Acknowledgment is also made to the Donors of The Petroleum Research Fund, administered by the ACS, for partial support of the symposium through a Type SE grant.

LISA SAUNDERS BOFFA
Corporate Research Laboratory
Exxon Research & Engineering Company
Route 22 East
Annandale, NJ 08801

BRUCE M. NOVAK
Department of Chemistry
North Carolina State University
Raleigh, NC 27695

Chapter 1

The Organometallic Polymerization of (Meth)acrylates: An Overview

Lisa Saunders Boffa

Corporate Research Laboratory, Exxon Research and Engineering Company, Route 22 East, Clinton Township, Annandale, NJ 08801

Metal-mediated (meth)acrylate polymerizations provide a good example of the versatility and importance of transition metal catalysis in polymer design. This chapter surveys several of these processes with respect to "living" behavior, experimental condition and monomer restrictions, and types of polymer architectures available. In addition to anionic methods, Group Transfer Polymerization (GTP) with silyl enolates or zirconocenes, organolanthanide- and aluminum porphyrin-initiated polymerization, and Atom Transfer Radical Polymerization (ATRP) are discussed.

Polymethacrylate and polyacrylate materials play an important role in our society. Almost two billion pounds of polymer products based on acrylic esters are produced each year for applications such as window plastics, dental materials, paints, contact lenses, fibers, and viscosity modifiers. Polymethacrylates--particularly poly(methyl methacrylate) (PMMA)--possess high strength, high impact resistance, excellent heat and chemical stability, and a highly amorphous nature which results in excellent optical clarity. For these reasons, they are used widely as building plastics (*Plexiglas, Lucite*). Polyacrylates are less rigid due to their lower glass transition temperatures, and as latex emulsions form the basis for durable automobile and house paints and coatings. Random copolymers of (meth)acrylates with vinylidene chloride (*Saran*), acrylonitrile, and ethylene find applications as clothing fibers, tubing, gaskets, disposable gloves, and heat- and oil-resistant automotive elastomers.

Despite the commercial utility of polymethacrylates and polyacrylates, refinements in synthetic methodology these macromolecules have lagged behind that of other polymers. The great majority of acrylic ester products in use today are still produced by radical polymerization, which allows little if any control over the fine points of the polymerization process. This is the result of the traditionally ill-controlled nature of acrylate polymerization. While chain termination and transfer side reactions may be largely eliminated for other monomers, such as styrene and dienes, through the use of controlled anionic polymerization, the (meth)acrylate ester

© 2000 American Chemical Society

group can participate in a number of side reactions during polymerization. For this reason the control necessary to produce well-defined acrylic macromolecules has been an elusive goal, and comparatively fewer special-architecture poly(meth)acrylate materials have been prepared.

The instances in which special-architecture acrylics have been studied indicate that these materials are both scientifically and technologically appealing. For example, telechelic polymethacrylates and -acrylates are useful toughening agents (1) and chain extenders (2) and are known to act as self-assembling rheothickeners (3). Diene-methyl methacrylate (MMA) diblock copolymers are excellent emulsifiers for filler blends, liquid dispersions, and high-impact polymeric matrices (4). Acrylics are predicted to be ideal components for styrene-butadiene-styrene-type triblock thermoplastic elastomers (5), and new copolymer combinations of (meth)acrylates and nonpolar monomers should produce interesting amphiphilic materials. A great deal of effort has thus been expended in the last two decades to develop polymerization techniques with the control necessary to produce well-defined poly(meth)acrylates of varying architectures. In addition to modified anionic methods, well-controlled or "living" pseudoanionic polymerizations (6-9) of (meth)acrylates based on organometallic initiators have provided a number of complimentary strategies for the synthesis of useful and novel materials. These processes, along with a discussion of the difficulties associated with traditional anionic polymerization, are discussed in this chapter.

Anionic Polymerization

Methacrylates and acrylates may be polymerized by traditional anionic initiators, including alkali metal alkyls and enolates, Grignard reagents, α-methylstyrene tetramer, and electron-transfer agents (sodium naphthalide, etc.). This process is extremely sensitive to both oxygen and active-hydrogen impurities. Two mechanisms are known to be operative (10). In weak or noncoordinating solvents such as toluene, the propagating enolate endgroup exists as an ion pair, and monomer precoordination to the counterion through the carbonyl group is favored. This requires a same-side configuration of the chain end and monomer ester groups, giving an isotactic unit upon addition (Scheme 1). In highly coordinating solvents--particularly chelating solvents such as dimethoxyethane (DME)--the incoming monomer unit is not able to displace the solvent molecule and addition takes place via a concerted process (Scheme 2). An opposite-side configuration of the monomer and endgroup esters is preferred for steric reasons and the resultant linkage is syndiotactic. This mechanism is most highly favored for lithium counterions, which interact very strongly with the solvent and chain end due to orbital overlap. When mixed solvents or larger, more weakly interacting cations are used, a mixture of both mechanisms results, giving largely atactic polymer. A concerted mechanism is also operable with cryptated cations, which cannot undergo monomer precoordination prior to addition, and in cases where free ions rather than ion pairs are the active species.

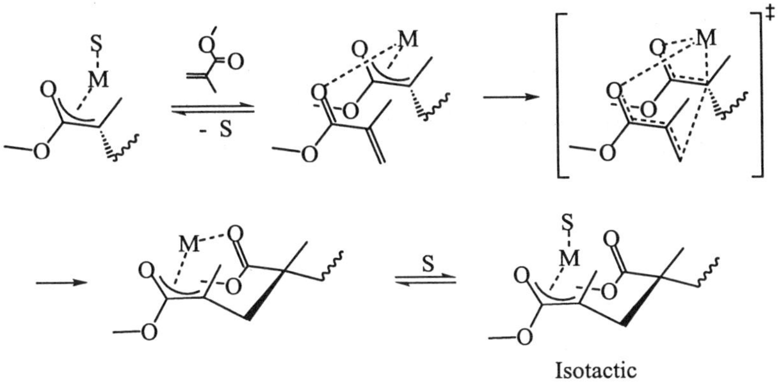

Scheme 1. Precoordination mechanism for anionic methacrylate polymerization (weakly coordinating solvents).

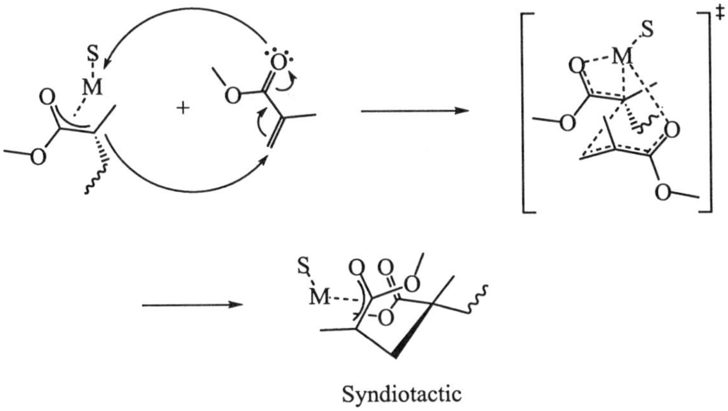

Scheme 2. Concerted mechanism for methacrylate polymerization (strongly coordinating solvents).

Side Reactions

Side reactions involving nucleophilic attack at the ester group render anionic (meth)acrylate polymerization nonliving under typical conditions (11-15). The most common process is intramolecular "backbiting" of the active endgroup to the antepenultimate ester unit of the polymer chain (Equation 1) (13, 16-18). This produces

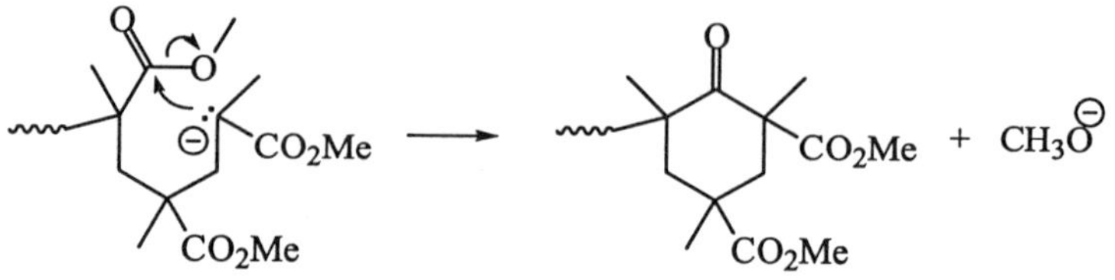

$$\text{(1)}$$

a nonreactive cyclic ketone endgroup and an alkoxide ion and results in lowered initiator efficiency and inflated molecular weights. Backbiting occurs mainly at the trimer stage of polymerization since additional chain units sterically inhibit attack at the carbonyl group (19). The alkoxide eliminated during cyclization may act as a weak initiator, and secondary polymerization of MMA by this species contributes to a lowering and broadening of the molecular weight distribution through chain transfer. The extent of initiation by alkoxide byproducts appears to be dependent on the identity of the counterion (10, 11, 20). A second side reaction occurring during early stages of the polymerization has been observed for alkyllithium and Grignard initiators (15, 21-26); this process involves 1,2- rather than 1,4-attack of the initiator on monomer, giving alkoxide and a vinyl ketone which may be added to a growing chain as a comonomer (Equation 2). The propagating enolate endgroup of the ketone

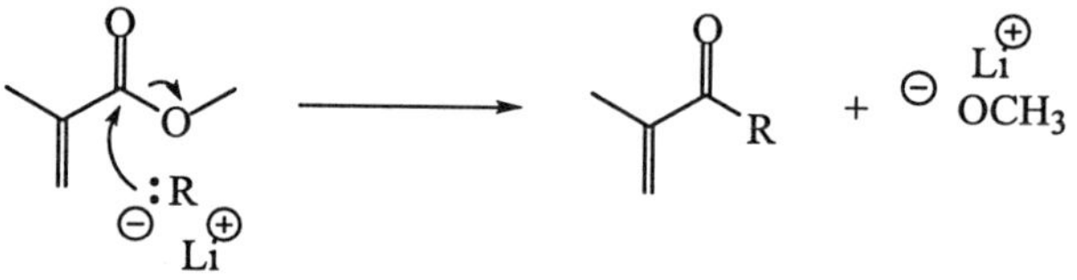

$$\text{(2)}$$

is not as reactive as the enolate anion of the (meth)acrylate, and also gives lowered and broadened molecular weight distributions since the overall polymerization rate is slowed. Schreiber (15) and Korotkov (27) and Löhr et al. (28) have proposed similar termination steps involving 1,2-addition at either the monomer or backbone carbonyl groups by the enolate propagating species; however, no evidence has been found for the existence of these mechanisms. Extremely bulky methacrylates, such as *tert*-butyl methacrylate (*t*BMA), are the only monomers which have been thought to be immune to terminating side reactions (29-31). However, recent evidence indicates that even for tBMA, oligomers resulting from side reactions are found in the polymerization mixture (32).

Side reactions involving chain transfer (other than secondary initiation from alkoxide byproducts) are not problematic for methacrylate polymerization. In contrast, chain transfer is facile for acrylates due to the presence of an acidic proton at the carbonyl α-position (10, 13, 33-36). Again, only bulky monomers such as isopropyl and t-butyl acrylate (tBuA), may be polymerized anionically to high conversion with even moderate degrees of control. Additionally, polymerizations of both methacrylates and acrylates carried out in noncoordinating solvents involve multiple active species due to the formation of chain end aggregates (10). Very broad, multimodal molecular weight distribution are obtained under these conditions, with polydispersities of up to 80 being reported for MMA polymerization.

Strategies for Living Polymerization

Several "ligated anionic polymerization" (LAP) approaches have been employed to suppress termination in anionic (meth)acrylate polymerization with the goal of obtaining "living" behavior (3, 37). Low temperatures ($\leq$ -75 °C), polar solvents (THF, DME), and large counterions are known to minimize terminating side reactions (10, 38, 39). The use of these conditions in conjunction with bulky, less nucleophilic initiators, such as alkali metal 1,1-diphenylalkanes (40, 41) and aromatics (42) and branched Grignard reagents (43), has allowed the preparation of relatively monodisperse PMMA (PDI < 1.20) with good molecular weight control. Alkali metal enolates, which mimic the active endgroup of the polymerization, are especially effective in this regard (18).

tert-Butylmagnesium bromide may be used to synthesize highly isotactic PMMA in noncoordinating solvents at low temperatures (44-46). The unusual control imparted by this reagent is thought to arise from an interaction between the initiating species, which has been postulated to exist as a complex aggregate bearing multiple tert-butyl groups due to the Schlenk equilibrium, and the active endgroup (47). This hypothesis is supported by the fact that the viscosity of the polymerization medium decreases upon termination. Similarly, the preparation of monodisperse, highly isotactic poly(isopropyl acrylate) was reported with a mesitylmagnesium bromide initiator in toluene (48).

A second strategy for imparting control in anionic methacrylate polymerization involves the addition of common-ion salts (49). In the absence of these additives, the active enolate chain ends exist in a slow equilibrium between free and associated species, which have differing activities and thus form polymer at different rates (10). When lithium chloride is added to the polymerization medium, a new, more rapid equilibrium involving μ-bridged endgroup-salt adducts results (Equation 3) (50-52). The faster interchange between active species results in a lowered polydispersity for

$$1/2 \; \sim\!\!\!\sim\!\! \overset{\overset{\displaystyle Li^{\oplus}}{\cdot\cdot}}{\underset{\underset{\displaystyle Li^{\oplus}}{\cdot\cdot}}{R^{\ominus}\;\;\;\;R^{\ominus}}}\!\!\sim\!\!\!\sim \; \rightleftharpoons \; \sim\!\!\!\sim R^{\ominus}\,Li^{\oplus} \; \underset{}{\overset{LiCl}{\rightleftharpoons}} \; \sim\!\!\!\sim\overset{\overset{\displaystyle Li^{\oplus}}{\cdot\cdot}}{\underset{\underset{\displaystyle Li^{\oplus}}{\cdot\cdot}}{R^{\ominus}\;\;\;\;Cl^{\ominus}}} \quad\quad (3)$$

the resultant polymer, and "protection" of the active center by the bridging groups serves to lessen side reactions. Interestingly, the addition of LiCl does not affect tacticity. Lithium *tert*-butoxide is also used as a μ-bridging additive although it is not as effective at promoting fast interchange as LiCl (53-55). Aluminum alkyls and phenoxides have also been employed as additives (46, 56). Recently, Zundel, Teyssié, and Jérôme have used lithium silanolates as μ-ligands with butyllithium initiators (57, 58). The resultant Bu(Me$_2$)SiOLi-stabilized alkyllithium species can produce high molecular weight, monodisperse, ≥ 90% isotactic PMMA in toluene at 0 °C. Well-defined poly(*n*-butyl acrylate) and PMMA-P*n*BA block copolymers can also be prepared at lower temperatures.

The use of a hindered lithium alkyl or enolate complex as an initiator, in conjunction with LiCl as an additive, has allowed Jérôme and Teyssié and coworkers to prepare PMMA with polydispersities as low as 1.02 and to achieve initiator efficiencies of over 95% at -78 °C in THF (51). Since the polymerization of *t*-butyl acrylate is also greatly improved by the addition of LiCl (PDI ≈ 1.2) (59), this methodology has been used to synthesize relatively monodisperse MMA-*t*BuA and styrene-*t*BuA block copolymers (60, 61).

A complementary approach, disruption of aggregation by complexation of the counterion rather than the anionic endgroup, has also been employed. Crown ethers, amine cryptands, and pyridine all promote the formation of narrow-PDI poly(meth)acrylates by eliminating chain-end aggregation and promoting the formation of monomeric, highly reactive species (2, 62-67). Large "metal free" organic counterions, including tetraalkylammoniums (68-70), tetraphenylphosphonium, and 'P$_4^+$,' (71) have also been employed to control polymerization at

P$_4^+$

LiOEEM

higher temperatures (0-20 °C) with reasonable success. Lithium 2-(2-methoxyethoxy) ethoxide (LiOEEM), which incorporates both cation-complexing and endgroup-bridging moieties, has recently been used as an additive to prepare monodisperse poly(*n*-butyl acrylate) (PDI = 1.15) and poly(2-ethylhexyl acrylate) (PDI = 1.07) and related block copolymers (72-74). Unfortunately, the presence of chain transfer prevents the synthesis of similarly well-defined polyacrylates with shorter primary ester chains.

Group Transfer Polymerization

Silyl Group Transfer Polymerization

The first strategy for achieving "living" (meth)acrylate polymerization through modification of the ionic nature of the endgroup was group transfer polymerization (GTP), developed by Webster and Sogah et al. at DuPont in 1983 (75-80). This process involves initiation with a silyl ketene acetal or related compound, producing a silyl enolate propagating species. GTP proceeds without significant interference from chain termination and transfer steps and is thus a "living" process. Since no highly charged species are involved, it is much more robust than anionic (meth)acrylate polymerization, and may be carried out at room temperature and in the presence of oxygen, although protic compounds still must be excluded. Solvents such as dimethylformamide (DMF) and chlorinated hydrocarbons, which are incompatible with anionic initiators, may also be used.

The mechanism of GTP monomer addition was originally postulated as associative, i.e. the silyl group remains attached to one particular polymer chain throughout the course of the polymerization and is simply "transferred" to each new monomer unit as it is added (Scheme 3) (81). However, studies by Quirk involving silyl endgroup exchange processes for "living" oligomers have indicated that a dissociative mechanism is more consistent (Scheme 4) (82-84). Viewed in this light, GTP is actually an anionic polymerization carried out via ester enolates, which are stabilized by rapid and reversible "termination" with silyl ketene acetals. Based on kinetic data, Müller has subsequently proposed an associative mechanism, at least for cases involving certain GTP catalyst components (see below) (85).

A small amount of either a nucleophilic or Lewis acid catalyst is needed in addition to the initiator to activate (or cleave, depending on which mechanism is actually operative) the silyl enolate endgroup for monomer addition. Nucleophilic catalysts function by promoting displacement at the silyl group through a hexavalent intermediate (for an associative process) and are needed only in low concentrations (< 0.1 mol % of initiator). A wide range of nucleophiles may be used, including fluoride, bifluoride (HF_2^-), azide, cyanide, cyanate, nitrite, carboxylates, phenolates, sulfinates, and sulfonamidates (75-77, 86-89). Lewis acid catalysts function by coordinating to the monomer carbonyl group and are thus needed in higher concentrations ($\geq$ 10 mol % of initiator). Zinc dihalides, dialkylaluminum halides, tetraalkylaluminoxanes, and more recently lanthanide triflates and HgI_2 have been used in this capacity (85, 90, 91). Lewis acid catalysts are preferred for acrylate polymerization, as nucleophilic catalysts have been found to give broader molecular weight distributions. GTP may also be performed without added catalysts under conditions of high pressure (1000-3000 atm) (92).

With the correct choices of initiator and catalyst, both methacrylates and acrylates may be polymerized quantitatively with GTP to give polymers with low polydispersities (1.03 $\leq$ PDI $\leq$ 1.4). Good control of molecular weight through the monomer to initator ratio is achieved. However, the synthesis of polymers with molecular weights in excess of 60 000 is difficult, especially with acrylates; it is thought that the silyl group possesses the ability to migrate to other locations along the polyacrylate backbone. Most reported examples of poly(meth)acrylates synthesized by GTP are of M_n < 20 000. Additionally, GTP does not produce

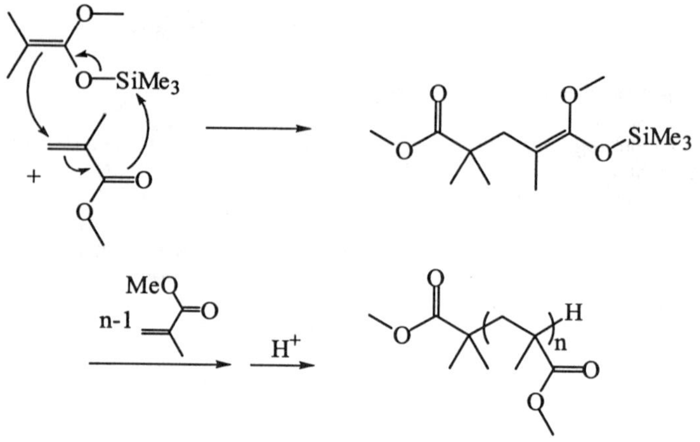

Scheme 3. Associative silyl Group Transfer Polymerization mechanism.

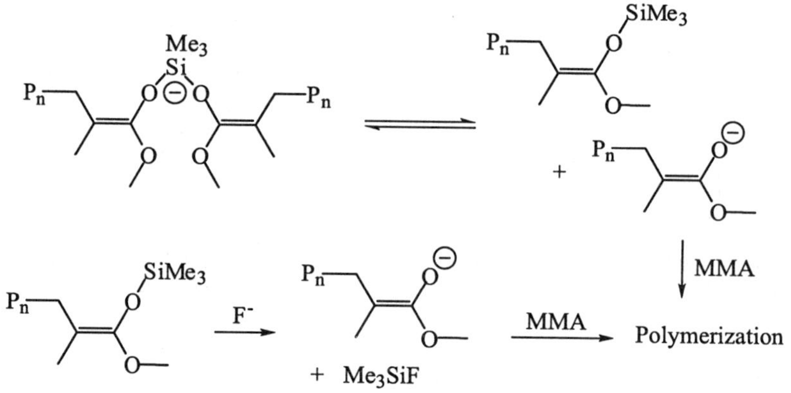

Scheme 4. Dissociative silyl Group Transfer Polymerization mechanism.

polymers having a high degree of stereoregularity. Similar to the trends seen with free-radical polymerization, syndiotacticities of over 80% are not achieved even at low temperatures (93).

The well-controlled nature of GTP has been exploited for the synthesis of a number of architecturally interesting polymers. Silyl initiators containing ether, vinyl, sulfide, cyanide, phosphonate, and phosphonium groups have been used to prepare end-functionalized poly(meth)acrylates (93, 94). Diblock, triblock, and random (meth)acrylate copolymers have been prepared with both monofunctional and bifunctional silyl initiators (79, 95-97); ladder polymers (92) have also been synthesized through GTP with bifunctional initiators. Cyclopolymerizations of dimethacrylates by GTP have recently been explored (98, 99). Telechelic, star, comb, and graft polymers have also been prepared through routes involving GTP (93).

Zirconocene Group Transfer Polymerization

Using a similar strategy, Collins has developed an associative group transfer-type polymerization for methyl methacrylate based on zirconocenes (100). A neutral zirconocene enolate serves as the initiating fragment while its conjugate zirconocene cation acts as a Lewis acid catalyst (Scheme 5). Each monomer addition step interconverts the two organometallic components. The PMMA obtained is predominantly syndiotactic (methylene chloride solvent, 0 °C), although isotactic PMMA has been obtained by using chiral indenyl zirconocenes in combination with non-zirconocene Lewis acid catalysts (101).

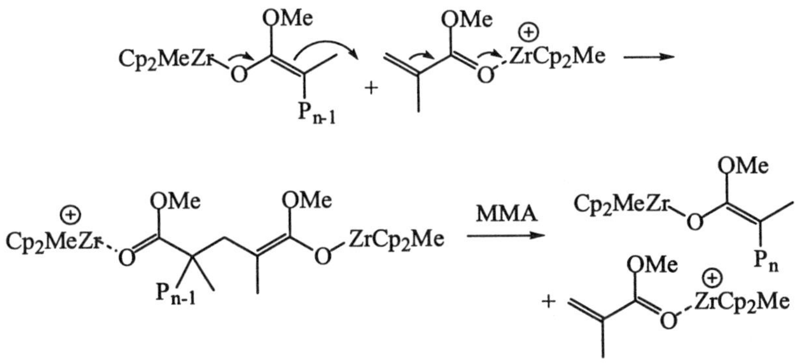

Scheme 5. Zirconocene Group Transfer Polymerization of methyl methacrylate.

Lanthanide(III) Initiators

In 1992, Yasuda and coworkers discovered that bis(cyclopentadienyl) lanthanide complexes could function as excellent initiators for the "living" polymerization of acrylates and methacrylates. Alkyl compounds such as $[Cp*_2SmH]_2$ ($Cp*$ = C_5Me_5), $Cp*_2LnMe$(ether), $Cp*_2LnAlMe_4$, and $[Cp_2LnMe]_2$ (Ln = Sm, Y, Yb, Lu; Cp = C_5H_5) were found to quantitatively and quickly polymerize MMA at temperatures from -115 to 40 °C, giving polymers with PDIs as low as 1.02. The PMMA is very highly syndiotactic when produced at lower temperatures (> 96% rr at -110 °C) (102-110); tacticity is still quite high under more reasonable conditions (77% rr at 40 °C). Due to the solubilizing effect of the cyclopentadienyl ligands, polymerization may be carried out in a wide range of solvents, although the catalysts are extremely sensitive to air and water. High-molecular weight, monodisperse *isotactic* PMMA can also be prepared through the use of certain nonmetallocene lanthanides, particularly a Yb[C(SiMe$_3$)$_3$]$_2$ precursor (110).

Lanthanide-initiated polymerization of (meth)acrylates is formally analogous to GTP. However, due to its large size and Lewis acidity, the lanthanide active center can simultaneously function as both an initiator and catalyst, and a second equivalent of Lewis acid is not needed. Isotopic labeling, active species hydrolysis, and crystallographic characterization of the 2:1 insertion product of MMA and [Cp*$_2$SmH]$_2$ were used to show that polymerization proceeds through an intramolecularly activated lanthanide(III) enolate (Equation 4) (102, 103). As

$$\text{(4)}$$

expected, the rate of polymerization increases with the size of the lanthanide and decreases with added steric hindrance around the metal center. Less-hindered initiators such as [Cp$_2$YMe]$_2$ show a greater loss of syndiotacticity at higher temperatures. Complexes with less nucleophilic alkoxide and halide initiating groups do not catalyze polymerization.

Lanthanidocenes also catalyze the living polymerization of other alkyl methacrylates, including those with secondary and tertiary ester groups. Interestingly, syndiotacticity *decreases* with increasing size of the methacrylate side chain (103, 105, 108). More impressively, living polymerizations of alkyl acrylates (including methyl and ethyl acrylate) have been achieved, although the propagating acrylate enolates are not quite as robust as their methacrylic analogues (111, 112). The resultant polyacrylates are only moderately syndiotactic (~ 60% rr).

Monodisperse (meth)acrylate-(meth)acrylate copolymers having both block and random structures have been prepared using Cp*$_2$Sm(III) catalysis (106, 107, 111). Since the reactivities of acrylates with these initiators is substantially higher than those of methacrylates, random copolymers are limited to those materials incorporating only one type of monomer. These materials are quite monodisperse (PDIs < 1.05). Diblock and triblock copolymers containing both methacrylate and acrylate segments exhibited elastomeric properties, such as elongations of 200% or more, at appropriate compositions.

A facile method for preparing these all-acrylic ABA triblocks involves electron-transfer initiation of (meth)acrylate polymerization with the divalent samarocenes Cp*$_2$Sm or Cp*$_2$Sm(THF)$_2$ (113-115). This process is analogous to the well-known initiation of anionic polymerization by sodium naphthalide: the samarocene carries out one-electron transfer to the (meth)acrylate monomer, forming radical anions which dimerize to give a bimetallic samarium(III) bis-enolate. This species goes on to grow "living" polymer in two directions simultaneously, circumventing activity restrictions for monomer crossover. Using post-polymerization deprotection methods, amphiphilic ABA triblocks with syndiotactic acid-ester segments have been prepared in this manner. A number of divalent lanthanidocenes in addition to samarocene have since been used to carry out similar electron-transfer-initiated (meth)acrylate homopolymerizations (116, 117).

Other specialized applications of the lanthanide(III) initiating system have allowed for the preparation of novel polymer architectures. Bis-Cp*$_2$Sm(III) initiators having functionalized, bridging initiating groups have been used to prepare "link-functionalized" PMMAs with double bonds at the center of the macromolecular backbone (113, 115, 118). PMMA oligomers quantitatively functionalized with hydroxyl groups were prepared via aldol condensation of a "living" samarium polymeric enolate with an endcapping agent, p-tolualdehyde, and subsequent hydrolysis (119). Lanthanide(III) initiators also polymerize other types of monomers with good control. This offers many possibilities for the synthesis of useful, well-defined hydrophilic / hydrophobic materials: diblock copolymers of ethylene and MMA, methyl acrylate, or ethyl acrylate prepared with Cp*$_2$Sm(III) initiators possessed superior dyeing capabilities (120). Diblock copolymers composed of (meth)acrylate and ε-caprolactone or δ-valerolactone segments were also synthesized in a controlled fashion for the first time with these catalysts (106, 111, 121).

Aluminum Porphyrin Initiators

A second main-group metal counterion strategy for living behavior was developed by Inoue and coworkers, who reported in 1987 that MMA could be polymerized to high yield and low polydispersity ($\leq$ 1.2) with (tetraphenylporphinato)aluminum methyl [(TPP)AlMe] (122, 123). This process utilizes visible light to activate the propagating aluminum-enolate bond through the porphyrin ring (Scheme 6), and was successfully employed for the preparation of block copolymers of MMA with n-butyl and t-butyl acrylate, methacrylonitrile, δ-valerolactone, and propylene oxide (122, 124, 125). Subsequently, it was found that aluminum-based Lewis acid additives could be used in place of visible light as activating agents (126). Methylaluminum-bis(2,4,6-trialkylphenolate) and -bis(1,1-diphenylalkanolate) complexes, when added in 1-3 equivalents per initiating center, accelerate the polymerization of MMA with aluminum porphyrins by factors of tens of thousands without any detrimental effects on yield or polydispersity. Amazingly, only three seconds was required to quantitatively produce PMMA of M_n = 25 500 (calculated M_n = 21 700) and PDI = 1.07 with (TPP)AlMe to which three equivalents of MeAl(OC$_6$H$_4$-2,4-t-Bu)$_2$ had been added (127). This conversion rate represents a 46 200-fold increase over the rate of the unaccelerated polymerization. At higher concentrations, Lewis Acid additives are also effective for the synthesis of ultra-high molecular weight, monodisperse PMMA (M_n = 1 017 000, PDI = 1.2) (128).

The "living" behavior seen with Lewis acid additives (evidenced by linear relationships between M_n and conversion and monomer to initiator ratio) is dependent on adequate steric protection of the active center. When porphyrins featuring

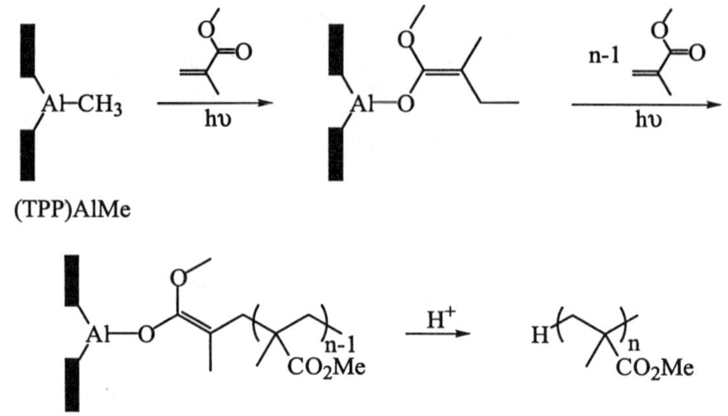

Scheme 6. *Polymerization of methyl methacrylate with photoactivated "immortal" aluminum porphyrin initiator.*

substituents smaller than phenyl groups or non-orthosubstituted aluminum phenolates are used, an exchange reaction occurs between the polymer chain and a Lewis acid substituent (Equation 5) (126, 127). This reaction is chain terminating since the

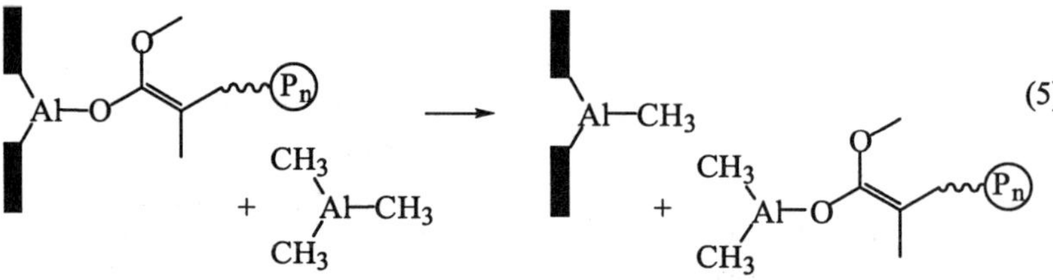

(5)

resulting aluminum enolate does not catalyze polymerization significantly. It may be suppressed at low temperature, however, and at -40 °C even trimethylaluminum may be used to promote MMA polymerization in a well-controlled fashion (125). Secondary methacrylates (benzyl, cyclohexyl, isopropyl) may also be successfully polymerized; less-hindered initiators and additives may be used due to the steric protection provided by the larger ester groups (126). With *tert*-butyl methacrylate, a unique polymerization mechanism is seen (127). This monomer is polymerized only to low conversion when bulky Lewis acid additives are used. With Me_3Al as an additive at room temperature, however, monodisperse P*t*BMA is produced in quantitative yield. The initiating species in this case was found to be the dimethylaluminum enolate resulting from ligand exchange between the porphyrin species and Me$_3$Al.

Polymerizations of both primary and branched acrylate monomers may also be carried out in a living fashion (129). Initiation with (TPP)AlMe is inefficient and slow; however, a more reactive propanethiolate complex undergoes initiation sufficiently rapidly to produce poly(*tert*-butyl acrylate) (P*t*BuA) with a PDI of 1.14 and an initiator efficiency of 0.67. When the initiating moiety is a living, preformed PMMA chain, initiation is rigorously quantitative. Thus, a PMMA-P*t*BuA diblock with an M_n of 24 000 and a PDI of 1.25 was prepared. A diblock copolymer featuring the opposite addition sequence (P*t*BuA → PMMA), a PMMA-P*t*BuA-PMMA triblock, and an ABC PMMA-P*t*BuA-poly(propylene oxide) copolymer were synthesized similarly. The reported molecular weights of polyacrylate materials prepared with aluminum porphyrins have been only moderate ($\leq$ 25 000).

Atom Transfer Radical Polymerization

While anionic polymerization is superior to radical polymerization in terms of "living" behavior, the latter technique offers certain advantages of practicality and generality. Radical initiators do not typically require as stringent air-free or low-temperature conditions as ionic polymerization, and are less subject to relative monomer reactivity constraints for copolymer synthesis. The development of copper-mediated Atom Transfer Radical Polymerization (ATRP) by Matyjaszewski and Wang in 1995 was thus an important advancement for the synthesis of well-defined polymer structures (130, 131). ATRP achieves control over highly reactive radical endgroups by *reversible* termination with a halogen radical derived from an alkyl halide initiator. A transition metal fragment which is able to reversibly abstract the halogen from the endgroup is added to the polymerization medium. The equilibrium between "terminated" and active radical endgroups heavily favors the inactive species, and since the concentration of active propagating sites is at any given time very small, side reactions are minimized (although not entirely eliminated). This technique has been used to polymerize both methacrylates and acrylates with high degrees of control.

ATRP typically employs an alkyl halide with a weak carbon-halogen bond as an initiator and a low-valent late metal complex, such as a copper(I) halide, as the halide abstractor (Scheme 7). Various ligands such as 2,2'-bipyridyl (bipy) are added to the reaction mixture or incorporated into the discrete metal components to solubilize and stabilize active species during polymerization. In addition to copper(I)-based systems, "living" radical polymerizations of (meth)acrylates can be carried out with complexes of ruthenium / aluminum oxide (132, 133), nickel (134), rhodium (135), palladium (136), iron (137, 138), and cobalt (139). The process may be carried out in solution, bulk, emulsion or dispersion in water, and on solid supports at a variety of temperatures. It should be noted that the final termination step is chemically similar to reversible termination during propagation and produces polymers with halogen endgroups. These species may be reactivated by addition of a second monomer for the synthesis of block copolymers.

MMA and various alkyl acrylates have been polymerized by ATRP with Cu^I(bipy) systems to high conversion and low or fairly low polydispersity (1.05 $\leq$

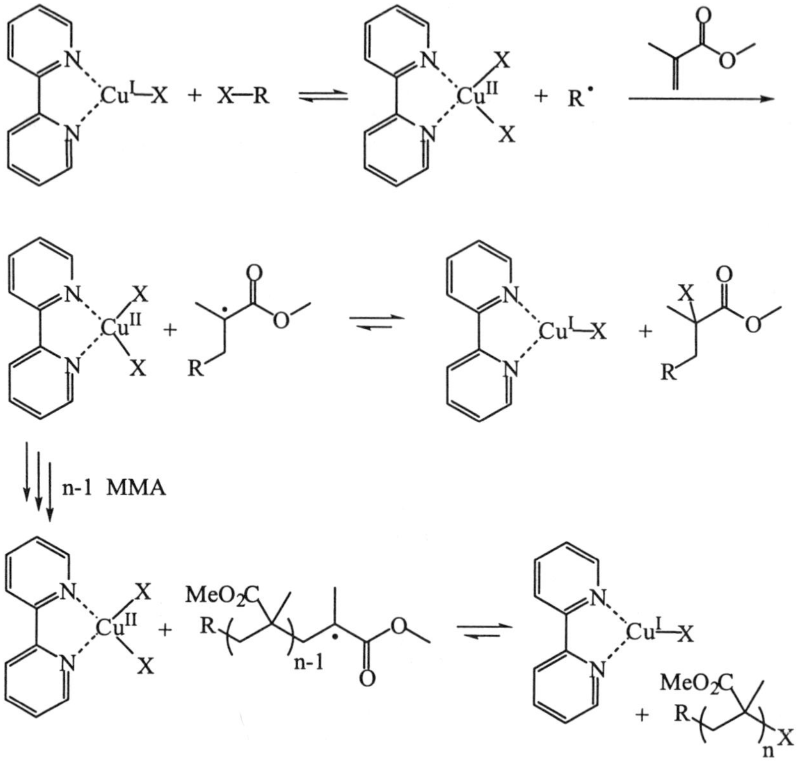

Scheme 7. Copper-mediated Atom Transfer Radical Polymerization (ATRP) of methyl methacrylate.

PDI ≤ 1.50) (131). Molecular weights of up to 180 000 for PMMA have been achieved with good control. Bromide initiators tend to give lower-PDI poly(meth)acrylates than chloride initiators, and tacticities are similar to those obtained through conventional radical polymerization. In direct contrast to pseudoanionic methods, small, primary acrylates are not especially problematic; poly(methyl acrylate)s with PDIs of 1.10 are available through ATRP. Monomers with alcohol, carboxylate, vinyl, and epoxide sidechains can also be polymerized (131, 140).

A number of useful structures such as block and random styrene / butyl acrylate copolymers (141), epoxide / sodium methacrylate block copolymers (140), all-acrylic gradient, random, and block copolymers (139, 142-145) and alternating copolymers of methyl acrylate and isobutylene or isobutyl vinyl ether (146) have been prepared using ATRP. Bifunctional, bis-initiating halide initiators based on small molecules or preformed telechelic oligomers have been employed for the synthesis of telechelics (147) and acrylate / acrylonitrile triblocks (148); tris-initiators have similarly been used to prepare stars (133). Halide initiators incorporating a variety of functional groups have been used to prepare end-functionalized PMMAs via ATRP, and the halogen endgroups resulting from termination have also been transformed into other useful moieties via azide reagents (131).

Conclusions

A wide variety of well-defined methacrylic and acrylic polymer architectures are becoming available through the use of "living" organometallic polymerization. Each of these inititating systems--silyl or zirconocene Group Transfer, lanthanide(III) metallocene, aluminum porphyrin, Atom Transfer Radical Polymerization, and ligated anionic polymerization--has unique advantages and limitations that must be considered when designing a macromolecular synthesis. Taken as a group, these methods illustrate the powerful contributions that organometallic catalysis can make towards rational macromolecular design. As these methods continue to be developed more thoroughly, we may anticipate the availability of even more unique and useful methacrylic materials.

References

1. Banthia, A. K.; Chaturvedi, P. N.; Jha, V.; Pendyala, V. N. S. in *Rubber-Toughened Plastics*; Riew, C. K., Ed.; Advances in Chemistry Series 222; American Chemical Society: Washington DC, 1989, pp 343-358.
2. Anderson, B. C.; Andrews, G. D.; Arthur, P. J.; Jacobson, H. W.; Melby, L. R.; Playtis, A. J.; Sharkey, W. H. *Macromolecules* **1981**, *14*, 1599-1601.
3. Teyssié, P.; Ph., B.; Jérôme, R.; Varshney, S. K.; Wang, J.-S. *Makromol. Chem., Macromol. Symp.* **1995**, *98*, 171-183 and references therein.
4. Teyssié, P.; Fayt, R.; Jacobs, C.; Jérôme, R.; Leemans, L.; Varshney, S. *Polym. Prepr., Am. Chem. Soc. Div. Polym. Chem.* **1988**, *29 (2)*, 52-53 and references therein.
5. Yu, Y. S.; Dubois, P.; Jérôme, R.; Teyssié, P. *Macromolecules* **1996**, *29*, 2738-2745.
6. Szwarc, M.; Levy, M.; Milkovich, R. *J. Am. Chem. Soc.* **1956**, *78*, 2656-2657.
7. Szwarc, M. *Nature* **1956**, *178*, 1168-1169.
8. Quirk, R. P.; Lee, B. *Polym. Int.* **1992**, *27*, 359-367.
9. Webster, O. W. *Science* **1991**, *251*, 887-893.
10. For a thorough discussion of and references pertaining to these mechanisms, see: (a) Müller, A. H. E. In *Comprehensive Polymer Science: The Synthesis, Characterization, Reactions, and Applications of Polymers*; Allen, G., Bevington, J. C., Eds.; Pergamon Press: Oxford, 1989; Vol. 3, pp 387-423. (b) Bywater, S. In *Comprehensive Polymer Science: The Synthesis, Characterization, Reactions, and Applications of Polymers*; Allen, G., Bevington, J. C., Eds.; Pergamon Press: Oxford, 1989; Vol. 3, pp 433-447.

11. Wiles, D. M. in *Structure and Mechanism in Vinyl Polymerization*; Tsuruta, T.; O'Driscoll, K. F., Ed.; Marcel Dekker: New York, 1969, pp 223-281.
12. Bywater, S. *Prog. Polym. Sci.* **1975**, *4*, 27-69.
13. Goode, W. E.; Owens, F. H.; Myers, W. L. *J. Polym. Sci.* **1960**, *47*, 75-89.
14. Rempp, P.; Volkov, V. I.; Parrod, J.; Sadron, C. *Bull. Soc. Chim. Fr.* **1960**, 919-926.
15. Schreiber, H. *Makromol. Chem.* **1960**, *36*, 86-88.
16. Glusker, D. L.; Lysloff, I.; Stiles, E. *J. Polym. Sci.* **1961**, *49*, 315-334.
17. Owens, F. H.; Myers, W. L.; Zimmerman, F. E. *J. Org. Chem.* **1961**, *26*, 2288-2298.
18. Lochmann, L.; Rodová, M.; Petránek, J.; Lím, D. *J. Polym. Sci., Polym. Chem. Ed.* **1974**, *12*, 2295-2304.
19. Müller, A. H. E.; Lochmann, L.; Trekoval, J. *Makromol. Chem.* **1986**, *187*, 1473-1482.
20. Webster, O. W. *Polym. Prepr., Am. Chem. Soc. Div. Polym. Chem.* **1996**, *37 (2)*, 682-683 and references therein.
21. Kawabata, N.; Tsuruta, T. *Makromol. Chem.* **1965**, *86*, 231-252.
22. Hatada, K.; Kitayama, T.; Fujikawa, K.; Ohta, K.; Yuki, H. *Polym. Bull. (Berlin)* **1978**, *1*, 103-108.
23. Hatada, K.; Kitayama, T.; Nagakura, M.; Yuki, H. *Polym. Bull. (Berlin)* **1980**, *2*, 125-129.
24. Hatada, K.; Kitayama, T.; Fumikawa, K.; Ohta, K.; Yuki, H. in *Anionic Polymerization: Kinetics, Mechanisms, and Synthesis*; McGrath, J. E., Ed.; ACS Symposium Series 166; American Chemical Society: Washington, DC, 1981, pp 327-341.
25. Wiles, D. M.; Bywater, S. *Chem. Ind. (London)* **1963**, 1209.
26. Bullock, A. T.; Cameron, G. G.; Elsom, J. M. *Eur. Polym. J.* **1977**, *13*, 751-756.
27. Korotkov, A. A.; Mitsengendler, S. P.; Krasulina, V. N. *J. Polym. Sci.* **1961**, *53*, 217-224.
28. Löhr, G.; Müller, A. H. E.; Warzelhan, V.; Schulz, G. V. *Makromol. Chem.* **1974**, *175*, 497-505.
29. Müller, A. H. E. in *Recent Advances in Anionic Polymerization*; Hogen-Esch, T. E.; Smid, J., Ed.; Elsevier: New York, 1987, pp 205-229.
30. Müller, A. H. E. *Makromol. Chem.* **1981**, *182*, 2863-2871.
31. Müller, A. H. E.; Jeuck, H.; Johann, C.; Kilz, P. *Polym. Prepr., Am. Chem. Soc. Div. Polym. Chem.* **1986**, *27 (1)*, 153-154.
32. Zune, C.; Dubois, P.; Jérôme, R. *Polym. Int.* **1999**, *48*, 565-570.
33. Busfield, W. K.; Methven, J. M. *Polymer* **1973**, *14*, 137-144.
34. Kitano, T.; Fujimoto, T.; Nagasawa, M. *Polym. J.* **1977**, *9*, 153-159.
35. Garrett, B. S.; Goode, W. E.; Gratch, S.; Kincaid, J. F.; Levesque, C. L.; Spell, A.; Stroupe, J. D.; Watanabe, W. H. *J. Am. Chem. Soc.* **1959**, *81*, 1007-1008.
36. Graham, R. K.; Panchak, J. R.; Kampf, M. J. *J. Polym. Sci.* **1960**, *44*, 411-419.
37. For a very recent overview of advances in controlled anionic polymerization of (meth)acrylates, see: Jérôme, R., Teyssié, Ph., Vuillemin, B.; Zundel, T.; Zune, C. *J. Polym. Sci., Polym. Chem. Ed.* **1999**, *37*, 1-10.

38. Müller, A. H. E. in *Anionic Polymerization: Kinetics, Mechanisms, and Synthesis*; McGrath, J. E., Ed.; ACS Symposium Series 166; American Chemical Society: Washington, DC, 1981, pp 441-461.
39. Müller, A. H. E.; Gerner, F. J.; Kraft, R.; Höcker, H.; Schulz, G. V. *Polym. Prepr., Am. Chem. Soc. Div. Polym. Chem.* **1980**, *21 (1)*, 36-37.
40. Cao, Z.-K.; Ute, K.; Kitayama, T.; Okamoto, Y.; Hatada, K. *Kobunshi Robunshu* **1986**, *43*, 435-441.
41. Long, T. E.; Allen, R. D.; McGrath, J. E. in *Recent Advances in Mechanistic and Synthetic Aspects of Polymerization*; Fontanille, M.; Guyot, A., Ed.; NATO Advanced Study Institute Series C 215; Plenum Press: New York, 1987, pp 79-100.
42. Leemans, L.; Fayt, R.; Teyssié, P. *Macromolecules* **1990**, *23*, 1554-1555.
43. Hatada, K.; Nakanishi, H.; Ute, K.; Kitayama, T. *Polym. J.* **1986**, *18*, 581-591.
44. Hatada, K.; Ute, K.; Tanaka, K.; Kitayama, T.; Okamoto, Y. *Polym. J.* **1985**, *8*, 977-980.
45. Hatada, K.; Ute, K.; Tanaka, K.; Kitayama, T.; Okamoto, Y. *Polym. Prepr., Am. Chem. Soc. Div. Polym. Chem.* **1986**, *27 (1)*, 151-152.
46. Hatada, K.; Ute, K.; Tanaka, K.; Okamoto, Y.; Kitayama, T. *Polym. J.* **1986**, *18*, 1037-1047.
47. Hatada, K.; Ute, K.; Tanaka, T.; Kitayama, T.; Okamoto, Y. in *Recent Advances in Anionic Polymerization*; Hogen-Esch, T. E.; Smid, J., Ed.; Elsevier: New York, 1987, pp 195-204.
48. Müeck, K. F.; Rolly, H.; Burg, K. *Makromol. Chem.* **1977**, *178*, 2773-2784.
49. Jérôme, R.; Forte, R.; Varshney, S. K.; Fayt, R.; Teyssié, in *Recent Advances in Mechanistic and Synthetic Aspects of Polymerization*; Fontanille, M.; Guyot, A., Ed.; NATO Advanced Study Institute Series C 215; Plenum Press: New York, 1987, pp 101-117.
50. Wang, J.-S.; Warin, R.; Jérôme, R.; Teyssié, P. *Macromolecules* **1993**, *26*, 6776-6781.
51. Varshney, S. K.; Hautekeer, J. P.; Fayt, R.; Jérôme, R.; Teyssié, P. *Macromolecules* **1990**, *23*, 2618-2622.
52. Kunkel, D.; Müller, A. H. E.; Janata, M.; Lochmann, L. *Polym. Prepr., Am. Chem. Soc. Div. Polym. Chem.* **1991**, *32 (1)*, 301-302.
53. Wiles, D. M.; Bywater, S. *J. Phys. Chem.* **1964**, *68*, 1983-1987.
54. Lochmann, L.; Janata, M.; Machová, L.; Vlcek, P.; Mitera, J.; Müller, A. H. E. *Polym. Prepr., Am. Chem. Soc. Div. Polym. Chem.* **1988**, *29 (2)*, 29-30 and references therein.
55. Wang, J.-S.; Jérôme, R.; Warin, R.; Teyssié, P. *Macromolecules* **1994**, *27*, 1691-1696.
56. Ballard, D. G. H.; Bowles, R. J.; Haddleton, D. M.; Richards, S. N.; Sellens, R.; Twose, D. L. *Macromolecules* **1992**, *25*, 5907-5913.
57. Zundel, T.; Teyssié, P.; Jérôme, R. *Macromolecules* **1998**, *31*, 2433-2439.
58. Zundel, T.; Teyssié, P.; Jérôme, R. *Macromolecules* **1998**, *31*, 5577-5581.
59. Fayt, R.; Forte, R.; Jacobs, C.; Jérôme, R.; Ouhadi, T.; Teyssié, P.; Varshney, S. K. *Macromolecules* **1987**, *20*, 1442-1444.
60. Varshney, S. K.; Jacobs, C.; Hautekeer, J.-P.; Bayard, P.; Jérôme, R.; Fayt, R.; Teyssié, P. *Macromolecules* **1991**, *24*, 4997-5000.
61. Hautekeer, J.-P.; Varshney, S. P.; Fayt, R.; Jacobs, C.; Jérôme, R.; Teyssié, P. *Macromolecules* **1990**, *23*, 3893-3898.

62. Wang, J. S.; Jérôme, R.; Warin, R.; Zhang, H.; Teyssié, P. *Macromolecules* **1994**, *27*, 3376-3382.
63. Wang, J.-S.; Jérôme, R.; Bayard, P.; Baylac, L.; Patin, M.; Teyssié, P. *Macromolecules* **1994**, *27*, 4615-4620.
64. Varshney, S. K.; Jérôme, R.; Bayard, P.; Jacobs, C.; Fayt, R.; Teyssié, P. *Macromolecules* **1992**, *25*, 4457-4463.
65. Huynh-ba, G.; McGrath, J. E. *Polym. Prepr., Am. Chem. Soc. Div. Polym. Chem.* **1986**, *27 (1)*, 179-180.
66. Huynh-ba, G.; McGrath, J. E. in *Recent Advances in Anionic Polymerization*; Hogen-Esch, T. E.; Smid, J., Ed.; Elsevier: New York, 1987, pp 173-184.
67. Janeczek, H.; Jedliñski, Z.; Bosek, I. *Macromolecules* **1999**, *32*, 4503-4507.
68. Raj, D. J. A.; Wadgaonkar, P. P.; Sivaram, S. *Macromolecules* **1992**, *25*, 2774-2776.
69. Reetz, M. T.; Minet, U.; Bingel, C.; Vogdanis, L. *Polym. Prepr., Am. Chem. Soc. Div. Polym. Chem.* **1991**, *32 (1)*, 296-297 and references therein.
70. Reetz, M. T.; Huette, S.; Goddard, R. *J. Phys. Org. Chem.* **1995**, *8*, 231-241.
71. Pietzonka, T.; Seebach, D. *Angew. Chem. Int. Ed. Engl.* **1993**, *32*, 716-717.
72. Wang, J.-S.; Bayard, P.; Jérôme, R.; Varshney, S. K.; Teyssié, P. *Macromolecules* **1994**, *27*, 4890-4895.
73. Wang, J.-S.; Jérôme, R.; Bayard, P.; Teyssié, P. *Macromolecules* **1994**, *27*, 4908-4913.
74. Bayard, P.; Jérôme, R.; Teyssié, P.; Varshney, S.; Wang, J. S. *Polym. Bull.* **1994**, *32*, 381-385.
75. Webster, O. W.; Hertler, W. R.; Sogah, D. Y.; Farnham, W. B.; RajanBabu, T. V. *J. Am. Chem. Soc.* **1983**, *105*, 5706-5708.
76. Webster, O. W. U.S. Patent 4 417 034, 1983; U.S. Patent 4 508 880, 1985.
77. Farnham, W. B.; Sogah, D. Y. U.S. Patent 4 414 372, 1983; U.S. Patent 4 581 428, 1986.
78. Webster, O. W.; Hertler, W. R.; Sogah, D. Y.; Farnham, W. B.; RajanBabu, T. V. *Polym. Prepr., Am. Chem. Soc. Div. Polym. Chem.* **1983**, *24 (2)*, 52-53.
79. Sogah, D. Y.; Hertler, W. R.; Webster, O. W.; Cohen, G. M. *Macromolecules* **1987**, *20*, 1473-1488.
80. Webster, O. W.; Hertler, W. R.; Sogah, D. Y.; Farnham, W. B.; RajanBabu, T. V. *J. Macromol. Sci., Chem.* **1984**, *A21*, 943-960.
81. Sogah, D. Y.; Farnham, W. B. in *Organosilicon and Bioorganosilicon Chemistry: Structure, Bonding, Reactivity and Synthetic Application*; Sakurai, H., Ed.; Halsted: New York, 1985, pp 219-230.
82. Quirk, R. P.; Ren, J. *Macromolecules* **1992**, *25*, 6612-6620.
83. Quirk, R. P.; Bidinger, G. P. *Polym. Bull.* **1989**, *22*, 63-70.
84. Quirk, R. P.; Ren, J.; Bidinger, G. *Makromol. Chem., Macromol. Symp.* **1993**, *67*, 351-363.
85. Müller, A. H. E. *Macromolecules* **1994**, *27*, 1685-1690.
86. Dicker, I. B.; Cohen, G. M.; Farnham, W. B.; Hertler, W. R.; Laganis, E. D.; Sogah, D. Y. *Macromolecules* **1990**, *23*, 4034-4041.
87. Dicker, I. B.; Farnham, W. B.; Hertler, W. R.; Laganis, E. D.; Sogah, D. Y.; Del Pesco, T. W.; Fitzgerald, P. H. U.S. Patent 4 588 795, 1986; *Chem. Abstr.* **1986**, *105*, 98315.
88. Schubert, W.; Sitz, H.-D.; Bandermann, F. *Makromol. Chem.* **1989**, *190*, 2193-2201.

89. Rodak, N. J.; Sogah, D. Y. *Polym. Prepr., Am. Chem. Soc. Div. Polym. Chem.* **1994**, *35 (2)*, 864-865.
90. Hertler, W. R.; Sogah, D. Y.; Webster, O. W.; Trost, B. M. *Macromolecules* **1984**, *17*, 1415-1417.
91. White, D.; Matyjaszewski, K. *Polym. Prepr., Am. Chem. Soc. Div. Polym. Chem.* **1995**, *36 (2)*, 286-287.
92. Sogah, D. Y.; Hertler, W. R.; Dicker, I. B.; DePra, P. A.; Butera, J. R. *Makromol. Chem., Macromol. Symp.* **1990**, *32*, 75-86.
93. Webster, O. W.; Sogah, D. Y. in *Comprehensive Polymer Science: The Synthesis, Characterization, Reactions, and Applications of Polymers*; Allen, G.; Bevington, J. C., Ed.; Pergamon Press: Oxford, 1989; Vol. 4, pp 163-169 and references therin.
94. Shen, W.; Zhu, W.; Yang, M.; Wang, L. *Makromol. Chem.* **1989**, *190*, 3061-3066.
95. Patrickios, C. S.; Lowe, A. B.; Armes, S. P.; Billingham, N. C. *J. Polym. Sci., Polym. Chem. Ed.* **1998**, *36*, 617-631.
96. Picard, J.-P.; Ekouya, A.; Dunogues, J.; Duffaut, N.; Calas, R. *J. Organomet. Chem.* **1972**, *93*, 51-70.
97. Picard, J.-P.; Dunogues, J.; Calas, R. *J. Organomet. Chem.* **1974**, *77*, 167-176.
98. Nakano, T.; Sogah, D. Y. *Polym. Prepr., Am. Chem. Soc. Div. Polym. Chem.* **1994**, *35 (2)*, 862-863.
99. Sogah, D. Y.; Nakano, S. T.; Rodak, N. J. *Polym. Prepr., Am. Chem. Soc. Div. Polym. Chem.* **1994**, *35 (2)*, 587-588.
100. Li, Y.; Ward, D. G.; Reddy, S. S.; Collins, S. *Macromolecules* **1997**, *30*, 1875-1883 and references therein.
101. Deng, H.; Shiono, T.; Soga, K. *Macromolecules* **1995**, *28*, 3067-3073.
102. Yasuda, H.; Yamamoto, H.; Yokota, K.; Miyake, S.; Nakamura, A. *J. Am. Chem. Soc.* **1992**, *114*, 4908-4910.
103. Yasuda, H.; Yamamoto, H.; Yamashita, M.; Yokota, K.; Nakamura, A.; Miyake, S.; Kai, Y.; Kanehisa, N. *Macromolecules* **1993**, *26*, 7134-7143.
104. Nakamura, A.; Yasuda, H. Jpn. Patent 02 258 808, 1990; *Chem. Abstr.* **1991**, *114*, 82748.
105. Yasuda, H.; Ihara, E. *Macromol. Chem. Phys.* **1995**, *196*, 2417-2441.
106. Yasuda, H.; Tamai, H. *Prog. Polym. Sci.* **1993**, *18*, 1097-1139.
107. Yasuda, H.; Yamamoto, H.; Takemoto, Y.; Yamashita, M.; Yokota, K.; Miyake, S.; Nakamura, A. *Makromol. Chem., Macromol. Symp.* **1993**, *67*, 187-201.
108. Yasuda, H.; Ihara, E.; Morimoto, M.; Nodono, M.; Yoshioka, S.; Furo, M. *Macromol. Symp.* **1995**, *95*, 203-206.
109. Yasuda, H. *Top. Organomet. Chem.* **1999**, *2*, 255-283.
110. Yasuda, H.; Ihara, E.; Nitto, Y.; Kakehi, T.; Morimot, M.; Nodono, M. in *Functional Polymers*; Patil, A. O.; Schulz, D. N.; Novak, B. M., Eds.; ACS Symposium Series 704; American Chemical Society: Washington, DC, 1998, pp 149-162.
111. Ihara, E.; Morimoto, M.; Yasuda, H. *Macromolecules* **1995**, *28*, 7886-7892.
112. Ihara, E.; Morimoto, M.; Yasuda, H. *Proc. Japan. Acad.* **1995**, *B71*, 126-131.
113. Boffa, L. S.; Novak, B. M. *Macromolecules* **1994**, *27*, 6993-6995.
114. Boffa, L. S.; Novak, B. M. *Tetrahedron* **1997**, *53*, 15367-15396.

115. Boffa, L. S. *J. Mol. Catal. A: Chem.* **1998**, *133*, 123-130.
116. Knjazhanski, S. Y.; Elizalde, L.; Cadenas, G.; Bulychev, B. M. *J. Polym. Sci., Polym. Chem. Ed.* **1998**, *36*, 1599-1606.
117. Mao, L.; Shen, Q.; Sun, J. *J. Organomet. Chem.* **1998**, *566*, 9-14 and references therein.
118. Boffa, L. S.; Novak, B. M. *Macromolecules* **1997**, *30*, 3494-3506.
119. Ihara, E.; Taguchi, M.; Yasuda, H. *Appl. Organomet. Chem.* **1995**, *9*, 427-429.
120. Yasuda, H.; Furo, M.; Yamamoto, H.; Nakamura, A.; Miyake, S.; Kibino, N. *Macromolecules* **1992**, *25*, 5115-5116.
121. Yasuda, H. Jpn. Patent 04 114 029, 1992; *Chem. Abstr.* **1992**, *117*, 213187.
122. Kuroki, M.; Aida, T.; Inoue, S. *J. Am. Chem. Soc.* **1987**, *109*, 4737-4738.
123. Sugimoto, H.; Inoue, S. *Adv. Polym. Sci.* **1999**, *146*, 39-119.
124. Kuroki, M.; Nashimoto, S.; Aida, T.; Inoue, S. *Macromolecules* **1988**, *21*, 3114-3115.
125. Inoue, S.; Aida, T.; Kuroki, M.; Hosokawa, Y. *Makromol. Chem., Macromol. Symp.* **1990**, *32*, 255-265.
126. Sugimoto, H.; Aida, T.; Inoue, S. *Macromolecules* **1994**, *27*, 3672-3674 and references therein.
127. Sugimoto, H.; Kuroki, M.; Watanabe, T.; Kawamura, C.; Aida, T.; Inoue, S. *Macromolecules* **1993**, *26*, 3403-3410.
128. Adachi, T.; Sugimoto, H.; Aida, T.; Inoue, S. *Macromolecules* **1992**, *25*, 2280-2281.
129. Hosokawa, Y.; Kuroki, M.; Aida, T.; Inoue, S. *Macromolecules* **1991**, *24*, 824-829 and references therein.
130. For a comprehensive review of copper-mediated ATRP, see: Patten, T. E.; Matyjaszewski, K. *Acc. Chem. Res.* **1999**, *32*, 895-903.
131. For recent general reviews of ATRP, see: (a) Sawamoto, M.; Kamigaito, M. *Chemtech* **1999**, *29*, 30-38. (b) Patten, T. E.; Matyjaszewski, K. *Adv. Mater.* **1998**, *10*, 901-915.
132. Takahashi, H.; Ando, T.; Kamigaito, M.; Sawamoto, M. *Macromolecules* **1999**, *32*, 3820-3823 and references therein.
133. Ueda, J.; Matsuyama, M.; Kamigaito, M.; Sawamoto, M. *Macromolecues* **1998**, *31*, 557-562.
134. Uegaki, H.; Kamigaito, M.; Sawamoto, M. *J. Polym. Sci., Polym. Chem. Ed.* **1999**, *37*, 3003-3009 and references therein.
135. Moineau, G.; Granel, C.; Dubois, P.; Jérôme, R.; Teyssié, P. *Macromolecules* **1998**, *31*, 542-544.
136. Lecomte, P.; Drapier, I.; Dubois, P.; Jérôme, R.; Teyssié, P. *Macromolecules* **1997**, *30*, 7631-7633.
137. Ando, T.; Kamigaito, M.; Sawamoto, M. *Macromolecules* **1997**, *30*, 4507-4510.
138. Matyjaszewski, K.; Wei, M.; Xia, J.; McDermott, N. E. *Macromolecules* **1997**, *30*, 8161-8164.
139. Wayland, B. B.; Basickes, L.; Mukerjee, S.; Wei, M.; Fryd, M. *Macromolecules* **1997**, *30*, 8109-8112.
140. Ashford, E. J.; Naldi, V.; O'Dell, R.; Billingham, N. C.; Armes, S. P. *Chem. Commun.* **1999**, 1285-1286.
141. Wang, J.-S.; Gaynor, S. G.; Matyjaszewski, K. *Polym. Prepr., Am. Chem. Soc. Div. Polym. Chem.* **1995**, *36 (1)*, 465-466.

142. Greszta, D.; Matyjaszewski, K. *Polym. Prepr., Am. Chem. Soc. Div. Polym. Chem.* **1996**, *37 (1)*, 569-570.
143. Shipp, D. A.; Wang, J.-L.; Matyjaszewski, K. *Macromolecules* **1998**, *31*, 8005-8008.
144. Granel, C.; Dubois, P.; Jérôme, R.; Teyssié, P. *Macromolecules* **1996**, *29*, 8576-8582.
145. Uegaki, H.; Kotani, Y.; Kamigaito, M.; Sawamoto, M. *Macromolecules* **1998**, *31*, 6756-6761.
146. Coca, S.; Matyjaszewski, K. *Polym. Prepr., Am. Chem. Soc. Div. Polym. Chem.* **1996**, *37 (1)*, 573-574.
147. Moineau, G.; Minet, M.; Dubois, P.; Teyssié, P.; Senninger, T.; Jérôme, R. *Macromolecules* **1999**, *32*, 27-35.
148. Jo, S. M.; Gaynor, S. C.; Matyjaszewski, K. *Polym. Prepr., Am. Chem. Soc. Div. Polym. Chem.* **1996**, *37 (2)*, 272-273.

New Chain-Building Mechanisms

Most transition metal-mediated polymerizations are formal analogues of traditional "organic" processes. Particularly, for 1,2-addition polymerizations of unsaturated monomers, the replacement of a "naked" ionic or radical propagating species with an organometallic fragment often provides great advantages in polymerization control. Examples include metallocene-catalyzed olefin polymerization (radical), Group 10 olefin-CO copolymerization (radical), and titanium-initiated polymerizations of heterocumulenes (anionic). Organometallic analogues for anionic ring-opening polymerizations of polar cyclic monomers also exist.

However, due to the great variety of organometallic reaction mechanisms, it is possible to develop unique metal-mediated polymerizations which are not based on previously known organic processes. These methods allow for the synthesis of macromolecules with architectures significantly differing from traditional main chain identity, substitution, and functionality patterns. Metathesis polymerization of olefins to give regularly unsaturated hydocarbon polymers (ROMP, ADMET) is a well-known example. Palladium- and nickel-mediated step-growth "cross-coupling" polymerizations of aryl halides with other aromatic reagents, giving poly(p-phenylenes) and other conjugated materials, is another.

This section presents papers focusing on the development of new types of polymerization mechanisms and polymer architectures. Weber et al. report on the use of ruthenium catalysts for step-growth polymerizations involving hydrosilylation and C-H addition processes to form carbosilanes. Extensive work in the area of alkyne cycloaddition polymerization is reviewed by Tsuda. This type of polymerization involves the metal-mediated cycloaddition of diynes with various small molecules to give polymers with heteroaromatic groups in their backbones.

We have also included two papers featuring "organic" polymerization processes that are directed towards the development of new mechanisms or architectures. Penelle reviews extensive work in the anionic polymerization of cyclopropanes, giving unusual carbon-chain polymers bearing substituents every *third*, rather than second, carbon. Miyamoto et al. report a new polymerization mechanism for methyl methacrylate, in which a Brønsted acid (rather than a Lewis acid) is used to activate the carbonyl group of the monomer.

Chapter 2

New Applications of "Carbonylbis(triphenylphosphine)ruthenium" Catalysis in Polymer Synthesis

William P. Weber, Jyri K. Paulasaari, Diyun Huang, Shashi Gupta, Timothy M. Londergan, Jonathan R. Sargent, and Joseph M. Mabry

D. P. and K. B. Loker Hydrocarbon Research Institute, Department of Chemistry, University of Southern California, Los Angeles, CA 90019–1661 (e-mail: wpweber@bcf.usc.edu)

A novel 1:1 complex of divinyldimethylsilane and $(PPh_3)_2RuCO$, catalyzes the step-growth copolymerization of aromatic ketones and *sym*-divinyltetramethyldisiloxane (I). The X-ray structure as well as the dynamic NMR behavior of this material is reported.

"$(Ph_3P)_2RuCO$" catalyzed reaction of thioxanthen-9-one with I fails to give copoly(1,8-thioxanthen-9-onylene/3,3,5,5-tetetra methyl-4-oxa-3,5-disila-1,7-heptanylene (II). Nevertheless, a reactive α,α'-difunctional aryl ketone monomer, 1,8-bis-(3,3,5,5,-pentamethyl-4-oxa-3,5-disilahexanyl)thioxanthen-9-one (III), has been prepared by "$(Ph_3P)_2RuCO$" catalyzed reaction of thioxanthen-9-one with excess vinylpentamethyldisiloxane (IV). III undergoes acid catalyzed siloxane equilibration polymerization to yield II. This approach, in which "$(Ph_3P)_2RuCO$" catalyzed reaction of less reactive aryl ketones with an excess of IV to yield reactive α,α-difunctional monomers which undergo acid catalyzed siloxane equilibration polymerization, has been explored.

Copolymers based on acetyl thiophene systems and I have been prepared by direct "$(Ph_3P)_2RuCO$" catalyzed copolymerization and by acid catalyzed siloxane equilibration polymerization.

Finally, "$(Ph_3P)_2RuCO$" catalysis permits hydrosilylation copolymerization between aromatic α,ω–diketones and α,ω-dihydrido siloxanes to yield aromatic poly(silyl ethers).

Introduction

Murai has reported that dihydridocarbonyltris(triphenylphosphine)ruthenium (V) [1] will catalyze the ortho-alkylation reaction of acetophenone with vinyltri-

© 2000 American Chemical Society

methylsilane [2,3]. This catalytic C-H activation reaction involves the regioselective anti-Markovnikov addition of an ortho C-H bond of acetophenone across the C-C double bond of vinyltrimethylsilane.

We have used this reaction to synthesize copolymers in which carbosilane or carbosilane/siloxane units alternate with aromatic ketone groups. For example, copolymerization of acetophenone and I catalyzed by V yields copoly(3,3,5,5-tetramethyl-4-oxa-3,5-disilaheptanylene/2-acetyl-1,3-phenylene) [4]. Initially, molecular weight of the copolymers was quite low. Exact balance of stoichiometry is essential to achieve high molecular weight in step-growth polymerizations [5]. The observation of Si-ethyl groups by NMR end group analysis suggested a stoichiometric imbalance due to hydrogenation of one of the C-C double bonds of I. This converts a difunctional monomer into a monofunctional end group. We have suggested that the hydrogen for this reduction came from the catalyst [6]. Prior treatment of V with a stoichiometric amount of styrene for a few min at 130 °C leads to a quantitative yield of ethyl benzene and an active catalyst. Loss of hydrogen from the catalyst is probably essential to create a site of coordinate unsaturation. The ^{31}P NMR of solutions of this catalyst show that only two of the three triphenylphosphine ligands are still bound to the ruthenium center. While, ruthenium carbonyl species are known to form carbonyl bridged dimers and trimers, we suggest that the active form of the catalyst may be the highly coordinately unsaturated species "$(Ph_3P)_2RuCO$". Addition of a 1:1 mixture of acetophenone and I to this activated catalyst leads to higher molecular weight copolymers [7].

Addition of ortho-acetylstyrene to a soluiton of V which has beeen activated with styrene gives a 1:1 complex of ortho-acetylstyrene and $(Ph_3P)_2RuCO$ whose structure has been determined by X-ray crystallography [8]. The ligand is coordinated to the Ru center via an O-Ru as well as by a π-bond between the vinyl group and Ru. Heating a solution of this complex with acetophenone and I results in high molecular weight copolymer. Recently, 1:1 complexes of $(Ph_3P)_2RuCO$ with 2-phenylpyridine or N-benzylideneaniline have been reported [9].

1:1 Complex of divinyldimethylsilane and $(Ph_3P)_2RuCO$.

We should like to report that addition of divinyldimethylsilane to a solution of "$(Ph_3P)_2RuCO$" yields a 1:1 complex whose X-ray structure has been determined at -100 °C. See figure 1 and table 1 for bond lengths and angles. The structure is unsymmetrical. Both the Ph_3P ligands and the C-C double bonds of the divinyldimethylsilane ligand are different. Bond lengths Ru-C_{29} and Ru-C_{30} are shorter 2.21 and 2.22 Å while Ru-C_{40} and Ru-C_{41} are longer 2.29 and 2.31 Å. The C_{29}-C_{30} double bond is more tightly bonded to Ru than the C_{40}-C_{41} double bond. Consistent with this, the C_{29}-C_{30} bond 1.41 Å is longer than the C_{40}-C_{41} bond 1.37 Å.

The ^{1}H and ^{31}P solution NMR of this complex change with temperature. At low temperature, two signals, which are coupled to one another, are detected in the ^{31}P NMR, while at higher temperature these resonances broaden, coalesce and finally sharpen to a single line (figure 2). Likewise at 20 °C in the ^{1}H NMR, the two Si-CH$_3$ groups give rise to two sharp singlets. At higher temperature, these broaden, collesce and finally sharpen to a single resonance. Similar dynamic behavior is observed in the ^{1}H NMR for the vinyl hydrogens. The ^{1}H NMR chemical shifts for the Ru complexed vinyl protons are at high field. Analysis based on exchange of two non-equivalent sites gives ΔG = 12.7 kcal/mol [10].

Addition acetophenone and I (1:1) to a solution of this 1:1 complex at 135 °C gives high molecular weight copolymer. Murai has suggested that coordination of the carbonyl oxygen to Ru followed by insertion of the coordinated Ru into an ortho C-H bond of acetophenone is critical for reaction. The fact that divinyldimethylsilane can be copolymerized with acetophenone and also form a 1:1 complex with (Ph$_3$P)$_2$RuCO suggests that π-complexation of the C-C double bonds of vinylsilane or vinyldisiloxane to Ru center may be important in the catalytic cycle.

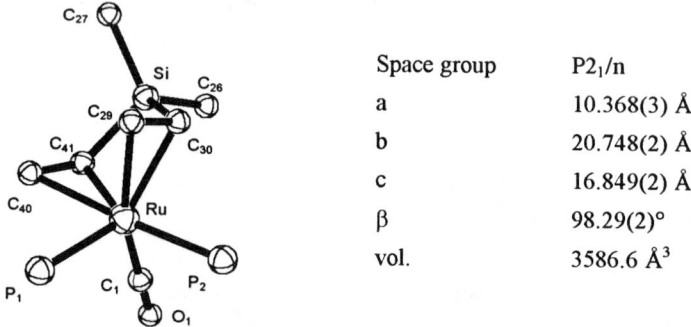

Space group	P2$_1$/n
a	10.368(3) Å
b	20.748(2) Å
c	16.849(2) Å
β	98.29(2)°
vol.	3586.6 Å^3

Figure 1. X-ray structure and dimensions of the complex, phenyl rings omitted.

Table 1. Bond distances (Å) and angles (deg) for the complex.

Bond Distances		Bond Angles			
Ru-C$_1$	1.848(12)	C$_1$-Ru-P$_1$	104.8(4)	P$_1$-Ru-P$_2$	98.93(12)
Ru-C$_{29}$	2.208(12)	C$_1$-Ru-P$_2$	87.1(4)	C$_{29}$-Ru-C$_{30}$	37.1(4)
Ru-C$_{30}$	2.220(11)	C$_1$-Ru-C$_{40}$	96.6(5)	C$_{40}$-Ru-C$_{41}$	34.7(4)
Ru-C$_{40}$	2.285(12)	C$_1$-Ru-C$_{41}$	80.1(5)	C$_{29}$-Ru-C$_{40}$	84.9(4)
Ru-C$_{41}$	2.308(12)	C$_1$-Ru-C$_{29}$	157.8(5)	C$_{30}$-Ru-C$_{41}$	74.7(4)
Ru-P$_1$	2.401(3)	C$_1$-Ru-C$_{30}$	120.8(5)	C$_{29}$-Ru-C$_{41}$	89.1(4)
Ru-P$_2$	2.353(3)	P$_1$-Ru-C$_{40}$	81.6(3)	C$_{30}$-Ru-C$_{40}$	91.8(4)
Si-C$_{41}$	1.838(12)	P$_1$-Ru-C$_{41}$	115.2(3)	C$_{30}$-Si-C$_{41}$	96.5(5)
Si-C$_{30}$	1.846(12)	P$_1$-Ru-C$_{29}$	97.3(3)	C$_{29}$-C$_{30}$-Si	119.2(9)
C$_{29}$-C$_{30}$	1.41(2)	P$_1$-Ru-C$_{30}$	134.4(3)	C$_{40}$-C$_{41}$-Si	125.3(9)
C$_{40}$-C$_{41}$	1.37(2)	P$_2$-Ru-C$_{40}$	176.0(3)	Ru-C$_{30}$-Si	94.7(5)
		P$_2$-Ru-C$_{41}$	145.5(3)	Ru-C$_{41}$-Si	92.0(5)
		P$_2$-Ru-C$_{29}$	91.1(3)	C$_{26}$-Si-C$_{27}$	108.1(7)
		P$_2$-Ru-C$_{30}$	85.0(3)		

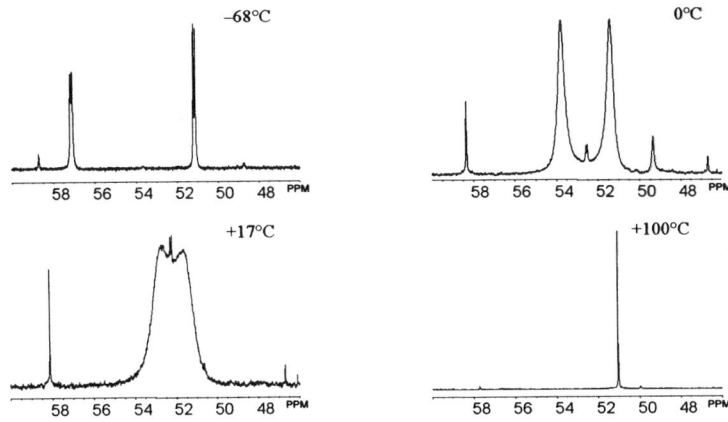

Figure 2. ^{31}P NMR of the 1:1 complex of divinyldimethylsilane and $(Ph_3)_2RuCO$.

"(Ph$_3$P)$_2$RuCO" synthesis of reactive α,α'-monomers for acid catalyzed siloxane equilibration polymerization.

The copolymerization reaction is successful with various substituted aromatic ketones [11-13] as well as with fluorenone, xanthone and anthrone [6]. Nevertheless, with "(Ph$_3$P)$_2$RuCO", benzophenone and I yield a mixture of irregular polymer and cyclic disiloxane monomer. Fortunately, acid catalyzed ring opening polymerization of cyclic disiloxane monomer gave regular copolymer whose T_g is higher [14].

This observation suggested that acid catalyzed siloxane equilibration polymerization might permit preparation of copolymers of aryl ketones and I for systems which do not undergo direct "(Ph$_3$P)$_2$RuCO" catalyzed copolymerization. For example, the "(Ph$_3$P)$_2$RuCO" catalyzed reaction of I with thioxanthen-9-one (1:1) fails. On the other hand, "(Ph$_3$P)$_2$RuCO" catalyzed reaction of thioxanthen-9-one

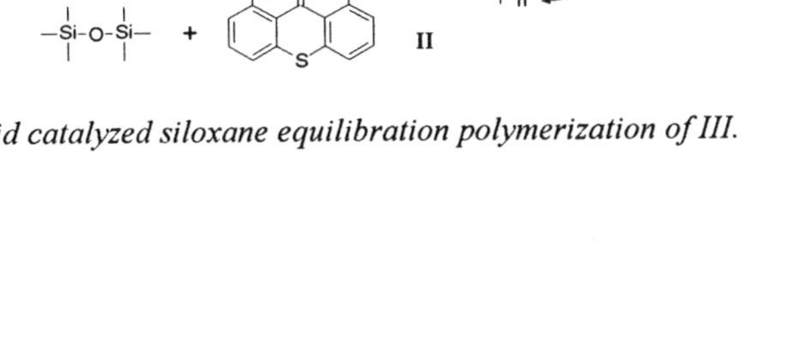

Figure 3. Acid catalyzed siloxane equilibration polymerization of III.

with an excess of IV gives III, which can undergo acid catalyzed siloxane equilibration polymerization to yield II (figure 3) [15].

We have been able to copolymerize 1,4-bis(5'-acetyl-2-thiophenyl) benzene (VI) and I to yield *alt*-copoly[3,3,5,5-tetramethyl-4-oxa-3,5-disila-1,7-heptanylene/2',2"-diacetyl-5',5"(1,4-benzene)-3',3"-dithiophenylene] (VII). However, "(Ph$_3$P)$_2$RuCO" catalyzed copolymerization of 1,3-bis(5'-acetyl-2'-thienyl)benzene (VIII) with I fails. We have solved this problem as follows. Thus, VIII undergoes "(Ph$_3$P)$_2$RuCO" catalyzed reaction with IV to yield 1,3-bis(4'-pentamethyldisiloxyethyl-5'-acetyl-2'-thienyl)benzene (IX) which undergoes triflic acid catalyzed equilibration polymerization to yield *alt*-copoly[3,3,5,5-tetramethyl-4-oxa-3,5-disila-1,7-heptanylene/2'2"-diacetyl-5',5"-(1,3-benzene)-3'3"-dithiophenylene] (X) (figure 4). Copolymers VII and X fluoresce. Sulfur containing compounds often poison Pt or Pd catalysts. Nevertheless, polymers containing thiophene and thioxanthone nuclei can be prepared by "(Ph$_3$P)$_2$RuCO". Murai has also reported reactions of 2 or 3-acetylthiophene and triethoxyvinylsilane catalyzed by V to yield 2-acetyl-3-(2'-triethoxysilylethyl)thiophene or 3-acetyl-2-(2'-triethoxysilylethyl)thiophene [3,16].

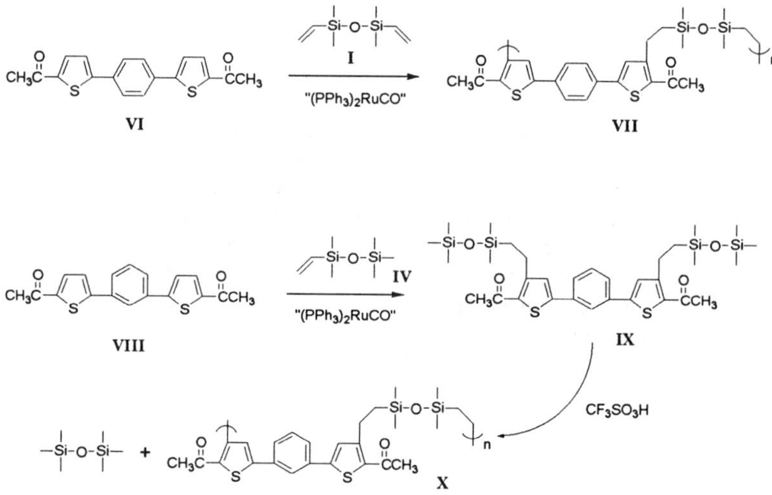

Figure 4. Synthesis of Copolymers VII and X.

A similar two step approach has been used to prepare *alt*-copoly[2,2'-bis(1,1'-ferrocene)benzoylene/3,3,5,5-tetramethyl-4-oxa-3,5-disila-1,7-heptanylene] [17].

"(Ph₃P)₂RuCO" catalyzed hydrosilylation polymerization of ketones and α,ω-bis(SiH) siloxanes.

"(Ph₃P)₂RuCO" also catalyzes the hydrosilylation polymerization of dimethylsilyloxy-aryl ketones as well as the copolymerization of aromatic α,ω-diketones and oligo-α,ω-dihydridodimethylsiloxanes to yield poly(silyl ethers). The Ru catalyzed addition of Si-H bonds across C-O double bonds to form CH-O-Si bonds is key to these reactions [18].

Symmetrical poly(silyl ethers) have been prepared by equilibration of dialkoxysilanes and α,ω-diols [19-21]. The quarternary alkyl ammonium or phosphonium salt catalyzed reaction of dichlorosilanes with bis-glycidyl ethers also yields symmetrical poly(silyl ethers) [22-24]. Unsymmetrical poly(silyl ethers) have been prepared by ring opening polymerization of 2-sila-1-oxacyclopentanes [25].

Reactive Si-O-C bonds in the backbone of poly(silyl ethers) are hydrolytically less stable than the Si-O-Si bonds of siloxanes [26]. Either acid or base can catalyze the hydrolysis of silyl ethers [27]. The hydrolytic stability of silyl ethers depends on the substituents bonded respectively to the silicon and carbon centers. Thus if the carbon of the Si-O-C bond is secondary, the silyl ether is considerably more resistant to hydrolysis than if it is primary [27]. The long term hydrolytic instability of poly(silyl ethers) may make these polymers attractive in a number of applications such as controlled release of drugs.

The hydrosilylation reaction usually involves the Pt catalyzed addition of a Si-H bond across a C-C double bond [28]. This reaction has been used to prepare poly(carbosilanes) [29-31] and copoly(carbosilane/siloxanes) [32-34]. While (Ph₃P)₃RhCl and (Ph₃P)₃RuCl₂ catalyzed hydrosilylation reactions between ketones or aldehydes and silanes to yield monomeric silyl ethers are well known [35-39], such reactions had not been applied to polymer synthesis.

Hydrosilylation polymerization of 4-dimethylsiloxyacetophenone (XI) catalyzed by "(Ph₃P)₂RuCO" yields high molecular weight *alt*-copy(1,4-phenylene/2,2,4-trimethyl-1,3-dioxa-2-silabutanylene) (XII). Likewise, "(Ph₃P)₂RuCO" catalyzed reaction of 4-dimethylsiloxybenzophenone (XIII) yields *alt*-copoly(1,4-phenylene/2,2-dimethyl-1,3-dioxa-4-phenyl-2-silabutanylene)(XIV). "(Ph₃P)₂RuCO" catalyzed hydrosilylation copolymerizations of 1,4-diacetylbenzene (XV), and *sym*-tetramethyl disiloxane (XVI), 1,1,3,3,5,5-hexamethyltrisiloxane (XVII), or 1,1,3,3,5,5,7,7-octamethyltetrasiloxane (XVIII) produce poly(silyl ethers). Thus, copolymerization of XV and XVI yields *alt*-copoly[1,4-phenylene/1,3,3,5,5,7-hexamethyl-2,4,6-trioxa-3,5-disilaheptanylene]. Similarly, "(Ph₃P)₂RuCO" catalyzed copolymerization of XV and XVII yields *alt*-copoly[1,4-phenylene/1,3,3,5,5,7,7,9-octamethyl-2,4,6,8-tetraoxa-3,5,7-tri- silanonanylene]. "(Ph₃P)₂RuCO" catalyzed reaction of XV and XVIII gives *alt*-copoly[1,4-phenylene/1,3,3,5,5,7,7,9,9,11-decamethyl-2,4,6,8,10-pentaoxa-3,5,7,9-tetrasila-undecanylene] (figure 5). Completion of the "(Ph₃P)₂RuCO" catalyzed hydrosilylation is usually indicated by a change in color of the reaction from brown to red which is associated with the initial activated "(Ph₃P)₂RuCO" catalyst. The T_g values for the copolymers decrease as the number of siloxane (Si-O) bonds between aromatic units increases. [40].

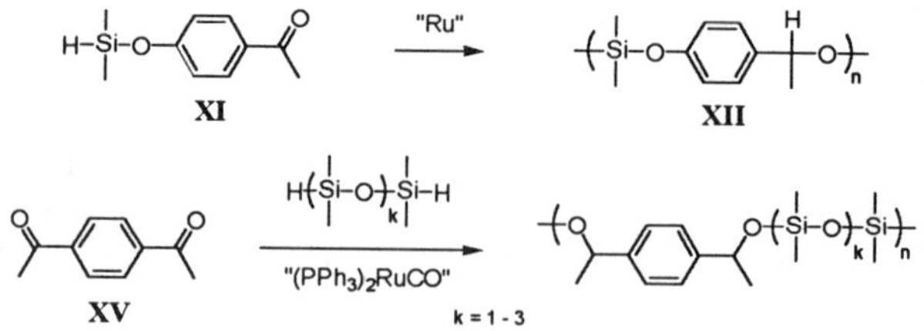

Figure 5. Poly(silyl ethers) via "(Ph$_3$P)$_2$RuCO" catalyzed hydrosilylation reaction.

The ^{1}H NMR spectra for these copolymers is complicated by the presence of two chiral centers associated with each polymer unit. The tetrasiloxane polymer contains two types of silicons. The Si-CH$_3$ groups bonded to each are affected only by the nearest chiral center. These can either be R or S in configuration, causing a large difference in the diastereotopic environment for the outer Si-CH$_3$ groups which are closest to a chiral center. The inner Si-CH$_3$ groups are further away from the chiral centers. The difference in the diastereotopic environment affecting them is smaller. In the case of the trisiloxane polymer, each unit contains three silicons. The central Si-CH$_3$ groups is affected by both chiral centers whose configurations can be RR, SS, RS, and SR. Of these, the RS and SR configurations are meso, and therefore equivalent. The result is three peaks in a 1:2:1 ratio.

Among the advantages of the "(Ph$_3$P)$_2$RuCO" catalyst is that the reaction does not equilibrate the oligodimethylsiloxane units. Equilibration is often observed in acid and base catalyzed reactions [23]. The reaction also generally gives high molecular weight copolymers in excellent chemical yield.

Experimental:

^{1}H and ^{13}C NMR spectra were obtained on a Bruker AMX-500 MHz spectrometer. Five percent w/v solutions in CDCl$_3$, toluene-d_8 or DMSO-d_6 were used to obtain NMR spectra. ^{13}C NMR was run with broad band proton decoupling. A heteronuclear gated ^{1}H decoupling pulse program (NONOE) with a 30-50 sec delay was used to acquire ^{29}Si NMR spectra. UV spectra were acquired on a Shimadzu UV-260 spectrometer. IR spectra of neat films on NaCl plates were recorded on a Perkin Elmer Spectrum 2000 FT-IR spectrometer. Raman spectra were measured on a Bruker Equinox 55 DPY 421-NII-OEM equipped with a Nd Yag laser 1064 nm. Photoluminescence spectra of degassed CH$_2$Cl$_2$ solutions were measured on a PTI instrument equipped with a model A1010 Xenon/Mercury lamp and a model 710 photomultiplier defraction detector.

GPC analysis of the molecular weight distribution of the polymers (M$_w$/M$_n$) was performed on a Waters system comprised of a U6K injector, a 510 HPLC pump, a R401 RI detector and a Millenium 2.15 Session Manager Control system. Two 7.8 mm x 300 mm Styragel columns HR4 and HR2 in series with THF at a flow rate of

0.8 mL/min were used for the analysis. The retention times were calibrated against monodisperse polystyrene standards (929,000; 212,400; 47,400; 13,700; and 794).

The T_gs of the polymers were determined on a Perkin-Elmer DSC-7 instrument. The DSC was calibrated from the melting points of indium (156.6 °C), 2,6-diisopropylphenol (18 °C) and the thermal transition of cyclohexane (-87.06 °C). The temperature program was begun at –100 °C. The temperature was increased at a rate of 10 °C per min to 200 °C.

TGA was carried out on a Shimadzu TGA-50 instrument with a nitrogen flow rate of 40 mL/min. The temperature was increased from 30 °C at a rate of 5 °C/min to 750 °C.

Satisfactory elemental analysis or exact mass measaurements were obtained for all compounds. Elemental analysis was carried out by Oneida Research Services Inc., Whitesboro, New York. Exact mass measurements by high resolution mass spectroscopy were carried out by peak matching against known masses of perfluorokerosene at the University of California, Riverside Mass Spectroscopy Facility on a VG-7070 EHF instrument with ammonia the chemical ionization agent.

All reaction are conducted in flame dried glass wear under an atmosphere of inert gas.

Reagents. Divinyldimethylsilane, hexamethyldisilazane, and vinylpentamethyldisiloxane, *sym*-divinyltetramethyldisiloxane (I), *sym*-tetramethyldisiloxane (XVI), 1,1,3,3,5,5-hexamethyltrisiloxane (XVII), and 1,1,3,3,5,5,7,7-octamethyltetrasiloxane (XVIII) were purchased from Gelest. Acetophenone, XV, 4-hydroxyacetophenone, 4-hydroxybenzophenone and styrene were obtained from Aldrich. Dihydridocarbonyltris(triphenylphosphine)ruthenium (V) was prepared from $RuCl_3$ hydrate [1].

"(Ph₃P)₂RuCO" V (65 μmol, 60 mg), toluene (1.0 mL), styrene (7.5 μL, 65 μmol, 6.8 mg) was placed in a 20 mL Ace pressure tube. The tube was sealed and heated at 135 °C for 3 min. The activated catalyst solution is red.

1:1 Complex of Divinyldimethylsilane and (PPh₃)₂RuCO. V (1.0 g, 1.1 mmol), toluene (2 mL) and styrene (150 mL, 1.3 mmol) were heated in a test tube, equipped with a Teflon covered magnetic stir bar and sealed with a rubber septum, at 120 °C for 10 min. A quantitative yield of ethylbenzene was formed. Divinyldimethylsilane (250 mL, 1.6 mmol) was then injected into the solution. After approximately 5 min, a precipitate formed. After cooling to rt, 10 mL of ether was added. The liquid phase was decanted and the solids were washed with ether and dried under vacuum for 5 h. In this way, 0.78 g, 93% yield of a light yellow powder was obtained. The powder was rapidly (5 min) dissolved in refluxing toluene (5-6 mL). The solution was rapidly cooled to rt. Transparent orange colored crystals were formed, mp = 187 °C, suitable for X-ray crystallography.

^{1}H NMR (at 107 °C, fast exchange) δ: 0.09 (s, 6H), 0.44 (d, 2H, J = 14.5 Hz), 1.87 (t, 2H, J = 12.5 Hz), 1.97 (d, 2H, J = 11 Hz), 6.99 (br. s, 18H), 7.44 (br. s, 12H). At –68 °C, slow exchange δ: -0.15 (br. s, 1H), -0.11 (s, 3H), 0.61 (br. s, 1H), 0.89 (s, 3H), 1.00 (br. s, 1H), 1.26 (br. s, 1H), 2.74 (br. s, 2H), 6.54 (t, 2H, J = 8 Hz), 6.62 (t, 2H, J = 7 Hz), 6.73 (t, 1H, J = 7 Hz), 6.83 (t, 2H, J = 7 Hz), 6.92 (s, 9H), 6.98

(t, 3H, J = 8 Hz), 7.14 (t, 2H, J = 7 Hz), 7.60 (br. s, 6 H), 7.68 (t, 2H, J = 8 Hz), 8.44 (t, 2H, J = 8 Hz). ^{13}C NMR δ: 213.61(t, C$_{C=O}$, J_{C-P} = 11.9 Hz), 139.07, 134.65, 134.25, 134.98, 128.80, 127.96, 127.82, 53.00, 29.75, 4.41, -1.93. ^{31}P NMR δ (at -68 °C slow exchange): 51.18 (d, 1P, J_{P-P} = 16 Hz), 57.12 (d, 1P, J_{P-P} = 16 Hz) (at 67 °C, fast exchange): 51.02 (s, 2P). ^{29}Si NMR δ: -4.30, -4.26, -4.23 (1:2:1). IR ν: 3056, 1884, 1585, 1479, 1433, 1310, 1241, 1184, 1086, 834, 743, 696, 555, 521 cm^{-1}. Raman ν: 3168, 3142, 3095, 3061, 3024, 3004, 2978, 2959, 2938, 2903, 2888, 1879, 1586, 1232, 1159, 1090, 1031, 1004, 685, 622, 609, 560, 467, 394, 360, 267, 160, 112, 99 cm^{-1}. UV λ $_{max}$ (ε): 337 (8,240), 248 (18,630). Anal. Calcd for C$_{43}$H$_{42}$OP$_2$RuSi: C, 67.42; H, 5.54; P, 8.09; Ru, 13.19. Found: C, 67.30; H, 5.44; P, 7.43; Ru, 13.17.

Copolymerization of Acetophenone and I by the 1:1 complex of divinyl-dimethylsilane and (PPh$_3$)$_2$RuCO.

Copoly(2-acetyl-1,3-phenylene/3,3,5,5-tetramethyl-4-oxa-3,5-disila-1,7-heptan-ylene) has been reported. For experimental details and characterization see [7].

X-ray Structure Analysis. A crystal of dimensions 0.2 x 0.35 x 0.45 mm^3 was mounted on a glass fiber and placed in a Siemens P$_4$/RA diffractometer for data collection at –100 °C. The annular settings of well-centered reflections indicated a monoclinic unit cell. Four monitor reflections indicated no significant decrease in intensity during data collection, with Cu Kα radiation up to a 2Θ maximum of 90°. The position of the Ru atom and the seven atoms attached to it were obtained from a Patterson map with crystallographic package SHELX-86 [41]. Other non-hydrogen atoms were subsequently located from a series of structure factor/difference Fourier calculations [42].

The X-ray analysis was completed with several cycles of full-matrix least-squares refinements, with anisotropic thermal parameters assigned to all atoms. The final agreement factor R(F) = 6 % for nonzero reflections with F > 4σ(F).

Synthesis of III. "(Ph$_3$P)$_2$RuCO" (27.5 mg, 0.03 mmol) 0.5 mL of toluene, thioxanthen-9-one (0.64 g, 3 mmol) and IV (1.23 g, 7 mmol) were placed in an Ace pressure tube. The tube was heated to 120 °C for 40 h. After cooling to rt, the volatiles were removed by evaporation under reduced pressure. The product was separated by chramogrraphy on a silica gel column with hexane/ethyl acetate (95:5). In this way, 0.96 g, 57% yield was obtained. ^{1}H NMR δ: 0.20 (s, 18 H), 0.26 (s, 12H), 1.08-1.15 (m, 4H), 3.16-3.22 (m, 4H), 7.31 (dd, 2H, J = 6.7 and 2.4 Hz), 7.37 (dd, 2H, J = 11 and 7.8 Hz), 7.39 (dd, 2H, J = 15 and 7.8 Hz). ^{13}C NMR δ: 0.51, 2.07, 21.17, 28.28, 123.03, 128.12, 130.40, 131.12, 136.30, 148.37, 186.62. ^{29}Si NMR δ: 7.44. Apparently the two silicon atoms fortuitously have identical chemical shifts. IR ν: 1649 cm^{-1}. C=O. UV λ $_{max}$ nm (ε): 374 (6300), 301 (7900), 359(39200).

Synthesis of II by acid catalyzed equilibration polymerization of III. III (0.56 g), mestylene (1 mL) and triflic acid (1 μL) were pleaced in a 5 mL round bottomed flasked equipped with a Teflon covered magnetic stir bar. The reaction was stirred at rt for 16 h. Hexamethyldisiloxane was continuously removed under vaccum. After neutralization with hexamethyldisilazane and an aquous work-up, the polymer was dissolved in a minimuym amount of THF and was purified by precipitation into ethanol. It was dried under vacuum. II, 0.25 g, 45% yield, M_w/M_n = 30000/21000, T_g = 17 °C was obtained. ^{1}H NMR δ: 0.16 (s, 12H), 0.98-1.05 (m, 4H), 3.03-3.09 (m, 4H), 7.17-7.20 (m, 2H), 7.26-7.29 (m, 4H). ^{13}C NMR δ: 0.65, 21.28, 28.37, 123.09, 128.23, 130.50, 131.10, 136.31, 148.39, 186.63. ^{29}Si NMR δ: 7.61. IR ν: 1643 cm^{-1}. C=O. UV λ $_{max}$ nm (ε): 375 (1720), 299(3140), 260 (14300).

Synthesis of VII. "(Ph$_3$P)$_2$RuCO" (8.44 mg, 9.2 μmol) and toluene (0.5 mL) were placed in an Ace pressure tube. VI (0.15 g, 0.46 mmol) [43] and I (106 mL, 0.46 mmol) were added. The tube was sealed and was heated at 135°C for 24 h. The solvent was removed by evaporation under reduced pressure. The solid residue was dissolved in a minimum amount of THF and the polymer was precipitated with methanol. The polymer was collected by centrifugation. In this way, 0.16 g, 68% yield, M_w/M_n = 44,860/15,400, T_g = 114 °C was obtained. ^{1}H NMR δ: 0.16 (s, 12H), 0.94 (m, 4H), 2.51 (s, 6H), 3.02 (m, 4H), 7.26 (s, 2H), 7.61(s, 4H). ^{13}C NMR δ: 0.26, 19.28, 24.32, 29.54, 126.51, 127.44, 133.52, 134.17, 146.92, 154.12, 190.67. ^{29}Si NMR δ: 6.9. IR ν: 1659, 1643 C=O cm^{-1}. UV λ$_{max}$ (ε): 250(11,240), 299(10,400), 371(40,490). When irradiated at either 318 or 410 nm (λ$_{max}$ in excitation scan), fluoresence was observed with a λ $_{max}$ at 438 nm.

1,3-bis(2'-Thiophenyl)benzene. 1,3-Diiodobenzene (3.37 g, 10.2 mmol), Pd(PPh$_3$)$_4$ (0.71 g, 0.61 mmol) and 1,2-dimethoxyethane (85 mL) were placed in a 250 mL flask. The solution was stirred for 10 min. 2-Thiophenylboronic acid (3.40 g, 26.6 mmol) and 1M NaHCO$_3$ solution (62 mL) were added and the reaction mixture was reluxed for 12 h with stirring. The mixture was filtered while warm. The organic layer was separated and dried over anhydrous MgSO$_4$, filtered and the solvent removed by evaporation under reduced pressure. The residue was purified by chromatography on a silica gel column with benzene as eluting solvent. The product, 1.9 g, 77% yield, mp 84-86 °C was obtained. ^{1}H NMR δ: 7.10 (dd, 2H, J = 5 and 3.5 Hz), 7.31 (dd, 2H, J = 5 and 1 Hz), 7.36 (dd, 2H, J = 4 and 1 Hz), 7.38 (t, 1H, J = 8 Hz), 7.52(dd, 2H, J = 7.5 and 1.5 Hz), 7.84(t, 1H, J = 1.5 Hz). ^{13}C NMR δ: 123.45, 123.52, 125.09, 128.02, 129.41, 135.03, 143.92. UV λ$_{max}$ (ε): 286 (28,700). When irradiated at 322 (λ $_{max}$ in exication scan) fluoresence was observed at 358 nm.

Synthesis of VIII. 1,3-Bis(2'-thiophenyl)benzene (1.43 g, 5.91 mmol) and acetic anhydride (115 mL) were placed in a 250 mL flask. Two drops of 70% HClO$_4$ was added. The reaction mixture was stirred at rt for 1 h and then poured into water. The precipitate was collected by filtration and was recrystallized from cyclohexane. The product 1.3 g, 67% yield, mp 147-149 °C was obtained. ^{1}H NMR δ: 2.55 (s, 6H), 7.54 (t, 1H, J = 8 Hz), 7.77 (dd, 2H, J = 8 and 1.5 Hz), 7.82 (d, 2H, J = 4 Hz), 7.99 (dd, 2H, J = 4 Hz), 8.11 (t, 1H, J = Hz). ^{13}C NMR δ: 26.21, 123.04, 125.98, 126.65, 130.40, 133.73, 135.13, 143.25, 150.30, 180.21. IR ν: 1664 C=O cm^{-1}. UV λ $_{max}$(ε); 329 (37,900). When irradiated at 376 nm (λ $_{max}$ in excitation scan) fluoresence was observed at 440 nm.

Synthesis of IX. VIII (0.2 g, 0.61 mmol), IV (1.4 mL, 6.1 mmol), "(Ph$_3$P)$_2$RuCO" (22.5 mg, 25 μmol) and toluene (0.5 mL) were placed in an Ace pressure tube. The reaction mixture was heated at 135 °C for 24 h. The solvent and excess IV were removed by evaporation under reduced pressure. The residue was purified by chromatography on a silica gel column with ethyl acetate and hexane (1:10). In this way, 0.27 g, 65% yield, mp 95.5-97 °C was obtained. ^{1}H NMR δ: 0.09 (s, 18 H), 0.12 (s, 12H), 0.92 (pentad, 4H, J = 4.5 Hz), 2.54 (s, 6H), 3.00 (pentad, 4H, J = 4.5 Hz), 7.27 (s, 2H), 7.42 (t, 1H, J = 8 Hz), 7.60 (d, 2H, J = 7.5 Hz), 7.85 (s, 1H). ^{13}C NMR δ: 0.21, 1.97, 19.33, 24.27, 29.67, 123.58, 126.31, 127.63, 129.73, 134.18, 134.40, 147.01, 153.98, 190.56. IR ν: 1634 C=O cm^{-1}. UV λ$_{max}$ (ε): 240(18,800), 327(43,300). When irradiated at 374 nm (λ$_{max}$ in exciation scan) fluoresence was observed at 436 nm.

Synthesis of X by acid catalyzed equilibration polymerization of IX. IX (184 mg, 0.27 mmol), 0.5 mL of mesitylene and a Teflon covered magnetic stir bar were placed in a 5 mL rb flask. Triflic acid 1 μL was added. The contents of the flask were stirred at rt for 10 h under vacuum (10 mm Hg). The reaction mixture was taken up in CHCl$_3$. The solution was extracted with aq. NaHCO$_3$, water, dried over anhydrous MgSO$_4$, filtered and the solvent removed under reduced pressure. The residue was taken up in a minimum amount of THF. The polymer was precipitated with methanol. In this way, 0.1 g, 70% yield, M$_w$/M$_n$ = 53,200/19,600, T$_g$ = 53 °C was obtained. ^{1}H NMR δ: 0.16 (s, 12H), 0.94 (m, 4H), 2.50 (s, 6H), 3.02 (m, 4H), 7.27 (s, 2H), 7.36 (t, 1H, J = 7 Hz), 7.55 (d, 2H, J = 6.5 Hz), 7.80 (s, 1H). ^{13}C NMR δ: 0.33, 19.33, 24.32, 29.69, 123.58, 126.27, 127.67, 129.69, 134.15, 134.30, 147.00, 154.00, 190.55. ^{29}Si NMR δ: 7.8. IR ν: 1661 cm^{-1}. UV λ$_{max}$ (ε): 1240 (15,250), 327 (33,900). When irradiated at 376 nm (λ$_{max}$ in exciation scan), fluoresence was observed at 450 nm.

Synthesis of XI was prepared by reaction of dimethylchlorosilane (11.4 g, 121 mmol) with 4-hydroxyacetophenone (12.7 g, 93 mmol) in the presence of Et$_3$N (11.5 g, 114 mmol) in THF (120 mL). After a non-aqueous work-up, the product was purified by fractional distillation. A fraction bp 75 °C/0.1 mm, 15 g, 82% yield was obtained. ^{1}H NMR δ: 0.37 (d, 6H, J = 3 Hz), 2.48 (s, 3H), 5.00 (sept. 1H, J = 3 Hz), 6.94 (dt, 2H, J = 9 and 2.5 Hz), 7.91 (dt, 2H, J = 9 and 2.5 Hz). ^{13}C NMR δ: -1.63, 26.17, 119.32, 130.90, 131.67, 159.97, 195.53. ^{29}Si NMR δ: 6.16. IR ν: 2139 Si-H, 1683 C=O cm^{-1} UV λ$_{max}$ nm(ε) 258 (300)

Synthesis of XIII was prepared by reaction of dimethylchlorosilane (7 g, 74 mmol) with 4-hydroxybenzophenone (11.3 g, 57 mmol) in the presence of Et$_3$N (11.5 g, 114 mmol). After filtration, the product was purified by fractional distillation. A fraction, 12 g, bp 141 °C/0.2 mm, 80% yield was isolated. ^{1}H NMR δ: 0.36 (d, 6H, J = 3 Hz), 5.02 (sept. 1H, J = 3 Hz), 6.99 (dt, 2H, J = 9 and 2.5 Hz), 7.41 (tt, 2H, J = 7.5 and 1.5 Hz), 7.50 (tt, 1H, J = 7.5 and 1.5 Hz), 7.73 (dt, 2H, J = 7 and 1.7 Hz), 7.77 (dt, 2H, J = 9 and 2.5 Hz). ^{13}C NMR δ: -1.63, 119.15, 128.51, 129.83, 131.29, 132.13, 132.65, 138.44, 159.67, 194.34. ^{29}Si NMR δ: 6.33. IR ν: 2139 Si-H, 1656 C=O cm^{-1}. UV λ$_{max}$ nm (ε): 269 (1430).

alt-**Copoly(1,4-phenylene/1,3-dioxa-2,2,4-trimethyl-2-silabutanylene (XII)** was prepared by an "(Ph$_3$P)$_2$RuCO" catalyzed reaction of XI (3.0 g, 15 mmol). M$_w$/M$_n$ = 119,000/57,800, T$_g$ = -10° C. ^{1}H NMR δ: 0.12 (s, 3H), 0.21 (s, 3H), 1.41 (d, 3H, J = 6 Hz), 5.07 (q, 1H, J = 6 Hz), 6.83 (d, 2H, J = 8 Hz), 7.21 (d, 2H, J = 8

Hz). ^{13}C NMR δ: -2.41, -1.94, 26.82, 71.05, 119.91, 127.01, 139.37, 154.12. ^{29}Si NMR δ: -5.67. UV λ $_{max}$ nm (ε): 272 (1250).

alt-Copoly(1,4-phenylene/1,3-dioxa-2,2-dimethyl-4-phenyl-2-silabutanylene) (XIV) was prepared by an "(Ph$_3$P)$_2$RuCO" catalyzed reaction of XIII (3.0 g, 12 mmol) as above. M$_w$/M$_n$ = 94,200/66,400, T$_g$ = 25 °C. ^{1}H NMR δ: -0.14 (s, 6H), 6.02 (s, 1H), 6.76 (d, 2H, J = 8 Hz), 7.11 (t, 1H, J = 7 Hz), 7.19 (d, 4H, J = 7 Hz), 7.35 (d, 12H, J = 7 Hz). ^{13}C NMR δ: -2.00, -1.98, 76.63, 120.06, 126.73, 127.52, 128.17, 128.64, 138.22, 145.00, 154.13. ^{29}Si NMR δ: -4.15. UV λ $_{max}$ nm (ε): 271 (1330).

alt-Copoly[1,4-phenylene/1,3,3,5,5,7-hexamethyl-2,4,6-trioxa-3,5-disila-1,7-heptanylene] was prepared by an "(Ph$_3$P)$_2$RuCO" catalyzed reaction of XV (2.415 g, 14.9 mmol) and XVI (2.0 g, 14.9 mmol). The color of the reacting solution was brown. After 12 h at 100 °C, color of the solution had become red. The copolymer was dissolved in a minimum amount of THF and was precipitated three times with methanol. In this way, 4.2 g, 95.1% yield, of polymer with M$_w$/M$_n$ = 46,200/19,650 = 2.35, T$_g$ = -55 °C was obtained. ^{1}H NMR δ: 0.026 (d, 6H, J = 9 Hz), 0.11 (d, 6H, J = 8 Hz), 1.47 (d, 6H, J = 7 Hz), 5.01 (q, 2H, J = 6 Hz), 7.32 (s, 4H). ^{13}C NMR δ: -0.78, -0.75, -0.61, -0.57, 26.69, 69.89, 69.90, 124.96, 144.68. ^{29}Si NMR δ: -13.63. UV λ$_{max}$ nm (ε): 247 (959). TGA: polymer is stable to 300 °C. Between 300 to 500 °C, catastrophic decomposition occurs; 90% of weight is lost.

alt-Copoly[1,4-phenylene/1,3,3,5,5,7,7,9-octamethyl-2,4,6,8-tetraoxa-3,5,7-trisilanonanylene] was prepared by an "(Ph$_3$P)$_2$RuCO" catalyzed reaction of XV (1.62 g, 10.0 mmol) and XVII (2.09 g, 10.0 mmol). After 12 h at 120 °C, the solution color had again become red. After purification by precipitation with methanol, 3.04 g, 82 % yield, M$_w$/M$_n$ = 89,200/40,100, Tg = -72 °C was obtained. ^{1}H NMR δ: -0.10 (s, 1.5H), -0.09 (s, 3H), -0.08 (s, 1.5H), 0.07 (s, 6H), 0.02 (s, 6H), 1.36 (d, 6H, J = 6.5 Hz), 4.90 (q, 2H, J = 6.5 Hz), 7.19 (s, 4H). ^{13}C NMR δ: -0.23, 0.95, 27.19, 70.48, 125.56, 145.46. ^{29}Si NMR δ: -14.03 (s, 2Si), -21.59 (s, 1Si). UV λ$_{max}$ nm(ε): 254 (1,081), 227 (1,478). TGA: Polymer is stable to 275 °C. Between 275 and 400 °C, catastrophic decomposition occurs; 90% of initial weight is lost.

alt-Copoly[1,4-phenylene/1,3,3,5,5,7,7,9,9,11-decamethyl-2,4,6,8,10-penta-oxa-3,5,7,9-tetrasila-undecanylene] was prepared by "(Ph$_3$P)$_2$RuCO" catalyzed reaction of XV (1.62 g, 10.0 mmol) and XVIII (2.83 g, 10.0 mmol). After 12 h at 120 °C, the color of the solution had become red. After purification by precipitation with methanol, 3.57 g, 80 % yield, M$_w$/M$_n$ = 168,000/75,700, T$_g$ = -81 °C was obtained. ^{1}H NMR δ: -0.07 (s, 6H), -0.05 (s, 6H), -0.04 (s, 6H), -0.02 (s, 6H), 1.36 (d, 6H, J = 6.5 Hz), 4.90 (q, 2H, J = 6.5 Hz), 7.19 (s, 4H). ^{13}C NMR δ: -0.17, 1.11, 27.25, 70.59, 125.58, 145.57. ^{29}Si NMR δ: -14.07(s, 2Si), -21.83(s, 2Si). UV λ$_{max}$ nm (ε) : 255(399), 227(832). TGA: Polymer is stable to 300 °C. Between 300 and 425 °C, catastrophic decomposition occurs; 90% of initial weight is lost.

Conclusion

New applications of the Murai - Ru catalyzed - reaction in polymer chemistry continue to be found. Siloxane copolymers which contain acetylthiophene and thioxanthone units are described . Further, X-ray crystallographic structural and dynamic NMR information of a 1:1 complex of divinyldimethylsilane and

($Ph_3P)_2RuCO$ is reported which may increase our understanding of the catalyst system. This may permit the design of even more efficient catalysts for the Murai Polymerization Reaction in the future.

Acknowledgment

This work was supported by the National Science Foundation and the Air Force Office of Scientific Research. Jyri K. Paulasaari thanks the Neste Foundation for fellowship support.

References

1. Levison, J. D.; Robinson, S. D. *J. Chem. Soc. A* **1970**, 2947.
2. Murai, S.; Kakiuchi, F.; Sekine, S.; Tanaka, Y.; Sonoda, M.; Chatani, N. *Nature* **1993**, *366*, 520.
3. Kakiuchi, F.; Sekine, S.; Tanaka, Y.; Kamatani, An; Sonoda, M.; Chatani, N.; Murai, S., *Bull. Chem. Soc. Jpn.* **1995**, *68*, 62.
4. Guo, H.; Weber, W. P. *Polym. Bull.* **1994**, *32*, 525.
5. Odian, G. *Principles of Polymerization*, J. Wiley & Sons, New York, NY, **1981**; p 82-96.
6. Guo, H.; Tapsak, M. A.; Weber, W. P. *Polymer Bull.* **1995**, *34*, 49.
7. Guo. H.; Wang, G.; Tapsak, M. A.; Weber, W. P. *Macromolecules* **1995**, *28*, 568.
8. Lu, P.; Paulasaari, J.; Jin, K.; Bau, R.; Weber, W. P. *Organometallics* **1998**, *17*, 584.
9. Hiraki, K.; Koizumi, M.; Kira, S. I.; Kawano, H. *Chem. Letters* **1998**, 47.
10. Bovey, F. A.; Hood III, F. P.; Anderson, E. W.; Kornegay, R. L. *J. Chem. Phys.* **1964**, *41*, 2042.
11. Guo, H.; Weber, W. P. *Polym. Bull.*, **1995**, *35*, 259.
12. Wang, G.; Guo, HJ.; Weber, W. P. *Polym. Bull.* **1996**, *37*, 169.
13. Guo, H.; Tapsak, M. A.; Wang, G.; Weber, W. P. in *Step-Growth Polymers for High Performance Materials, New Synthetic Methods*, Hedrick, J. L,; Labadie, J. W., Eds.; ACS Symposium Series 624, American Chemical Society, Washington, DC **1996**, p 99-112.
14. Kepler, C. L.; Londergan, T. M.; Lu, J.; Paulasaari, J.; Weber, W. P. *Polymer* **1998**, *40*, 765.
15. Gupta, S. K.; Weber, W. P. *Polym. Prepr.* **1999**, *40-I*, 62.
16. Kakiuchi, F.; Sato, T.; Tsujimoto, T.; Yamauchi, M.; Chatani, N.; Murai, S. *Chem. Letters* **1998**, 1053.
17. Sargent, J. R.; Weber, W. P. *Polym. Prept.* **1998**, *39-I*, 274.
18. Paulasaari, J. K.; Weber, W. P. *Macromolecules* **1998**, *31*, 7105.
19. Bailey, D. L.; O'Connor, F. M. *Brit. Pat.* 880022, 5-22-58, CA 57: P1087G **(1962)**.
20. Koepnick, H.; Delfs, D.; Simmler, W. *German Pat.* 1108917, 6-15-61, CA 56: P2576F **(1962)**.

21. Bailey, D. L.; O'Connor, F. M. *German Pat.* 1012602, 7-25-57, CA 53: P15647I (1957).
22. Itoh, H.; Kameyama, A.; Nishikubo, T. *J. Polym. Sci. A, Polym. Chem.* **1997**, *35*, 3217.
23. Liaw, D. J. *Polymer* **1997**, *38*, 5217.
24. Nishikubo, T.; Kameyama, A.; Kimura, Y.; Nakamura, T. *Macromolecules* **1996**, *29*, 5529.
25. Mironov, V. F.; Kozlikov, V. L.; Fedotov, N. S. *Zhur. Obshch. Chim.* **1969**, *39*, 966.
26. Voronkov, M. G.; Mileshkevich, V. P.; Yuzhelevskii, Yu, A. *The Siloxane Bond*, Consultants Bureau, New York, N.Y, **1978;** Si-O-Si p 146, Si-O-C p 323.
27. Burger, C.; Freuzer, F. H. *Polysiloxane and Polymers Containing Siloxane Groups in Silicon Polymer Synthesis*, Ed. Kricheldorf, H. R., Springer-Verlag: Berlin, German, **1996;** p 139.
28. Ojima, I., *The Chemistry of Organic Silicon Compounds*, Patai, S.; Rappoport, Z., Eds.; J. Wiley & Sons: New York, NY, **1989;** p 1479-1526.
29. Curry, J. W. *J. Am. Chem. Soc.* **1956**, *78*,1686.
30. Boury, B.; Carpenter, L S.; Curriu, R. J. P. *Angew Chem. Int. Ed. Engl.* **1990**, *29*, 785.
31. Boury, B.; Curriu, R. J. P.; Leclercq, D.; Mutin, P. H.; Paneix, J. M. an Vioux, A. *Organometallics* **1991**, *10*, 1457.
32. Mathias, L. J.; Lewis, C. M. *Macromolecules* **1993**, *26*, 4070.
33. Dvornic, P. R.; Gerov, V. V.; Govedarica, M. N. *Macromolecules* **1994**, *27*, 7575.
34. Hu, Y.; Son, D. Y. *Polym. Preprints* **1998**, *39-I*, 567.
35. Frainnet, E.; Martel-Siegfried, V.; Borousee, E.; Dedier, J. *J. Organometal. Chem.* **1975**, *85*, 297.
36. Ojima, I.; Kogura, T.; Nihonyanagi, M.; Nagai Y. *Bull. Chem. Soc. Jpn.* **1972**, *45*, 3506.
37. Corriu, R. J. P.; Moreau, J. J. E. *J. Chem. Soc., Chem. Commun.* **1973**, 38.
38. Eaborn, C.; Odell, K.; Pidcock, A. *J. Organometal. Chem.* **1973**, *63*, 93.
39. Semmelhack, M. F.; Misra, R. N. *J. Org. Chem.* **1982**, *47*, 2469.
40. Noll, W. *The Chemistry and Technology of Silicone*, Academic Press: New York, NY, **1968**, p 326.
41. Sheldrick, G. M. SHELX-86; University of Göttingen: Göttingen, German, 1986.
42. Shedlrick, G. M. SHELX-93; University of Göttingen: Göttingen, Germany 1993.
43. Ribereau, P.; Pastour, P. *Bull. Soc. Chim. Fr.* **1969**, *6*, 2076.

Chapter 3

Transition Metal-Catalyzed Alkyne Cycloaddition Polymerization

Tetsuo Tsuda

Department of Polymer Chemistry, Graduate School of Engineering, Kyoto University, Yoshida, Kyoto 606–8501, Japan

The scope and feature of the polymer synthesis by three types of transition metal-catalyzed alkyne cycloaddition polymerizations, i.e., 1) a diyne cycloaddition copolymerization, 2) monoyne and diyne double-cycloaddition copolymerizations, and 3) a diyne cycloaddition terpolymerization, together with a diyne cycloaddition copolymerization with a low-valent transition metal complex without a transition metal catalyst are summarized. These polymerizations afforded poly(2-pyrone)s, poly(2-pyridone)s, poly(pyridine)s, poly(thiophene)s, poly(bicyclo[2.2.2]oct-7-ene)s, poly(imide)s, and poly(enone)s together with transition metal-containing polymers. These results indicate the effectiveness of the transition metal-catalyzed alkyne cycloaddition polymerization as a new method of polymer synthesis.

Introduction

Polymer synthesis is closely related to organic synthesis because a polymer is formed by repetition of a selective organic reaction. Consequently, mutual feedback of organic synthesis and polymer synthesis may be a useful strategy for finding new organic and polymerization reactions. Organic synthesis utilizing transition metal catalyst's characteristic organic reactions has recently attracted much attention and produced excellent results. Among these organic reactions, there is the transition metal-catalyzed alkyne cycloaddition reaction (1,2). It is interesting therefore to develop a new method of polymer synthesis using the transition metal-catalyzed alkyne cycloaddition reaction as a polymer-forming elementary reaction.

The author has recently developed a transition metal-catalyzed alkyne cycloaddition polymerization as a new method of polymer synthesis (3-6). This review article summarizes the scope and feature of the polymer synthesis by three types of transition metal-catalyzed alkyne (monoyne and diyne) cycloaddition polymerizations (7): 1) a diyne cycloaddition copolymerization, 2) monoyne and diyne double-cycloaddition copolymerizations, and 3) a diyne cycloaddition terpolymerization. Synthesis of a transition metal-containing polymer by the diyne cycloaddition copolymerization with a low-valent transition metal complex without a transition metal catalyst is also described.

About twenty-five years ago, the copolymerization of a mixture of 1,3- and 1,4-diethynylbenzenes with phenylacetylene, which is classified into a diyne/monoyne cycloaddition copolymerization, was examined by using a $TiCl_4/AlEt_2Cl$ catalyst, but a highly branched poly(phenylene) with wide molecular weight distribution was produced due to uncontrollable and inevitable diyne cyclotrimerization (*vide infra*) (8). The related study of the preparation of thermosetting resins via branched and cross-linked prepolymers formed by the cobalt-catalyzed diyne cyclotrimerization polymerization and copolymerization was reported in a patent (9). Thus, these diyne cyloaddition polymerizations cannot afford linear cycloaddition polymers.

As for a thermally induced cycloaddition polymerization, the Diels-Alder polymerization is well-known. The Diels-Alder polymerizations using a benzocyclobutene functional group (*10*) and producing π-conjugated ladder polymers (*11*) were reported recently.

Diyne Cycloaddition Copolymerization

On the basis of a new conception, that is, the transition metal-catalyzed intermolecular diyne cycloaddition in the presence of an effective cycloaddition component (Z), the author developed recently a transition metal-catalyzed diyne cycloaddition copolymerization, which is schematically expressed in Scheme 1. The diyne, which does not undergo intramolecular cycloaddition efficiently, is intermolecularly connected by the cycloaddition component in the presence of a transition metal catalyst to produce a copolymer containing rings.

One important side reaction of the diyne cycloaddition copolymerization is the diyne cyclotrimerization without participation of the cycloaddition component to form a benzenoid unit containing a pendant carbon-carbon triple bond, which may cause copolymer branching and/or cross-linking to give an insoluble copolymer (*vide ante*) (Scheme 2). Suppression of the side diyne cyclotrimerization is indispensable for the efficient diyne cycloaddition copolymerization.

On the basis of the finding (*12*) that 3,10-tridecadiyne (m = 5, R = Et) gave the bicyclic 2-pyrone only in low yield by the unfavorable intramolecular diyne cycloaddition in comparison to diynes with m = 3, 4 and R = Et, which undergo the intramolecular diyne cycloaddition efficiently (eq 1), the reported nickel(0)-catalyzed monoyne/CO_2 cycloaddition to form a 2-pyrone (*13*) was found to be an effective polymer-forming elementary reaction to produce a poly(2-pyrone) by the diyne/CO_2 copolymerization. Further, a poly(2-pyridone) and a poly(pyridine) were prepared by the transition metal-catalyzed diyne cycloaddition copolymerization on the basis of the nickel(0)-catalyzed monoyne/isocyanate cycloaddition to form a 2-pyridone (*14*) and the cobalt-catalyzed monoyne/nitrile cycloaddition to form a pyridine (*15*) as polymer-forming elementary reactions, respectively. The mechanism of the 2-pyrone (*16*) and 2-pyridone (*17*) formations was proposed to involve nickelacyclopentene intermediate **1**, not nickelacyclopentadiene intermediate **2a** (Scheme 3). In contrast, the pyridine formation was proposed to proceed via cobaltacyclopentadiene intermediate **2b** (*15*).

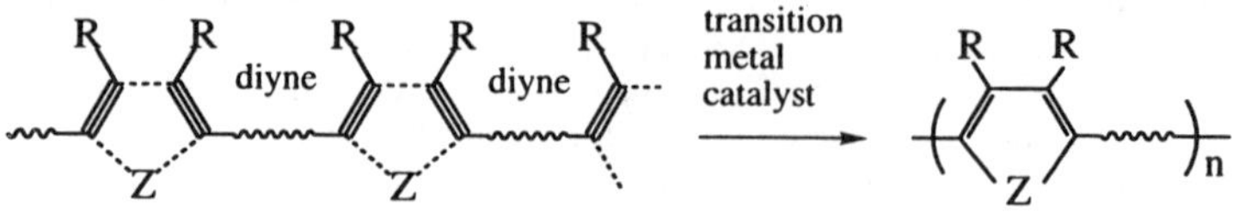

Scheme 1. *Diyne cycloaddition copolymerization*

Z: cycloaddition component

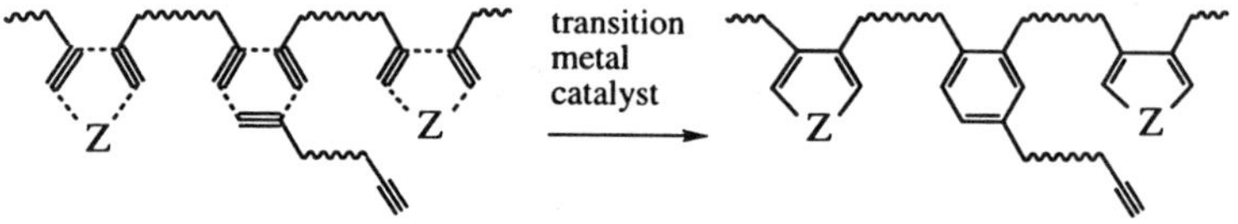

Scheme 2. *Side diyne cyclotrimerization in the diyne cycloaddition copolymerization*

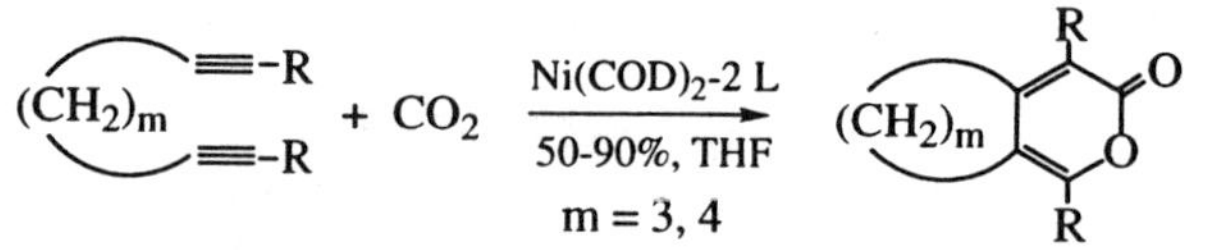

$$(CH_2)_m \begin{matrix} \equiv\text{-R} \\ \equiv\text{-R} \end{matrix} + CO_2 \xrightarrow[\substack{50-90\%,\ THF \\ m = 3, 4}]{Ni(COD)_2\text{-}2\ L} (CH_2)_m \begin{matrix} R \\ \diagdown \\ O \\ \diagup \\ R \end{matrix} \qquad (1)$$

R = H, Me, Et, Pr^i, Bu, Bu^{sec}, SiMe$_3$ L: P(c-C$_6$H$_{11}$)$_3$, P(C$_8$H$_{17}$)$_3$, Bu$_2$PCH$_2$CH$_2$-(2-pyridyl)

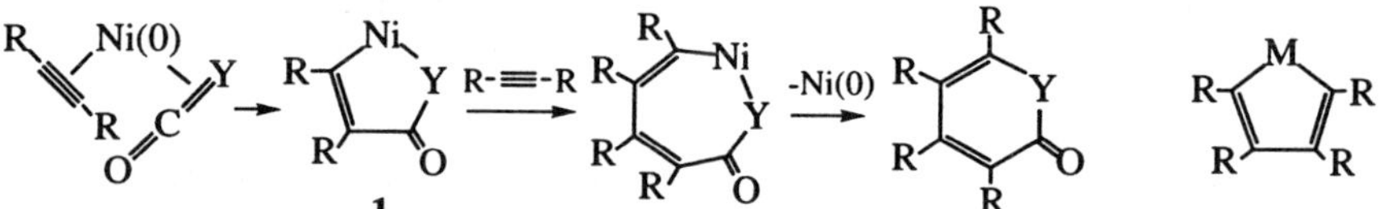

Scheme 3. *Formations of 2-pyrone (Y = O) and 2-pyridone (Y = NR)*

M = Ni (**2a**), Co (**2b**)

Nickel-Catalyzed Diyne/CO_2 Cycloaddition Copolymerization to Form Poly(2-pyrone)s

3,11-Tetradecadiyne (**3a**) underwent a diyne/CO_2 cycloaddition copolymerization in THF/MeCN [1/1 (v/v)] in the presence of a nickel(0) catalyst generated from commercially available bis(1,5-cyclooctadiene)nickel (Ni(COD)$_2$) and 2 equiv of a tri-n-alkylphosphine ligand such as triethyl- or trioctylphosphine to form poly(2-pyrone) **4a**, which was isolated by the dichloromethane/ether combination (eq 2) (*7, 18, 19*). A tricyclohexylphosphine ligand did not produce **4a**.

Comparison of the ^{13}C NMR spectrum of **4a** with those of 2:1 and 3:2 cooligomers **5a,b**, which are model compounds of a repeat unit of **4a**, was a decisive method for the structure determination of **4a**. This method was also very useful for the structural identification of other cycloaddition polymers prepared by the transition metal-catalyzed alkyne cycloaddition polymerizations (*7*). The cooligomers **5a,b** were prepared by shortening the reaction time to 20 min. The isolation of **5a,b** indicates that the copolymerization proceeds via a step-growth mechanism. The cooligomer **5a** consisted of four regioisomeric 2-pyrones. This result indicates that the formation of the 2-pyrone ring of **4a** is not regioselective (*19*). Homopolymerization of **3a** gave a low molecular weight homopolymer, which is a model compound of the diyne cyclotrimerization unit in **4a**.

The ^{13}C NMR C=O and C=C signals of **4a** prepared at 90 °C in THF/MeCN were similar to those of **5a,b**. This result confirms the formation of **4a**. No signal of the C=C group of the diyne homopolymer together with the C≡C group of an end diyne moiety was observed. In contrast, **4a** prepared at 110 °C in THF/MeCN and at 90 °C in THF alone showed the weak C=C signals of the diyne homopolymer. Thus, the lower reaction temperature of 90 °C and the MeCN cosolvent suppressed the side diyne cyclotrimerization reaction.

2,6-Octadiyne (**3b**) (*20*), 1,4-di(2-hexynyl)benzene (**3c**) (*21*), and ether diyne MeC≡C(CH$_2$)$_2$O(CH$_2$)$_4$O(CH$_2$)$_2$C≡CMe (**3d**) (*22*) also formed the corresponding poly(2-pyrone)s **4b-d**. Thermogravimetric analysis (TGA) of **4a,c** showed a rapid weight loss around 420 °C under nitrogen while **4b,d** had the lower thermal stability.

The poly(2-pyrone) formation with C (comonomer)-C (CO_2) bond formation is interesting because the catalytic CO_2 copolymerization is attractive for chemical utilization of CO_2 (*3*) and the zinc-catalyzed epoxide/CO_2 copolymerization with O (comonomer)-C (CO_2) bond formation to form a polycarbonate is the only precedent of the efficient CO_2 copolymerization (*23*).

Nickel-Catalyzed Diyne/Isocyanate Cycloaddition Copolymerization to Form Poly(2-pyridone)s

Aliphatic diynes such as **3a,b** underwent a nickel(0)-catalyzed diyne cycloaddition copolymerization with phenyl (**6a**), octyl (**6b**), and cyclohexyl (**6c**) isocyanates to afford poly(2-pyridone)s **7** with GPC molecular weights of 10,000-40,000 in good yield (eq 3) (*19, 24*). For examples, **7aa** from **3a** and **6a** and **7ab** from **3a** and **6b** were isolated by the dichloromethane/ether and methanol combinations (*7*), respectively. The diyne/isocyanate copolymerization showed different ligand and solvent effects from those of the diyne/CO_2 copolymerization: the tricyclohexylphosphine ligand was effective and the acetonitrile cosolvent retarded the copolymerization. An excess amount of the isocyanate raised the copolymer yield and molecular weight. Ether diynes **3d,e** also copolymerized with **6b** to produce poly(2-pyridone)s with M_n = 11000-18000 (*25*).

1,4-Bis(phenylethynyl)benzene (**3f**), which does not copolymerize with CO_2, copolymerized with equimolar amounts of aryl isocyanates (eq 4) (*19, 26*). This result shows the higher copolymerizability of the isocyanate than CO_2. The equimolar **3f/6b** copolymerization produced a copolymer containing a considerable amount of the diyne cyclotrimerization unit, but a 5-fold excess of **6b** suppressed the diyne cyclotrimerization to afford a 1:1 copolymer, which was isolated by the dichloromethane/methanol combination (*7*). Poly(2-pyridone) **7fa** obtained from **3f** and **6a** had the high thermal stability to show a rapid weight loss around 500 °C in air.

The copolymerization of 1,4-diethynylbenzene (**3g**) with the isocyanate provided an extreme case of the diyne cycloaddition copolymerization due to its enhanced cyclotrimerization reactivity (*27*). A soluble **3g/6d** copolymer with M_n = 3300 was obtained only under the selected condition (**6d/3g** = 5, 5 min, and the increased solvent amount). The **3g/6b** copolymerizations (**6b/3g** = 7, 120 min and 20 h) produced soluble poly(2-pyridone)s **7gb** with high M_w/M_n values of 6.1 and 18, respectively. Progress of a bimodal character of the GPC peak of **7gb** with the reaction time was observed. These results may be explained as follows: the low molecular weight poly(2-pyridone)s containing diyne cyclotrimerization units are formed first and then the higher molecular weight poly(2-pyridone)s having longer poly(2-pyridone) branches with the higher M_w/M_n value are produced. Thus, the M_w/M_n value of a copolymer may be a criterion of the formation of a branched copolymer in the transition metal-catalyzed diyne cycloaddition copolymerization.

The 2-pyrone and 2-pyridone rings undergo a variety of chemical reactions (*28, 29*). The poly(2-pyrone) and the poly(2-pyridone) therefore may be useful as a reactive polymer with a functional group. The poly(2-pyridone)s were obtained by the reaction of the poly(2-pyrone) with amines including ammonia (eq 5) (*30*).

Cobalt-Catalyzed Diyne/Nitrile Cycloaddition Copolymerization to Form Poly(pyridine)s

Poly(pyridine)s were prepared by a 1,11-dodecadiyne/nitrile cycloaddition copolymerization using a cobaltocene catalyst, which is commercially available (eq 6) (*19, 31*). An excess amount of the nitrile suppressed the diyne cyclotrimerization and formed a soluble poly(pyridine). Poly(pyridine)s **9a,c** were isolated by the toluene/hexane combination and **9b** was isolated by the toluene/methanol combination (*7*). This poly(pyridine) formation is the first example of catalytic synthesis of a poly(pyridine) having a pyridine ring in a polymer chain (*3*). The poly(pyridine) formation at 80 °C by the $Ph_4C_4Co(Cp)(PPh_3)$-catalyzed copolymerization of diynes containing a 4-(2-propynyloxy)phenyl group with 4-methoxybenzonitrile was reported (*32*).

Palladium-Catalyzed Diyne/Elemental Sulfur Cycloaddition Copolymerization to Form Poly(thiophene)s

Elemental sulfur S_8 acts as a carbene-type cycloaddition component, which donates an unshared electron pair to the cycloaddition. The thermal cycloaddition of

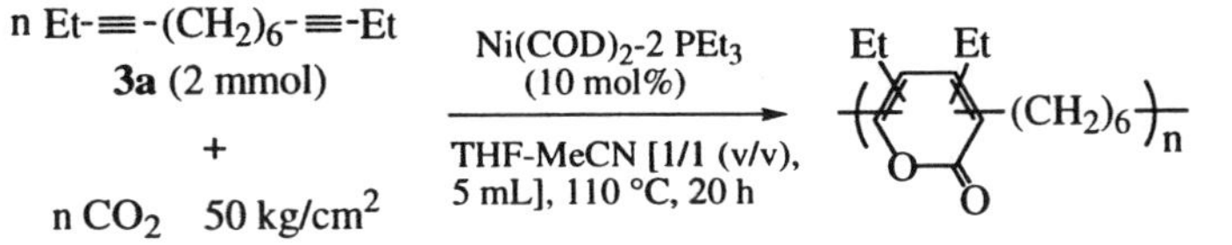

n Et-≡-(CH$_2$)$_6$-≡-Et
3a (2 mmol)

\+

n CO$_2$ 50 kg/cm^2

$\xrightarrow[\text{THF-MeCN [1/1 (v/v),}\\ \text{5 mL], 110 °C, 20 h}]{\text{Ni(COD)}_2\text{-2 PEt}_3 \\ \text{(10 mol\%)}}$

(2)

4a, 86%, M_n = 13300, M_w/M_n = 2.3

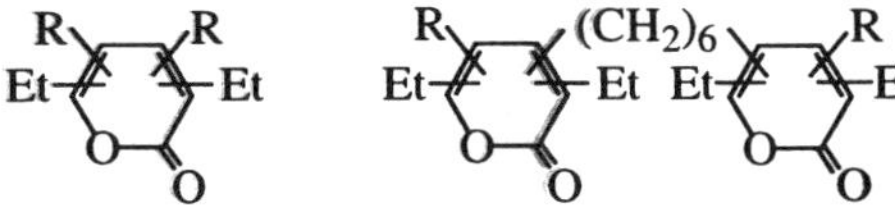

2:1 Cooligomers **5a**   3:2 Cooligomers **5b**   etc. R = (CH$_2$)$_6$-C≡C-Et

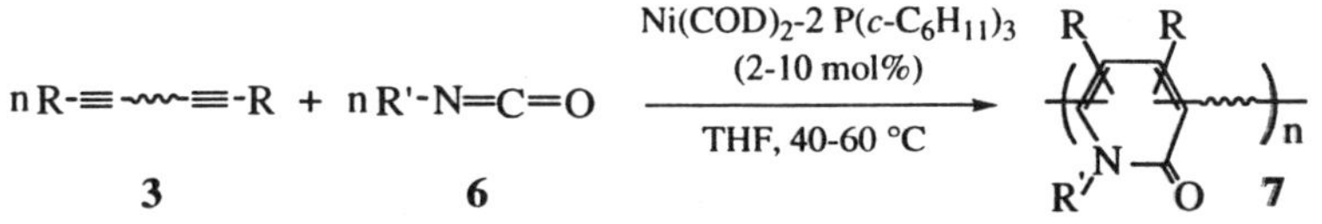

nR-≡-∼∼∼-≡-R + nR'-N=C=O $\xrightarrow[\text{THF, 40-60 °C}]{\text{Ni(COD)}_2\text{-2 P(}c\text{-C}_6\text{H}_{11}\text{)}_3 \\ \text{(2-10 mol\%)}}$ **7**   (3)

3   **6**

R-≡-(CH$_2$)$_m$-≡-R: R = Et, m=6 (**3a**), R = Me, m=2 (**3b**)
R-≡-(CH$_2$)$_2$O(CH$_2$)$_4$O(CH$_2$)$_2$-≡-R: R = Me (**3d**), Et (**3e**)

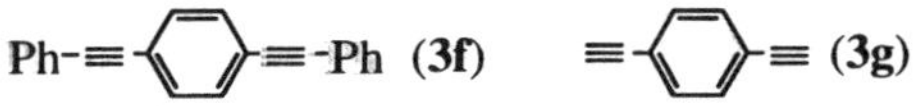

Ph-≡-⟨⟩-≡-Ph (**3f**)   ≡-⟨⟩-≡ (**3g**)

R' = C$_6$H$_5$ (**a**), C$_8$H$_{17}$ (**b**), c-C$_6$H$_{11}$ (**c**)

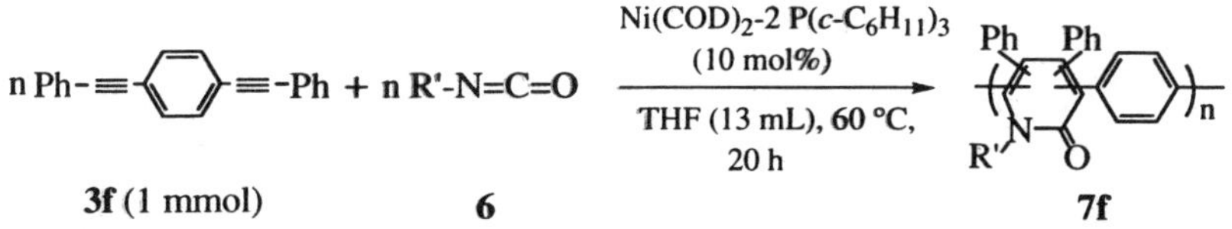

n Ph-≡-⟨⟩-≡-Ph + n R'-N=C=O $\xrightarrow[\text{THF (13 mL), 60 °C,}\\ \text{20 h}]{\text{Ni(COD)}_2\text{-2 P(}c\text{-C}_6\text{H}_{11}\text{)}_3 \\ \text{(10 mol\%)}}$ **7f**   (4)

3f (1 mmol)   **6**

R' = C$_6$H$_5$ (**a**), C$_8$H$_{17}$ (**b**), 4-MeC$_6$H$_4$ (**d**), 3-MeC$_6$H$_4$ (**e**)

7fa, fd, fe (**6a, d, e**/**3f** = 1): 50-70%, M_n = 20000-46000, M_w/M_n = 1.9-4.0
7fb (**6b**/**3f** = 5): 51%, M_n = 12300, M_w/M_n = 1.5

diphenylacetylene (**10a**) with S$_8$ was reported to take place at 200-210 °C in benzene to produce 2,3,4,5-tetraphenylthiophene (**11**) (*33*). Pd(OAc)$_2$ was found to catalyze this cycloaddition (eq 7) (*34*). The Pd(OAc)$_2$-catalyzed **10a**/S$_8$ reactions at 160 and 140 °C formed **11** in good yield while the corresponding thermal reactions gave **11** only in low yield. This result is noteworthy because the reported transition metal-catalyzed monoyne/S$_8$ cycloaddition to form a thiophene is limited to activated monoynes such as dimethyl acetylenedicarboxylate (*35*).

Pd(OAc)$_2$ catalyzed a cycloaddition copolymerization of 1,4-bis(4'-hexylphenylethynyl)benzene (**3i**) with S$_8$ to form poly(thiophene) **12**, which was isolated by the dichloromethane/ether combination (eq 8) (*7, 19, 34*). Two hexyl groups improved the poly(thiophene) solubility. Palladium catalysis was not efficient, but was evident at 160 °C because the thermal copolymerization at 160 °C gave almost no copolymer. Prolonging the reaction time and decreasing the solvent amount raised the copolymer yield and/or molecular weight. The poly(thiophene) **12** showed a λ_{max} value at 355 nm, which is similar to that of terthiophene and indicates that π-conjugation of **12** is not extensive due to the nonregioselective formation of the thiophene ring.

Nickel-Catalyzed Cyclic Diyne/CO$_2$ and Isocyanate Cycloaddition Copolymerizations to Form Ladder Poly(2-pyrone)s and Poly(2-pyridone)s

Use of a structurally unique diyne in the diyne cycloaddition copolymerization affords a structurally unique cycloaddition copolymer. Thus, a cyclic diyne cycloaddition copolymerization produces a ladder polymer. The nickel(0)-catalyzed cycloaddition copolymerization of cyclic diynes **3j-l** with isocyanates **6a-c** afforded soluble ladder poly(2-pyridone)s **7ja-lc**, which were isolated by the dichloromethane/ether combination, with molecular weights of ca. 15000-65000 in high yield (eq 9, Table I) (*7, 36*). The equimolar copolymerizations of **3l**, which has the largest ring size among three cyclic diynes, with **6a-c** gave insoluble copolymers (*37*). A 2-fold excess of **6a-c** suppressed the diyne cyclotrimerization to afford soluble ladder poly(2-pyridone)s **7la-c**. The ladder poly(2-pyridone) **7la** had the thermal stability similar to the poly(2-pyridone) **7aa** to show a rapid weight loss around 420 °C under nitrogen.

Relationship of the copolymer yield and molecular weight with the reaction time was examined in the **3j/6c** copolymerization: comonomers were rapidly and quantitatively converted into cooligomers and low molecular weight copolymers, which then grew to the high molecular weight ladder poly(2-pyridone)s. This result confirms a stepwise polyaddition mechanism.

The **3j**/CO$_2$ copolymerization in THF/MeCN [1/1 (v/v)] formed ladder poly(2-pyrone) **4j** (eq 10) (*18, 38*), which was precipitated during the copolymerization reaction due to its limited solubility and was isolated by the dichloromethane/ether combination (*7*). The poly(2-pyrone) **4j** showed a rapid weight loss around 420 °C under nitrogen. The CO$_2$ copolymerizations of **3k,l**, which have the larger ring sizes than **3j**, gave partly insoluble ladder poly(2-pyrone)s (*37*). Thus, the transition metal-catalyzed cycloaddition copolymerization of a cyclic diyne with a heterocumulene provides a new synthetic method of a ladder polymer (*11*).

$$\left(\begin{array}{c}\text{Et} \quad \text{Et}\\ \text{—}\!\!\left(\!\!\begin{array}{c}\\ \text{O}\end{array}\!\!\right)\!\!\text{—(CH}_2)_6\!\!-\!\!\right)_n + n\,RNH_2 \xrightarrow[-n\,H_2O]{170\text{-}180\ °C,\ 20\ h} \left(\!\!\begin{array}{c}\text{Et}\quad \text{Et}\\ \text{—}\!\!\left(\!\!\begin{array}{c}\\ R'\ \ O\end{array}\!\!\right)\!\!\text{—(CH}_2)_6\!\!-\!\!\right)_n$$ (5)

R = H, Bu, C_8H_{17}, $HOCH_2CH_2$, H_2N-⟨⟩-CH_2CH_2

$$n\ H\text{-}\!\!\equiv\!\!-(CH_2)_8\text{-}\!\!\equiv\!\!-H + n\,RC\!\!\equiv\!\!N \xrightarrow[\text{toluene (5 mL),}\ 150\ °C,\ 20\ h]{Co(C_5H_5)_2\ (10\ mol\%)} \left(\!\!\begin{array}{c}(CH_2)_8\\ N\quad R\end{array}\!\!\right)_n$$ (6)

3h (1 mmol)　　　　**8**　　　　　　　　　　　　　　　　**9**

9a: R = Me (**a**), 52%, $M_n = 15900$, $M_w/M_n = 1.3$ (**8a/3h** = 10)

9b: R = C_8H_{17} (**b**), 44%, $M_n = 3700$, $M_w/M_n = 1.4$ (**8b/3h** = 30)

9c: R = Me-⟨⟩- (**c**), 72%, $M_n = 9500$, $M_w/M_n = 1.7$ (**8c/3h** = 2)

$$2\ Ph\text{-}\!\!\equiv\!\!-Ph + \frac{1}{8}S_8 \xrightarrow[\text{toluene (2 mL),}\ 6\ h]{Pd(OAc)_2\ (20\ mol\%)} \begin{array}{c}Ph\quad Ph\\ Ph\text{-}\!\!\langle S \rangle\!\!\text{-}Ph\end{array}$$ (7)

10a (0.5 mmol)　　　　　　　　　　　　　　　　**11**

71% (160 °C, S/**10a** = 4)
64% (140 °C, S/**10a** = 4)

$$n\ R\text{-}\langle\rangle\!\!\equiv\!\!\langle\rangle\!\!\equiv\!\!\langle\rangle\text{-}R + \frac{n}{8}S_8 \xrightarrow[\text{toluene (4 mL),}\ 20\ h]{Pd(OAc)_2\ (20\ mol\%)} \left(\!\!\begin{array}{c}R\text{-}\langle\rangle\quad \langle\rangle\text{-}R\\ \langle S \rangle\!\!-\!\!\langle\rangle\end{array}\!\!\right)_n$$ (8)

3i (R = C_6H_{13}, 0.2 mmol)　　　　　　　　　　　　**12**

12: 32%, $M_n = 9800$, $M_w/M_n = 1.5$ (S/**3i** = 50, 160 °C)

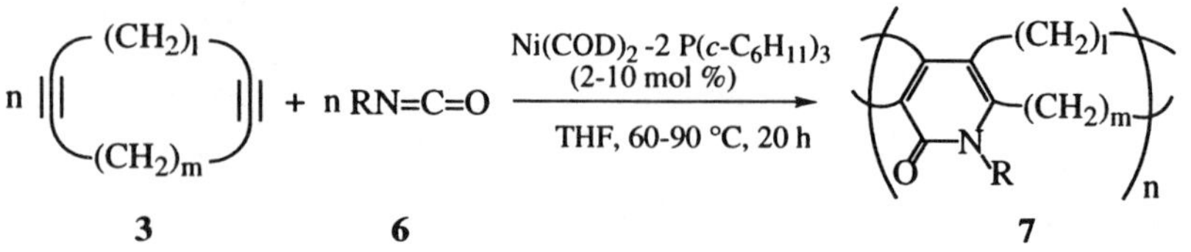

$$n \left(\begin{array}{c} -(CH_2)_l- \\ \| \\ -(CH_2)_m- \end{array} \right) + n\ RN=C=O \xrightarrow[\text{THF, 60-90 °C, 20 h}]{\text{Ni(COD)}_2 \cdot 2\ P(c\text{-}C_6H_{11})_3 \ (2\text{-}10\ \text{mol}\%)} \left(\begin{array}{c} -(CH_2)_l- \\ \text{pyridone ring} \\ -(CH_2)_m- \end{array} \right)_n \quad (9)$$

         3   6   7

l, m = 4, 5 (**j**), 4, 6 (**k**), 5, 6 (**l**)

R = Ph (**a**), C_8H_{17} (**b**), c-C_6H_{11} (**c**)

Table I. Ladder Poly(2-pyridone) Formation (eq 9)[a]

3	6	6/3	7				
				yield (%)	M_n	M_w/M_n	remarks
j:	a	1	**ja**	80	36800	4.6	soluble copolymer
l, m = 4, 5	b	1	**jb**	94	26200	4.9	
	c	1	**jc**	92	17900	3.9	
	c	2	**jc**[b]	99	43600	4.2	
k:	a	1	**ka**				insoluble copolymer
l, m = 4, 6	a	5	**ka**	81	63100	3.5	soluble copolymer
	b	1	**kb**	98	26000	6.9	
	c	1	**kc**	93	16400	6.0	
l:	a-c	1	**la-lc**				insoluble copolymers
l, m = 5, 6	a	2	**la**	90	39800	1.9	soluble copolymer
	b	2	**lb**	89	19100	2.0	
	c	2	**lc**	91	23800	5.4	

[a] **3**, 1 mmol; Ni(COD)$_2$/**3** = 0.10; solvent, 5 mL; temperature, 60 °C.
[b] Ni(COD)$_2$/**3j** = 0.02; solvent, 2.5 mL; temperature, 90 °C.

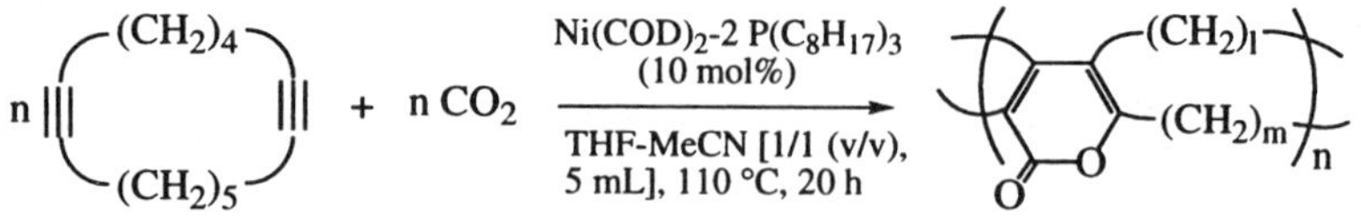

$$n \left(\begin{array}{c} -(CH_2)_4- \\ \| \\ -(CH_2)_5- \end{array} \right) + n\ CO_2 \xrightarrow[\text{THF-MeCN [1/1 (v/v), 5 mL], 110 °C, 20 h}]{\text{Ni(COD)}_2 \cdot 2\ P(C_8H_{17})_3 \ (10\ \text{mol}\%)} \left(\begin{array}{c} -(CH_2)_l- \\ \text{pyranone ring} \\ -(CH_2)_m- \end{array} \right)_n \quad (10)$$

3j (1 mmol)   50 kg/cm^2 **4j** (l, m = 4, 5) :
                                                               100%, M_n = 7500, M_w/M_n = 2.0

Diyne/Transition Metal Complex Cycloaddition Copolymerization to Form Transition Metal-Containing Polymers

A new synthetic method of a transition metal-containing polymer was developed by using the principle of the transition metal-catalyzed diyne cycloaddition copolymerization. A CpCo(PPh$_3$)$_2$ complex as a cycloaddition component connected diyne **3m** intermolecularly by oxidative addition without a transition metal catalyst to form cobaltacyclopentadiene-containing polymer **13** (eq 11). The insoluble copolymer **13a** exhibited a λ_{max} value at 480 nm and was proposed to have a conjugated structure (q/p = 0) formed by 2,5-regioselectivity (*39*). The soluble 4,4'-bis(phenylethynyl)-biphenyl/CpCo(PPh$_3$)$_2$ copolymer **13b**, which is stable in air and showed a λ_{max} value at 312 nm, contained a side (η^4-cyclobutadiene)cobalt moiety generated by thermal rearrangement of the cobaltacyclopentadiene repeat unit during the copolymerization (*19, 40*). Among various diynes, 1,10-diphenyl-1,9-decadiyne and 1,4-C$_8$H$_{17}$C≡CCO-C$_6$H$_4$COC≡CC$_8$H$_{17}$ formed the corresponding cobaltacyclopentadiene repeat units chemoselectively with favored 2,5-regioselectivity. The polymer **13b** underwent various chemical reactions characteristic of a cobaltacyclopentadiene ring such as the thermal rearrangement to a (η^4-cyclobutadiene)cobalt moiety (*41*) and the conversion to 2-pyridone and pyridine rings by the reactions with an isocyanate (*42*) and a nitrile (*43*), respectively.

A Cp$_2$Zr complex generated *in situ* from Cp$_2$ZrCl$_2$ and BuLi in THF reacted regiospecifically with silyl diyne **3n** to form 2,5-zirconacyclopentadiene-containing polymer **14** (eq 12) (*44*). Macrocyclic Cp$_2$Zr/diyne trimer **15** was obtained quantitatively by heating **14** in refluxing THF. The formation of **15** is due to the depolymerization of **14** by reversible Cp$_2$Zr/alkyne cycloaddition. In contrast, the reaction of 1,4-Me$_3$SiC≡CC$_6$H$_4$C$_6$H$_4$C≡CSiMe$_3$ with Cp$_2$Zr formed a mixture of macrocyclic Cp$_2$Zr/diyne trimer and tetramer. The latter was converted quantitatively into the thermodynamically stable trimer in refluxing THF (*45*). The cyclic diyne/Cp$_2$Zr reaction produced a polymer consisting of macrocyclic rings connected by zirconacyclopentadiene rings (*46*).

Monoyne and Diyne Double-Cycloaddition Copolymerizations

Nickel-Catalyzed Monoyne/Maleimide Double-Cycloaddition to Form Bicyclo[2.2.2]oct-7-enes

It is interesting to prepare a cycloaddition copolymer having a carbon backbone by a diyne/alkene cycloaddition copolymerization. Examples of an efficient transition metal-catalyzed intermolecular monoyne/alkene cycloaddition, which is a polymer-forming elementary reaction of the diyne/alkene cycloaddition copolymerization, however, are very limited (*2*). The interesting, but unsatisfactory monoyne/maleimide cycloaddition catalyzed by Ni(CO)$_2$(PPh$_3$)$_2$ (*47*) in benzene to form derivatives of a 1,3-cyclohexadiene and/or a bicyclo[2.2.2]oct-7-ene was reported about twenty-five years ago (*48*). If the efficient bicyclo[2.2.2]oct-7-ene synthesis by the monoyne/maleimide double-cycloaddition is available, it is possible to develop a diyne/maleimide double-cycloaddition copolymerization to form poly(bicyclo[2.2.2]oct-7-ene) **17** (Scheme 4).

An efficient chemo- and stereoselective nickel(0)-catalyzed monoyne/maleimide double-cycloaddition to form bicyclo[2.2.2]oct-7-ene **18** was found by using highly soluble *N*-octylmaleimide (**16b**) along with *N*-phenyl- and *N*-ethylmaleimides (**16a, c**), an excess amount of the maleimide, and a THF solvent (eq 13) (*49*). A variety of bicyclo[2.2.2]oct-7-enes with an *exo,exo*-stereochemistry were prepared in high to excellent yields. Two examples are shown in eqs 14 and 15. The reaction using **16b/10b** = 2 formed **18bb** chemo-, regio- and stereoselectively in excellent yield while the equimolar reaction gave **18bb** only in 53% yield. In this equimolar reaction, benzene derivative **19** formed by dehydrogenation of the 1,3-cyclohexadiene intermediate was detected in 8% yield together with the formation of **18bb** in 53% yield, which indicates that an excess amount of the maleimide suppressed the side dehydrogenation reaction and promoted the bicyclo[2.2.2]oct-7-ene formation. The regioselective formation of **18bb** containing a 1,8-diphenyl-substituted bicyclo[2.2.2]oct-7-ene ring suggests that the reaction proceeds via initial monoyne/alkene coupling to generate a nickelacyclopentene intermediate, not via initial monoyne/monoyne coupling to give a nickelacyclopentadiene intermediate (Scheme 3, *1, 49*). The reaction using **16b/10c** = 2 produced two regioisomeric bicyclo[2.2.2]oct-7-enes **18cbA** and **18cbB** chemoselectively in excellent yield. Sterically bulky tri-*o*-tolylphosphine favored the formation of **18cbA** and the reaction without the ligand formed **18cbB** regioselectively. X-ray structural analysis of **18dc** confirmed an *exo,exo*-stereochemistry.

The present bicyclo[2.2.2]oct-7-ene synthesis (eq 13) is the first example of an efficient transition metal-catalyzed intermolecular monoyne/alkene double-cycloaddition reaction involving a 1,3-cyclohexadiene intermediate and was found to be an excellent polymer-forming elementary reaction to provide the following three types of alkyne double-cycloaddition copolymerizations, which demonstrates the usefulness of the strategy of mutual feedback of organic synthesis and polymer synthesis (*vide ante*).

Nickel-Catalyzed Diyne/Maleimide Double-Cycloaddition Copolymerization to Form Poly(bicyclo[2.2.2]oct-7-ene)s

1,4-Diethynylbenzene (**3g**) and 1,11-dodecadiyne (**3h**) copolymerized with *N*-octylmaleimide (**16b**) in the presence of the $Ni(CO)_2(PPh_3)_2$ or $Ni(COD)_2/PPh_3$ catalyst to afford soluble poly(bicyclo[2.2.2]oct-7-ene)s **17gb,hb**, which were isolated by the dichloromethane/ethyl acetate and dichloromethane/hexane combinations, respectively (eqs 16, 17) (*7, 19, 49, 50*), but copolymerizations of **3g** with *N*-phenyl- and *N*-ethylmaleimides (**16a, c**) gave insoluble copolymers. $Ni(COD)_2$ alone gave an insoluble **3g/16b** copolymer, which indicates that the PPh_3 ligand suppressed the diyne cyclotrimerization. T_{10} (the temperature causing a 10% weight loss by TGA) of **17gb** was 322 °C. In contrast to the maleimide, maleic anhydride and ester gave only low molecular weight products. The poly(bicyclo[2.2.2]oct-7-ene)s **17gb,hb** containing two octyl groups per repeat unit have unique rigid and flexible fence-like structures **20**,

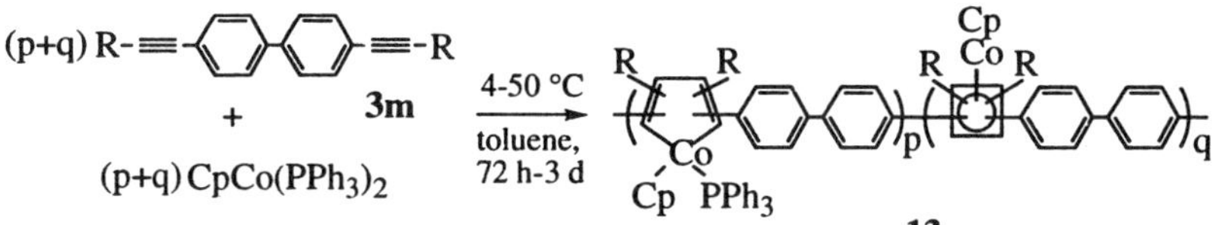

$(p+q) R-\equiv-\langle\bigcirc\rangle-\langle\bigcirc\rangle-\equiv-R$
+ **3m** $\xrightarrow[\text{toluene,}]{\text{4-50 °C}}$ **13** (11)
$(p+q) CpCo(PPh_3)_2$ 72 h-3 d

13a (R = H): 61%
13b (R = Ph): 94%, $M_n = 9600$, $M_w/M_n = 2.1$, p/q = 84/16

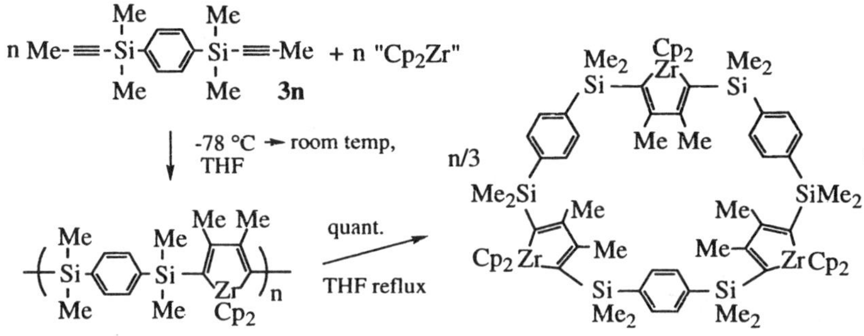

n Me-≡-Si-⟨◯⟩-Si-≡-Me + n "Cp$_2$Zr" → **15** (12)

3n

↓ -78 °C → room temp, THF

14: 90%, $M_n = 4600$, $M_w/M_n = 2.8$ →(n/3, THF reflux)→ **15**

2 R-≡-〜-≡-R + **16** $\xrightarrow{Ni(0)}$

3

→ → **17**

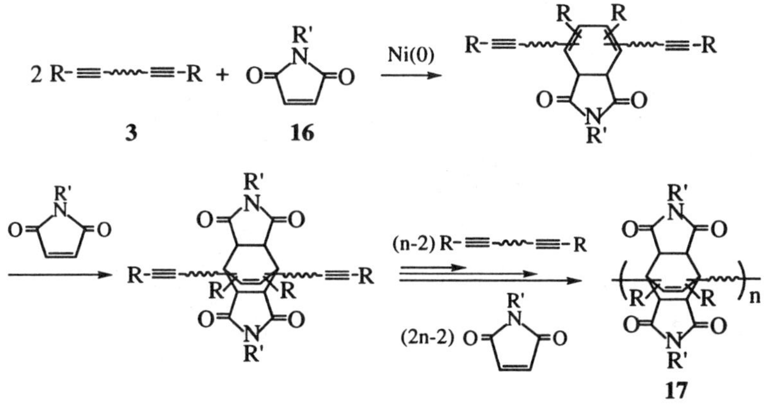

〜 = $(CH_2)_6$, R = Et (**3a**); 〜 = 1,4-C_6H_4, R = H (**3g**);
〜 = $(CH_2)_8$, R = H (**3h**)
R' = C_6H_5 (**a**), C_8H_{17} (**b**), Et (**c**)

Scheme 4. Diyne/maleimide double-cycloaddition copolymerization

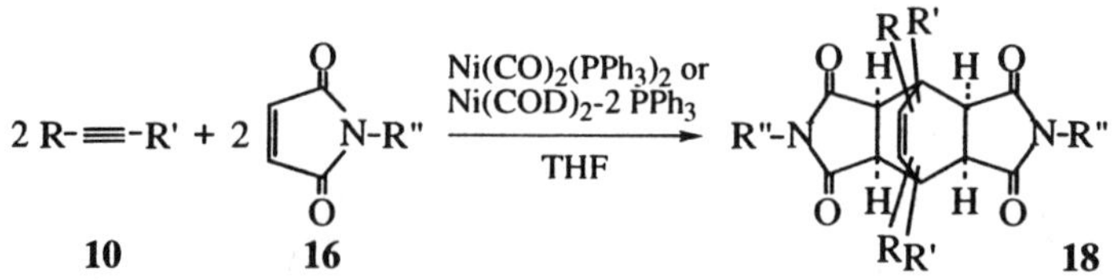

$$2\text{ R}\!\!-\!\!\equiv\!\!-\text{R}' + 2 \underset{16}{\underset{10}{\overset{O}{\underset{O}{\bigcup}}}}\text{N-R"} \xrightarrow[\text{THF}]{\text{Ni(CO)}_2(\text{PPh}_3)_2 \text{ or} \atop \text{Ni(COD)}_2\text{-2 PPh}_3} \underset{18}{\text{R"-N}\cdots\text{N-R"}} \quad (13)$$

R = Ph, R' = H (**b**); R = Pr, R' = H (**c**); R = R' = Et (**d**)
R" = C_6H_5 (**a**), C_8H_{17} (**b**), Et (**c**)

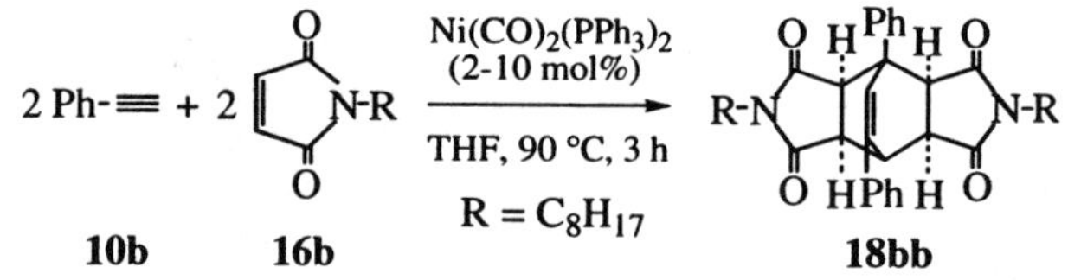

$$2\text{ Ph-}\!\!\equiv + 2 \underset{\mathbf{16b}}{\overset{O}{\underset{O}{\bigcup}}}\text{N-R} \xrightarrow[\text{THF, 90 °C, 3 h} \atop R = C_8H_{17}]{\text{Ni(CO)}_2(\text{PPh}_3)_2 \atop (2\text{-}10\text{ mol\%})} \underset{\mathbf{18bb} \atop 88\text{-}97\%\ (53\%)}{\text{R-N}\cdots\text{N-R}} \quad (14)$$

10b   **16b**
16b/10b = 2 (**16b/10b** = 1)

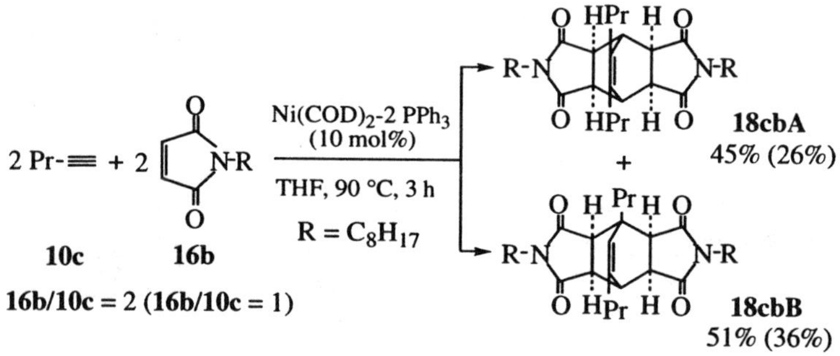

$$2\text{ Pr-}\!\!\equiv + 2 \underset{\mathbf{16b}}{\overset{O}{\underset{O}{\bigcup}}}\text{N-R} \xrightarrow[\text{THF, 90 °C, 3 h} \atop R = C_8H_{17}]{\text{Ni(COD)}_2\text{-2 PPh}_3 \atop (10\text{ mol\%})} \begin{array}{l} \text{R-N}\cdots\text{N-R} \quad \mathbf{18cbA} \\ \quad + \quad 45\%\ (26\%) \\ \text{R-N}\cdots\text{N-R} \\ \quad \mathbf{18cbB} \\ \quad 51\%\ (36\%) \end{array} \quad (15)$$

10c   **16b**
16b/10c = 2 (**16b/10c** = 1)

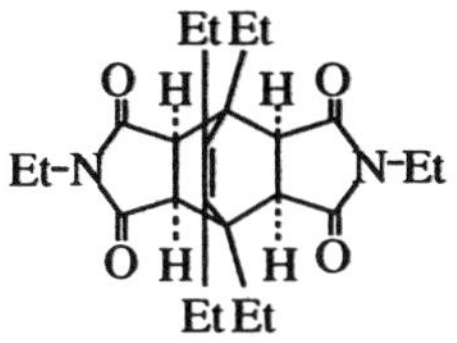

18dc

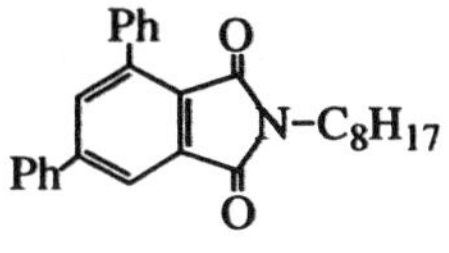

19

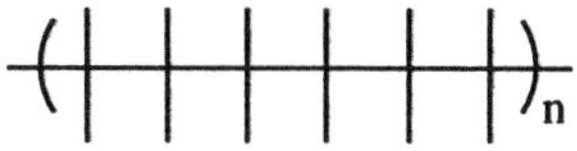

fence-like polymer **20**

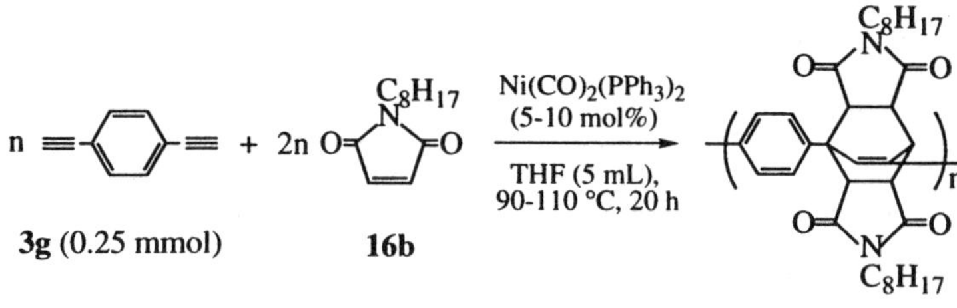

$$n \equiv\!\!-\!\!\bigcirc\!\!-\!\!\equiv\ +\ 2n\ \text{16b} \xrightarrow[\text{THF (5 mL)},\ 90\text{-}110\ °C,\ 20\ h]{\text{Ni(CO)}_2(\text{PPh}_3)_2\ (5\text{-}10\ \text{mol\%})} \text{17gb} \quad (16)$$

3g (0.25 mmol)   **16b**   **17gb**

17gb: 51%, $M_n = 27200$, $M_w/M_n = 2.1$ (**16b/3g** = 2);
50%, $M_n = 26800$, $M_w/M_n = 3.5$ (**16b/3g** = 2, Ni/**3g** = 0.05);
50%, $M_n = 19800$, $M_w/M_n = 1.9$ (**16b/3g** = 4);
60%, $M_n = 34600$, $M_w/M_n = 1.9$ (**16b/3g** = 2, 110 °C)

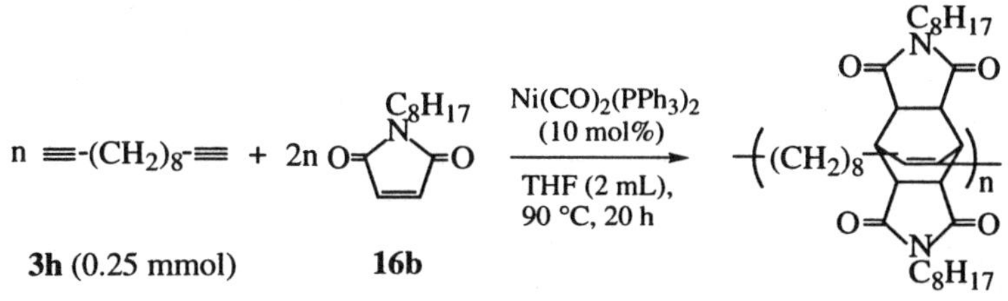

$$n \equiv\!\!-(CH_2)_8\!\!\equiv\ +\ 2n\ \text{16b} \xrightarrow[\text{THF (2 mL)},\ 90\ °C,\ 20\ h]{\text{Ni(CO)}_2(\text{PPh}_3)_2\ (10\ \text{mol\%})} \text{17hb} \quad (17)$$

3h (0.25 mmol)   **16b**   **17hb**

17hb: 70%, $M_n = 12400$, $M_w/M_n = 2.4$ (**16b/3h** = 2);
96%, $M_n = 9300$, $M_w/M_n = 1.8$ (**16b/3h** = 4)

respectively, and are potentially interesting, because structurally related comb-like polymers have attracted considerable attention on account of their specific properties (*51*).

The ^{13}C NMR C=O and C=C signals of **17hb** prepared using **16b/3h** = 2 and 4 were identical and were a superposition of those of two model compounds **18cbA** and **18cbB**. These results confirm the copolymer structure and also indicate that the monomer feed ratio did not influence the copolymer structure. No ^{13}C NMR signal of the C≡C group of an end diyne moiety was observed. In contrast to the bicyclo-[2.2.2]oct-7-ene formation reaction (eq 13), in the diyne/maleimide double-cycloaddition copolymerization, there was no remarkable influence of the monomer feed ratio upon the copolymer yield and molecular weight together with the copolymer structure. This finding indicates that the side aromatization reaction of the intermediate 1,3-cyclohexadiene moiety such as the formation of **19** is indifferent to copolymer growth and is not significant in the poly(bicyclo[2.2.2]oct-7-ene) formation. These results may be explained by the sterically suppressed dehydrogenation of the internal 1,3-cyclohexadiene unit in the copolymer.

Nickel-Catalyzed Monoyne/Dimaleimide Double-Cycloaddition Copolymerization to Form Poly(imide)s

Further application of the bicyclo[2.2.2]oct-7-ene synthesis (eq 13) to the polymer synthesis is a nickel(0)-catalyzed monoyne/dimaleimide double-cycloaddition copolymerization to form a poly(imide) (eq 18, Table II) (*19, 52*). This copolymerization provides a new synthetic method of a poly(imide) and affords a variety of new poly(imide)s with complex structures by various combinations of the monoyne and the dimaleimide.

The introduction of relatively long alkyl substituents such as octyl and hexyl groups into the monoyne or a relatively long alkylene chain such as a hexylene group into the dimaleimide was necessary for the preparation of a soluble poly(imide): the **10b/21a,d** and **10c,d/21a** copolymerizations produced totally and partly insoluble copolymers, respectively. In contrast to 1-decyne (**10e**), ethoxyacetylene (**10g**) copolymerized with **21b** to produce **22gb** containing a 1,8-diethoxy-substituted bicyclo[2.2.2]oct-7-ene ring. Thus, the presence of an oxygen functional group in the monoyne plays an important role in the regioselective bicyclo[2.2.2]oct-7-ene ring formation. T_{10} was 341, 362, and 343 °C for **22ea**, **22ed**, and **22ee**, respectively.

Nickel-Catalyzed Diyne/Dimaleimide Double-Cycloaddition Copolymerization to Form a Poly(imide) with a Pendant Carbocycle

1,7-Octadiyne (**3p**) acted as two monoyne molecules to produce soluble poly(imide) **23** containing a pendant cyclohexene ring formed by the intramolecular diyne cyclization (eq 19) (*52*). The poly(imide) **23** was isolated by the dichloromethane/ether combination (*7*).

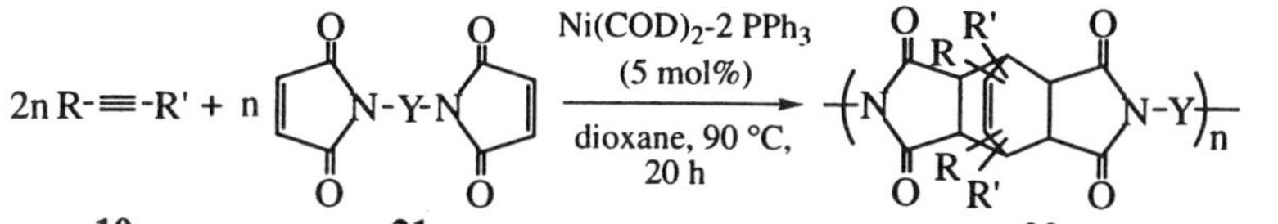

$$2n\ R\text{-}{\equiv}\text{-}R' + n\ \underset{\mathbf{21}}{\overset{\mathbf{10}}{\text{(maleimide)}}} \xrightarrow[\substack{\text{dioxane, 90 °C,}\\ \text{20 h}}]{\substack{\text{Ni(COD)}_2\text{-2 PPh}_3\\ (5\ \text{mol\%})}} \mathbf{22} \quad (18)$$

R = Ph, R' = H (**b**); R = C_8H_{17}, R' = H (**e**); R = R' = C_6H_{13} (**f**); R = EtO, R' = H (**g**)

Y = 1,1',4,4'-$C_6H_4CH_2C_6H_4$ (**a**), $(CH_2)_6$ (**b**), 1,1',4,4'-$C_6H_4OC_6H_4$ (**c**), 1,4-C_6H_4 (**d**), 1,3-C_6H_4 (**e**)

Table II. Poly(imide) Formation (eq 18)[a]

10	21	22	Yield (%)	M_n	M_w/M_n	10	21	22	Yield (%)	M_n	M_w/M_n
e	a	ea	68	8500	1.9	e	c	ec	92	7400	2.3
f	a	fa	49[b]	12100	2.2	e	d	ed	71	8800	1.7
b	b	bb	33	4500	1.9	f	d	fd	23[b]	7700	1.8
e	b	eb	53	15000	1.4	e	e	ee	91	8300	1.9
g	b	gb	41	5000	1.7						

[a] **10** = 1 mmol; **10/21** = 2; solvent, 10 ml.
[b] Temperature, 150 °C.

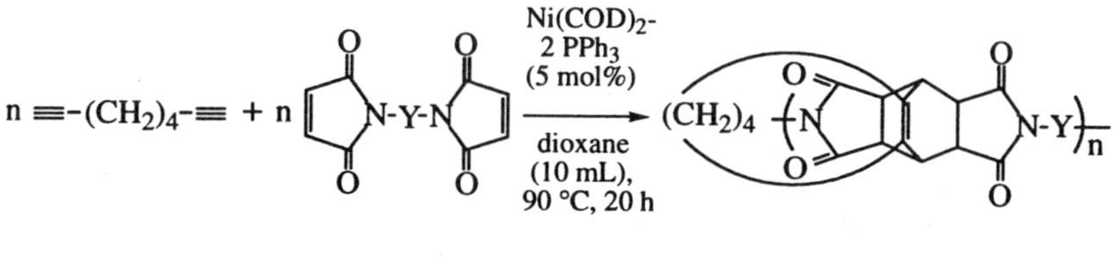

$$n\ {\equiv}\text{-}(CH_2)_4\text{-}{\equiv} + n\ \underset{\mathbf{21b}\ (0.5\ \text{mmol})}{\text{(maleimide-Y)}} \xrightarrow[\substack{\text{dioxane}\\ (10\ \text{mL}),\\ 90\ °C,\ 20\ h}]{\substack{\text{Ni(COD)}_2\text{-}\\ 2\ \text{PPh}_3\\ (5\ \text{mol\%})}} \mathbf{23} \quad (19)$$

3p (0.5 mmol)   **21b** (0.5 mmol)   **23** (Y = $(CH_2)_6$): 69%, M_n = 4400, M_w/M_n = 1.7

Diyne Cycloaddition Terpolymerization

Cobalt-Catalyzed Diyne/Norbornadiene/CO Cycloaddition Terpolymerization to Form Poly(enone)s

The development of a new and efficient catalyst for the known transition metal-catalyzed alkyne cycloaddition reaction generates a new alkyne cycloaddition polymerization. The Pauson-Khand alkyne/alkene/CO cycloaddition to form an enone usually employs a stoichiometric amount of a cobalt complex. Co(indenyl)(COD) was recently reported to catalyze efficiently the intermolecular Pauson-Khand reaction (53). It effected a diyne/norbornadiene/CO cycloaddition terpolymerization to form poly(enone) **25**, which was isolated by the dichloromethane/ether combination (eq 20) (7, 54). The monomer feed ratio influenced the terpolymerization, which suggests that some side reactions consuming the diyne took place during the terpolymerization. T_{10} of **25h** was 395 °C. 1,4-Diethynylbenzene (**3g**) gave an insoluble poly(enone).

This diyne cycloaddition terpolymerization involves two cycloaddition components and is a new variation of the diyne cycloaddition copolymerization. Considering the previously reported poly(ketone) synthesis by the palladium-catalyzed CO/alkene polyaddition reaction (55) and poly(amide) or poly(ester) synthesis (56, 57) from CO, dihalides, and diamines or diols by the palladium-catalyzed polycondensation reaction, the Co(indenyl)(COD)-catalyzed poly(enone) formation provides a new type of transition metal-catalyzed CO incorporation into a polymer.

Summary and Conclusions

The three types of transition-metal catalyzed alkyne cycloaddition polymerizations, i.e., 1) the diyne cycloaddition copolymerization, 2) the alkyne (monoyne and diyne) double-cycloaddition copolymerization, and 3) the diyne cycloaddition terpolymerization, were developed as a new method of polymer synthesis and a variety of cycloaddition polymers were prepared by using various combinations of alkynes (monoynes and diynes) and cycloaddition components such as heterocumulenes (CO_2 and the isocyanate), unsaturated compounds (the nitrile, the maleimide, and the dimaleimide), and carbene-type compounds (elemental sulfur and CO). Among these cycloaddition polymers, the poly(2-pyrone) including the ladder poly(2-pyrone), the poly(2-pyridone) including the ladder poly(2-pyridone), the poly(bicyclo[2.2.2]oct-7-ene), and the poly(enone) are new polymers, which cannot easily be prepared by other existing methods of polymer synthesis. The poly(pyridine), the poly(thiophene), and the poly(imide) prepared here have new structures. These polymers may be interesting as a conjugated polymer, a reactive polymer with a functional group, a molecule-recognizing polymer, a fence-like polymer, a rigid polymer, a ladder polymer, etc. In addition, the new synthetic method of the transition metal-containing polymer was developed by using the principle of the transition-metal catalyzed diyne cycloaddition copolymerization. These results indicate the effectiveness of the transition-metal catalyzed alkyne cycloaddition polymerization as a new method of polymer synthesis.

A remarkable feature of the transition metal-catalyzed alkyne cycloaddition polymerization is the one-step construction of a polymer repeat unit with a complex structure by the cycloaddition reaction connecting three molecules. Another feature in the diyne cycloaddition copolymerization is the use of an excess amount of the cycloaddition component, i.e., a comonomer, which promotes the copolymerization and suppresses the

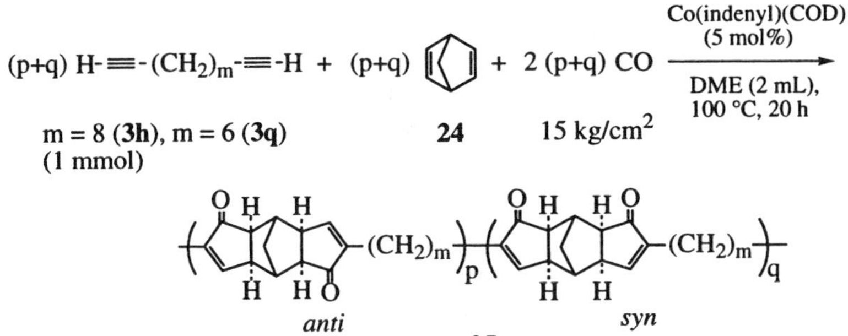

$$(p+q)\ H\text{-}\equiv\text{-}(CH_2)_m\text{-}\equiv\text{-}H\ +\ (p+q)\ \text{[norbornadiene]}\ +\ 2(p+q)\ CO\ \xrightarrow[\text{DME (2 mL), 100 °C, 20 h}]{\text{Co(indenyl)(COD) (5 mol\%)}}$$

m = 8 (**3h**), m = 6 (**3q**)   **24**   15 kg/cm^2
(1 mmol)

[anti and syn structures of **25**]   (20)

25h (m = 8): 54%, M_n = 6700, M_w/M_n = 2.9 (**24**/**3h** = 1);

43%, M_n = 9000, M_w/M_n = 2.0, p/q = 1.8 (**24**/**3h** = 0.8)

25q (m = 6): 30%, M_n = 5700, M_w/M_n = 1.5 (**24**/**3q** = 0.8)

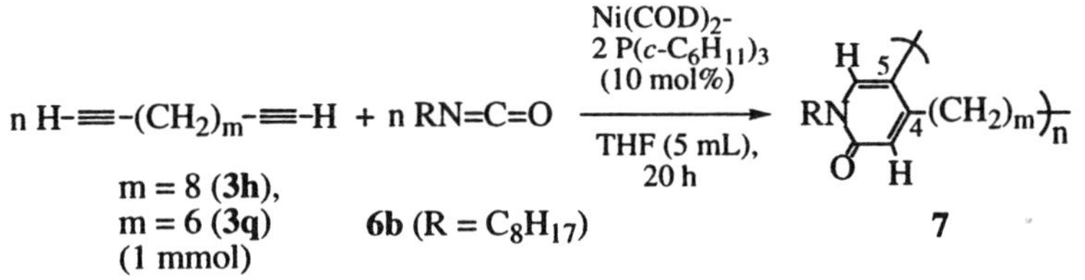

$$n\ H\text{-}\equiv\text{-}(CH_2)_m\text{-}\equiv\text{-}H\ +\ n\ RN=C=O\ \xrightarrow[\text{THF (5 mL), 20 h}]{\text{Ni(COD)}_2\text{-}2\text{P}(c\text{-}C_6H_{11})_3\ (10\text{ mol\%})}\ \text{[pyridone polymer 7]}$$

m = 8 (**3h**),
m = 6 (**3q**)   **6b** (R = C$_8$H$_{17}$)   **7**   (21)
(1 mmol)

7hb (m = 8): 67%, M_n = 9200, M_w/M_n = 2.0 (**6b**/**3h** = 5, 60 °C)

7qb (m = 6): 58%, M_n = 8500, M_w/M_n = 2.1 (**6b**/**3q** = 5, 90 °C)

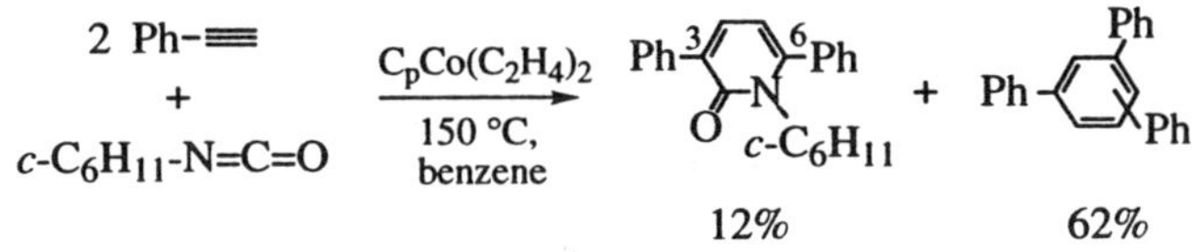

$$2\ Ph\text{-}\equiv\ +\ c\text{-}C_6H_{11}\text{-}N=C=O\ \xrightarrow[\text{benzene}]{Cp Co(C_2H_4)_2,\ 150\ °C}\ \text{[2-pyridone]}\ +\ \text{[benzene derivative]}$$

(22)

12%   62%

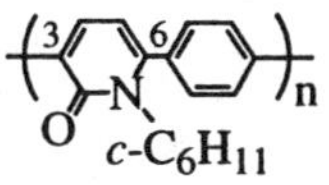

rigid-rod conjugated
poly(2-pyridone) **26**

side diyne cyclotrimerization reaction without its homopolymerization. The reaction temperature and solvent along with the ligand also suppresses the side diyne cyclotrimerization.

The transition metal-catalyzed alkyne cycloaddition polymerization demonstrated the usefulness of the transition metal-catalyzed monoyne cycloaddition reaction as a polymer-forming elementary reaction. The exploitation of a new transition metal-catalyzed monoyne cycloaddition reaction along with the improvement of the catalysis efficiency and regiocontrol of the reported monoyne cycloaddition reaction therefore is expected to bring new aspects of the transition-metal catalyzed alkyne cycloaddition polymerization.

The regiocontrol of the transition metal-catalyzed monoyne cycloaddition has not well been studied and may be approached by changing a transition metal, a monoyne structure, and a ligand. In contrast to the nickel(0)-catalyzed poly(2-pyridone) synthesis using the internal diynes (eqs 3, 4), the nickel(0)-catalyzed poly(2-pyridone) synthesis with the regioselective 4,5-substituted 2-pyridone ring formation was found very recently by using terminal diynes **3h,q** (eq 21) (*58*). The observed regiochemistry is favorable to the Diels-Alder reaction at 3,6-positions (*29*). On the other hand, a cobalt catalyst effects the regioselective 3,6-substituted 2-pyridone ring formation (eq 22) (*59*), which may permit the synthesis of rigid-rod conjugated poly(2-pyridone) **26**. These results indicate that the regiocontrol of the ring formation is an important problem in the transition-metal catalyzed alkyne cycloaddition polymerization. The regiocontrol using a sterically well-designed ligand may be one promising approach.

References

1. Grotjahn, D. B. In *Comprehensive Organometallic Chemistry II*; Abel, E. W.; Stone, F. G. A.; Wilkinson, G.; Hegedus, L. S., Eds.; Pergamon: Oxford, 1995; Vol. 12, pp 741-770.
2. Lautens, M.; Klute, W.; Tam, W. *Chem. Rev.* **1996**, *96*, 49-92.
3. Tsuda, T. *Gazz. Chim. Ital.* **1995**, *125*, 101-110.
4. Tsuda, T. In *The Polymeric Materials Encyclopedia*; Salamone, J. C., Ed.; CRC: Boca Raton, FL, 1996; Vol. 3, pp 1905-1915.
5. Tsuda, T. In *The Polymeric Materials Encyclopedia*; Salamone, J. C., Ed.; CRC: Boca Raton, FL, 1996; Vol. 5, pp 3525-3530.
6. Tsuda, T. *Recent Research Developments in Macromolecules Research* **1997**, *2*, 23-32.
7. Detailed reaction conditions of representative polymer syntheses are indicated in the equations and tables in the text and representative experimental details for the polymerizations discussed in this paper can be found in references (*20*), (*26*), (*31*), and (*49*). General experimental procedures of the transition metal-catalyzed alkyne cycloaddition polymerization are as follows. The polymerization was carried out under nitrogen. In a 50 mL stainless steel autoclave, a reaction solvent, a catalyst solution, a ligand when necessary, a cycloaddition component, and a monoyne or diyne were placed in this order under magnetic stirring at ambient temperature. In the diyne/CO_2 copolymerization, CO_2 was added finally. In the Pd(OAc)$_2$-catalyzed diyne/S$_8$ copolymerization or the Ni(CO)$_2$(PPh$_3$)$_2$-catalyzed diyne/maleimide double-cycloaddition

copolymerization, the catalyst and the cycloaddition component were placed first. In the Ni(COD)$_2$/PPh$_3$-catalyzed monoyne/dimaleimide double-cycloaddition copolymerization, the dimaleimide, PPh$_3$, and Ni(COD)$_2$ were placed first. In the cobalt-catalyzed diyne/norbornadiene/CO terpolymerization, the solvent, the diyne, norbornadiene, the catalyst solution, and CO were added in this order. The reaction mixture was heated at an appropriate temperature for 20 h. After the reaction mixture was cooled by ice water, it was concentrated under vacuum. Addition of a polymer-insolubilizing solvent precipitated a cycloaddition polymer, which was isolated by dissolution in a small amount of a polymer-dissolving solvent and subsequent precipitation by addition of a polymer-insolubilizing solvent. The combination of the polymer-dissolving and polymer-insolubilizing solvents is indicated in the text. The cycloaddition polymer was identified by IR, ^{1}H, and ^{13}C NMR spectroscopy. The structure of the cycloaddition polymer was further confirmed by comparison of the ^{13}C NMR spectrum of the cycloaddition polymer with that of a cooligomer, a heterocyclic compound, or a carbocyclic compound, which is a model compound of a repeat unit of the cycloaddition polymer. A molecular weight of the cycloaddition polymer except for the poly(pyridine) was determined by GPC in chloroform with polystyrene standards. The molecular weight of the poly(pyridine) was determined by GPC in THF with polystyrene standards.

8. Chalk, A. J.; Gilbert, A. R. *J. Polym. Sci. A-1* **1972**, *10*, 2033-2043.
9. Korshak, V. V.; Volpin, M. E.; Sergeev, A. A.; Shitikov, V. K.; Kolomnikov, I. S. U.S. Patent 3,705,131, 1972.
10. Kirchhoff, R. A.; Bruza, K. J. In *The Polymeric Materials Encyclopedia*; Salamone, J. C., Ed.; CRC: Boca Raton, FL, 1996; Vol. 1, pp 474-484.
11. Schlüter, A-D.; Löffler, M.; Schlicke, B.; Schirmer, H. In *The Polymeric Materials Encyclopedia*; Salamone, J. C., Ed.; CRC: Boca Raton, FL, 1996; Vol. 5, pp 3538-3543.
12. Tsuda, T.; Morikawa, S.; Sumiya, R.: Saegusa, T. *J. Org. Chem.* **1988**, *53*, 3140-3145.
13. Walther, D.; Schönberg, H.; Dinjus, E.; Sieler, J. *J. Organomet. Chem.* **1987**, *334*, 377-388.
14. Hoberg, H.; Oster, B. N. *Synthesis* **1982**, 324-325.
15. Wakatsuki, Y.; Yamazaki, H. *Synthesis* **1976**, 26-28.
16. Hoberg, H.; Schaeffer, D.; Burkhurt, G.; Krüger, C.; Romao, M. J. *J. Organomet. Chem.* **1984**, *266*, 203-224.
17. Hoberg, H.; Oster, B. N. *J. Organomet. Chem.* **1983**, *252*, 359-364.
18. Tsuda, T.; Maruta, K.; Kitaike, Y. *J. Am. Chem. Soc.* **1992**, *114*, 1498-1499.
19. The expression of a repeat unit of the poly(2-pyrone) **4a** in eq 2 is based on the result that the formation of the 2-pyrone ring of **4a** is not regioselective. For example, one ethyl substituent takes the 3- or 4-position of the 2-pyrone ring and the methylene chain exists at the 4- or 3-position, respectively. The similar expression of a repeat unit of other cycloaddition polymer also means the nonregioselective ring formation in the transition metal-catalyzed alkyne cycloaddition polymerization.
20. Tsuda, T.; Maruta, K. *Macromolecules* **1992**, *25*, 6102-6105.
21. Tsuda, T.; Kitaike, Y.; Ooi, O. *Macromolecules* **1993**, *26*, 4956-4960.
22. Tsuda, T.; Yasukawa, H.; Komori, K. *Macromolecules* **1995**, *28*, 1356-1359.

23. Inoue, S.; Koinuma, H.; Tsuruta, T. *J. Polym. Sci., Polym. Lett. Ed.* **1969**, *7*, 287-292.
24. Tsuda, T.; Hokazono, H. *Macromolecules* **1993**, *26*, 1796-1797.
25. Tsuda, T.; Miyoshi, J. unpublished results.
26. Tsuda, T.; Tobisawa, A. *Macromolecules* **1994**, *27*, 5943-5947.
27. Tsuda, T.; Tobisawa, A. *Macromolecules* **1995**, *28*, 1360-1363.
28. Ellis, G. P. In *Comprehensive Heterocyclic Chemistry*; Boulton, A. J.; Mckillop, A., Eds.; Pergamon: Oxford, 1984; Vol. 3, pp 675-689.
29. Scriven, E. F. In *Comprehensive Heterocyclic Chemistry*; Boulton, A. J.; Mckillop, A., Eds.; Pergamon: Oxford, 1984; Vol. 2, pp 165-314.
30. Tsuda, T.; Hokazono, H. *Macromolecules* **1994**, *27*, 1289-1290.
31. Tsuda, T.; Maehara, H. *Macromolecules* **1996**, *29*, 4544-4548.
32. Tomita, I.; Lee, J-C.; Endo, T. *Macromolecules* **1995**, *28*, 5688-5690.
33. Nakayama, J.; Yomoda, R.; Hoshino, M. *Heterocycles* **1987**, *26*, 2215-2222.
34. Tsuda, T.; Takeda, A. *J. Chem. Soc., Chem. Commun.* **1996**, 1317-1318.
35. Kajitani, M.; Suetsugu, T.; Wakabayashi, R.; Igarashi, A.; Akiyama, T.; Sugimori, A. *J. Organomet. Chem.* **1985**, *293*, C15-C18.
36. Tsuda, T.; Hokazono, H. *Macromolecules* **1993**, *26*, 5528-5529.
37. Tsuda, T.; Ooi, O.; Maruta, K. *Macromolecules* **1993**, *26*, 4840-4844.
38. Tsuda, T.; Yasukawa, H.; Hokazono, H.; Kitaike, Y. *Macromolecules* **1995**, *28*, 1312-1315.
39. Ohkubo, A.; Aramaki, K.; Nishihara, H. *Chem. Lett.* **1993**, 271-274.
40. Lee, J-C.; Nishio, A.; Tomita, I.; Endo, T. *Macromolecules* **1997**, *30*, 5205-5212.
41. Tomita, I.; Nishio, A.; Endo, T. *Macromolecules* **1994**, *27*, 7009-7010.
42. Tomita, I.; Nishio, A.; Endo, T. *Macromolecules* **1995**, *28*, 3042-3047.
43. Lee, J-C.; Tomita, I.; Endo, T. *Polym. Bull.* **1997**, *39*, 415-422.
44. Mao, S. S. H.; Tilley, T. D. *J. Am. Chem. Soc.* **1995**, *117*, 5365-5366.
45. Mao, S. S. H.; Tilley, T. D. *J. Am. Chem. Soc.* **1995**, *117*, 7031-7032.
46. Mao, S. S. H.; Tilley, T. D. *Macromolecules* **1996**, *29*, 6362-6364.
47. Tanaka, K.; Kawata, Y.; Tanaka, T. *Chem. Lett.* **1974**, 831-832.
48. Chalk, A. J. *J. Am. Chem. Soc.* **1972**, *94*, 5928-5929.
49. Tsuda, T.; Mizuno, H.; Takeda, A.; Tobisawa, A. *Organometallics* **1997**, *16*, 932-941.
50. Tsuda, T.; Tobisawa, A.; Mizuno, H.; Takeda, A. *J. Chem. Soc., Chem. Commun.* **1997**, 201-202.
51. Shibasaki, Y. In *The Polymeric Materials Encyclopedia*; Salamone, J. C., Ed.; CRC: Boca Raton, FL, 1996; Vol. 2, pp 1336-1342.
52. Tsuda, T.; Shimada, M.; Mizuno, H. *J. Chem. Soc., Chem. Commun.* **1996**, 2371-2372.
53. Lee, B. Y.; Chung, Y. K.; Jeong, N.; Lee, Y.; Hwang, S. H. *J. Am. Chem. Soc.* **1994**, *116*, 8793-8794.
54. Tsuda, T.; Tsugawa, F. *J. Chem. Soc., Chem. Commun.* **1996**, 907-908.
55. Drent, E.; Budzelaar, P. H. M. *Chem. Rev.* **1996**, *96*, 663-681.
56. Perry, R. J.; Wilson, B. D. *Macromolecules* **1994**, *27*, 40-44.
57. Yoneyama, M.; Kakimoto, M.; Imai, Y. *Macromolecules* **1989**, *22*, 2593-2596.
58. Tsuda, T.; Yamanishi, M. unpublished results.
59. Diversi, P.; Ingrosso, G.; Lucherini, A.; Malquori, S. *J. Mol. Catal.* **1987**, *40*, 267-280.

Chapter 4

Synthetic Control over Substituent Location on Carbon-Chain Polymers Using Ring-Opening Polymerization of Small Cycloalkanes

J. Penelle [1]

Département de Chimie, Université Catholique de Louvain,
Place L. Pasteur 1, B–1348 Louvain-la-Neuve, Belgium

Placement of two esters, carboxylate salts and/or cyano substituents on every third carbon of a carbon-chain polymer can be easily achieved by direct anionic ring-opening polymerization of cyclopropanes bearing cyano and ester substituents on geminal position or by modification of the polymers obtained therefrom. Cyclopropyl monomers are activated by electron-withdrawing ester and cyano substituents via the stabilization of the propagating carbanion. Under appropriate conditions, living polymerization, with full control over the nature of the end-groups at both ends, molecular weight distribution, and number-average degree of polymerization, can be obtained. Preliminary investigations have shown that cyclobutanes bearing identical substituents do not polymerize under otherwise identical experimental conditions. Under basic conditions, quantitative hydrolysis of the pendant esters allows access to a new class of highly charged, semicrystalline polyelectrolytes containing carboxylate ions on every third atom along the chain. Preliminary morphological and spectroscopic studies demonstrate that the unusual location of the polycyclopropane substituents along the backbone (quaternary carbon centers separated by CH_2-CH_2 spacers), impacts backbone local conformation and global rigidity via minimization of non-bonded repulsive interactions between next neighbors. The increased stiffness imparted by the substitution pattern dramatically changes ultimate materials properties. It provides higher glass transition temperatures and improved thermal stability.

[1]Current address: Department of Polymer Science and Engineering, University of Massachusetts, Amherst, MA 01003-4530.

Introduction

The unfolding of structure-property relationships linking macromolecular structures to their properties will most certainly remain as one of the most successful scientific accomplishments of this century. Through the collective efforts of chemists in both academia and industry, many polymer structures have been obtained and characterized. These efforts have made possible to gain both a very diverse set of materials properties and a reasonable understanding of some of the structure-property relationships that govern polymeric materials. Ultimately they have made widespread, everyday use of 'plastic' materials possible in our modern society.

Today, after more than seventy years of research and the successful synthesis of many original polymer structures, this area is more active than ever. The title of this symposium section: "Beyond Tacticity: The Next Level of Control" conveys this resurgent interest. In particular, structural design of polymers has recently gone through a renaissance, as an accrued worldwide interest in discovering methodologies for building specific structures has been noticed. There is a need for more 'intelligent' materials inspired from Nature modes of action, capable of incorporating a wide variety of side-substituents, controlling local conformations and long distances interactions between functional groups, and encoding targeted materials properties into the repeating units of the backbone. A key aspect in the design of these well-controlled architectures resides in the discovery of efficient strategies to obtain highly functionalized backbones whose conformation can be controlled by the side substituents. In order to reach that objective, Nature makes use of confor-mationally rigid amide bonds (in proteins) or pentose phosphate subunits (in DNA and RNA). Various synthetic macromolecules characterized by conformationally rigid helical structures have been proposed as possible mimics (for recent examples, see[1-13]).

In the past few years, our group has been working on alternate strategies for the design of polymers with well-controlled architectures. We have looked in particular at highly functionalized carbon-chain polymers substituted on every third or fourth atom rather than on every second carbon or at random positions alongside the backbone.[14-17] This program was fulfilled with two main objectives in mind. First, to understand how to minimize strong repulsive interactions between side substituents in carbon-chain polymers (in particular for applications such as piezoelectricity, non-linear optics, monomolecular electron or ion transport, that require specific alignments of the side-substituents). Secondly, to exploit these interactions to control local conformations of the backbone and possibly fine-tune the stiffness of the entire macromolecule. In this short review, we will discuss how some of these polymers can be synthesized and how the obtained substitution pattern exerts a strong control over local conformations and ultimate properties. The discussion will focus primarily on carbon-chain polymers substituted by ester and/or cyano substituents placed on every third carbon. Some preliminary results on identical polymers substituted on every fourth carbon have been included, however. The general structure for these substituted poly(trimethylene)s and poly(tetramethylene)s can be schematized by the structural formulas $(CH_2CH_2(CXY))_n$ and $(CH_2CH_2CH_2(CXY))_n$ with X,Y= COOR,COOR; CN,COOR; and CN,CN.

Polytrimethylene Synthesis and Characterization

Synthetic Strategy

An obvious synthetic route to substituted polytrimethylenes (i.e., carbon-chain polymers substituted on every third atom) involves the polymerization of cyclopropanes bearing the two substituents of interest, in analogy to the well-known ring-opening polymerizations of three-membered heterocyclic oxiranes and episulfides (eq 1).

$$\triangle\begin{matrix}X\\Y\end{matrix} \longrightarrow \left(\begin{matrix}X & Y\\ & \end{matrix}\right)_n \qquad (eq.1)$$

By direct structural analogy, it can be expected that the large ring strain typical of these small cycles can be used as a driving force for the ring-opening of the cyclopropane ring, and favors polymerization. Simple thermochemical calculations indicate that, for unsubstituted systems, the theoretical free energy of polymerization at room temperature is considerably more negative for cyclopropane than for ethylene (-22.1 vs. -15.6 kcal.mol^{-1}).[18-20] Recent ab initio calculations showed equally favorable numbers for the polymerization of functionalized cyclopropanes.[21] Despite this favorable thermodynamic driving force, cyclopropanes are traditionally considered very poor monomers. The classical review paper by Snow and Hall on ring-opening polymerizations via carbon-carbon σ-bond cleavage suggests that if further strain is not added to an all-carbon three-membered ring, only drastic conditions will allow polymerization (oligomerization) to occur.[22] No example of an efficient polymerization has ever been reported for the unsubstituted cyclopropane (a contrary claim proved later to be incorrect[23-26]).

We can thus conclude that cyclopropane polymerization is thermodynamically feasible but kinetically difficult, and that further improvement is required to facilitate the kinetics of the propagation (and possibly initiation) step. At the beginning of this project, we reasoned that for an anionic polymerization to occur efficiently, it might be better to generate a highly stabilized carbanion during the ring-opening step. We looked in particular for a stabilities matching those of aliphatic alcoholates or thiolates, in analogy again to the ring-opening of oxirane and episulfide. Based on the pK$_a$ scale as a measure of anion stability, such carbanions must be stabilized by at least two strong electron-withdrawing groups, like esters or nitriles (pK$_a$(CH$_3$O-H) = 29.0 (DMSO) or 15.5 (water), pK$_a$(RS-H) = 10-11 (water), pK$_a$((NC)$_2$CH-H) = 11.0 (DMSO), and pK$_a$((EtOOC)$_2$CH-H) = 16.4 (DMSO)).[27]

This crude reasoning proved to be correct, as detailed below. We found that cyclopropanes geminally substituted by two esters, two nitriles or a combination of an ester and a nitrile, do indeed polymerize under anionic conditions. Recently, Alder et coll. discovered that a fluorenyl moiety (pKa = 22.6 (DMSO))[27] also allows some oligomerization to occur (starting from spiro[cyclopropane-1,9'-fluorene], eq

2).[28] The poor solubility of the polymer prevented high molecular weights to be obtained.

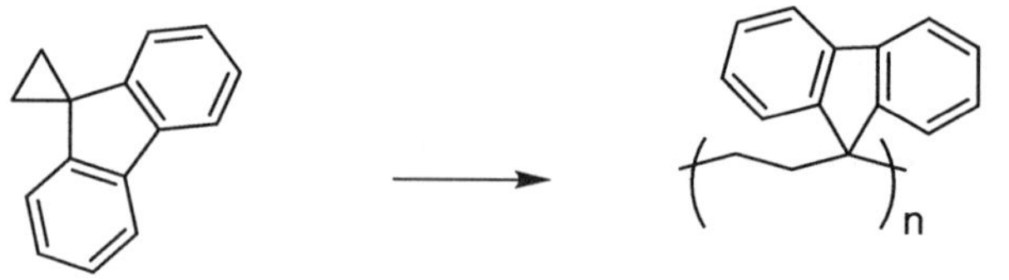

(eq. 2)

The ring-opening polymerization of cyclopropane-1,1-dicarboxylates and 1,1-dicyanocyclopropane further activated by a third substituent (vinyl, phenyl, or electron-donating substituent) on one of the adjacent carbons had been observed earlier by Cho et coll.[29-31]

Choice of Initiators and Reaction Conditions

In most of the polymerization experiments described below, thiophenolate salts **3** were used as initiators. These organic salts can be generated *in situ* by mixing the right amount of thiophenol with an excess of metallic sodium in the presence of the liquid monomer at room temperature. They can also be synthesized and isolated using a synthetic procedure described in one of our recent papers.[16] In the latter case, some care has to be exercized in order to minimize the very rapid aerobic oxidation of the thiolate salt, as the products of the oxidation reaction are not at all active as initiators. ^{1}H-NMR provides a very convenient way to assess the purity of **3** and its suitability as an initiator.[16] Use of commercial samples of **3** is not recommended for the same reason.

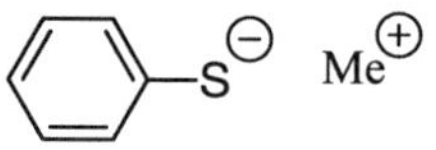

3

Me = Li, Na, K and complexes thereof

The presence of traces of water, a classical difficulty in anionic polymerization, is not a major problem in the polymerizations of these activated cyclopropanes. Polymerization is possible even if a small amount of water is present in the reaction mixture. This lack of sensitivity results from the large stabilization of the carbanions involved ($pK_a((NC)_2CH-H)$ = 11.0 (DMSO), and $pK_a((EtOOC)_2CH-H)$ = 16.4 (DMSO)) and the very low acidity of water in polar solvents (for example $pK_a(H_2O)$ = 32 in DMSO).[27,32] A similar reactivity pattern is known for activated vinylidene monomers like α-cyanoacrylates (super glue), whose ability to polymerize in humid environment is well established.[33]

Though convenient, the choice of initiators for the polymerization is not restricted to thiophenolates. The ring-opening reaction (not polymerization) of cyclopropane-1,1-dicarboxylates 1 is relatively well known.[34] The first experiments date back to the end of the 19th century when Perkin reported that $(EtOOC)_2CH^-Na^+$ the sodium salt of diethyl malonate, reacts mostly quantitatively with diethyl cyclopropane-1,1-diacarboxylate 1b in refluxing ethanol.[35] From the work of Danishefsky and others, it is also known that cyclopropanes activated by two strong electron-withdrawing groups ring-open, when reacted with soft nucleophiles like thiolates, azide, organocuprates.[34,36-43] Hard nucleophiles on the other hand, which include a large fraction of traditional initiators used in anionic polymerization (butyl lithium or Grignard reagent for example), attack on the carbonyls of the ester groups rather than on the cyclopropyl ring itself.[40] Benefits of using thiophenolate initiators include their simple synthesis, easy storage for extended periods of time (in the absence of oxygen), wide range of chemical structures and counterions available, and simplicity to dry. It is also known from the literature that sodium thiophenolate opens cyclopropane-1,1-dicarboxylate 1b quantitatively and rapidly,[43] ensuring that a clean initiation mechanism is available for that monomer.

In a typical polymerization experiment, a weighed amount of the thiophenolate salt is dissolved in a minimal amount of dimethyl sulfoxide (DMSO), typically around 0.35 mmol of thiophenolate per milliliter of DMSO. The initiating solution is then added to the monomer, and the final solution is heated rapidly to the final reaction temperature. A direct addition of the solid thiophenolate to the liquid monomer was attempted several times, but the solubility of most of the thiophenolates investigated is much too low at room temperature. Heating the suspension results in a polymerization, but an induction period was often observed, reproducibility was poor, and no living polymerization could be obtained under experimental conditions that were operative under the regular procedure. A few experiments were carried out using the *in situ* generation technique (thiophenol + metallic sodium) described above. This approach does not work very well with dicyanocyclopropane 6, whose reactivity at room temperature is too high.

Ring-Opening Polymerization of Cyclopropane-1,1-Dicarboxylates

Though several dialkyl cyclopropane-1,1-dicarboxylates are commercially available, samples obtained from several commercial sources were found to be unsuitable for polymerization. Analysis of commercial 1b indicated the presence of up to 3 mol-% of diethyl malonate.[14] This contaminant probably results from some unreacted reagent during the synthesis of the cyclopropane, as cyclopropane-1,1-dicarboxylates are traditionally synthesized from a 1,2-bisalkylating agent, like 1,2-dibromoethane, and the corresponding malonate under basic conditions. Under the polymerization conditions, the contaminating malonate acts as a chain-transfer agent and effectively limits the degree of polymerization. A procedure described in the literature[44] and slightly improved in our group[14,16] allows to obtain very pure cyclopropane-1,1-dicarboxylates in excellent yields. It requires the appropriate malonate, 1,2-dibromoethane, dimethylsulfoxide, and potassium carbonate. Direct separation of the cyclopropane from the mixture is also possible in some cases, but is difficult and leads to severe drops in isolated yield.

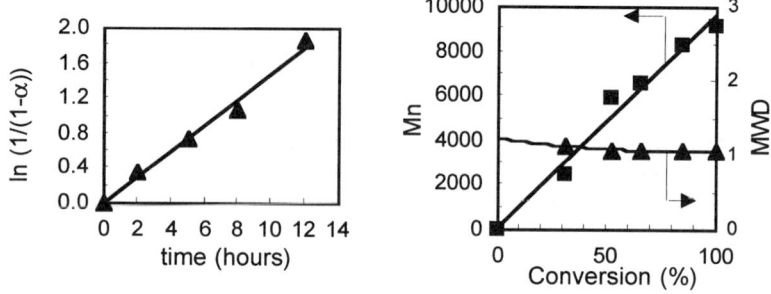

(eq. 3)

R = Me (**a**), Et (**b**), n-Pr (**c**), n-Hex (**d**), i-Pr (**e**), t-Bu (**f**)

If the final polymer is soluble enough under the reaction conditions, polymerization (eq. 3) occurs livingly as demonstrated in the case of cyclopropane **1e** (Fig. 1). A first-order kinetic behavior can be observed over the full conversion range (Fig. 1, left). Very narrow distributions of molecular weights are obtained (Fig. 1, right), and molecular weights, measured by absolute techniques (VPO, end-group analysis by quantitative ^{1}H-NMR, and/or GPC coupled to multiangle light scattering), increase linearly with conversion, reaching the theoretical number at full conversion (Fig. 1, right).

*Figure 1. First-order kinetic plot (left figure, α: conversion) and evolution of number-average molecular weight and polydispercity index with conversion (right figure, theoretical $\overline{M}_n$ at full conversion is 9,400) for **1e** polymerization (general conditions: T=140°C, 22.2 mmol **1e** in 1.6 mL DMSO, 2.3 mol-% PhSNa)*

Characterization of the polymers by several techniques confirms the absence of side reactions. No unexpected functionality can be identified by IR, ^{1}H- or ^{13}C-NMR, even on low molecular weight polymers. A typical MALDI-ToF mass spectrum is provided in Figure 2. The major distribution corresponds to the expected polymer cationized by a potassium ion and end-terminated on one side by a phenylthio group (M_r = 109) and on the other side by a hydrogen (polymerizations were terminated by pouring the reaction mixture into a HCl/CHCl$_3$ mixture). Other distributions of much lower intensities can also be observed. They correspond to cationization by proton, lithium, or sodium ions. Here again, no side reactions can be identified from the spectrum. Particular attention was paid to some potential side reactions that, due to some precedents in the organic chemistry literature, appeared to be possible in particular at high temperarures. They include the Krapcho reaction (alkylation-decarboxylation of malonate esters by good nucleophiles), the Dieckmann condensation (backbiting of the malonate carbanion on the ester of the neighboring unit with the release of a alcoholate anion), and the formation of a thioester by direct attack of the thiophenolate initiator on an ester group of the monomer or polymer. The last reaction is not a very probable side reaction here as quantitative and rapid reaction of the thiophenolate initiator with the cyclopropane is observed in the initiation step. This rapid disappearance of the thiolate, which is independently confirmed by the perfect linearity observed when plotting conversion vs. degree of polymerization (Fig. 1, left), does not leave any thiophenolate available for further (side) reactions.

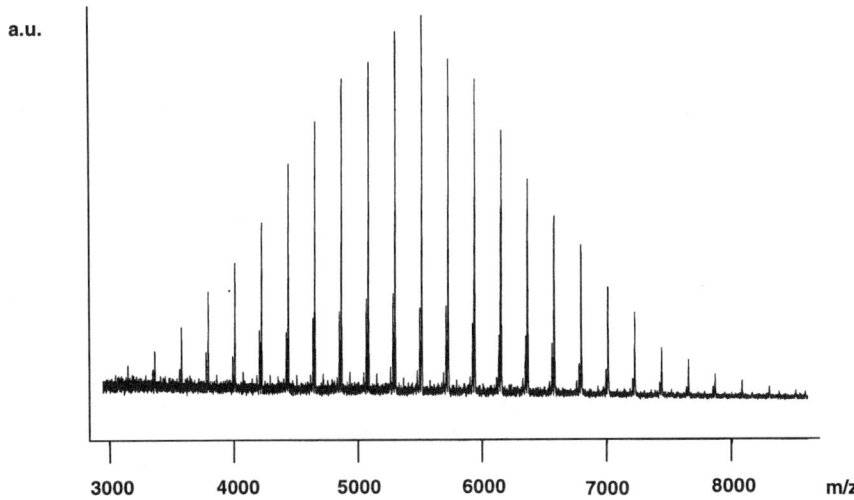

Figure 2. MALDI-ToF mass spectrumof poly(diisopropyl cyclopropane-1,1-dicarboxylate) 2e

Polymerization of cyclopropanes **1** occurs at elevated temperatures (above 80°C) and maintains its living character at temperatures as high as 190°C. A classical Arrhenius behavior is observed in this temperature range (Fig. 3). The activation energy for the reaction is 21.3 kcal.mol^{-1}. At temperatures higher than 200°C, a progressive deviation from the linear Arrhenius behavior can be noticed, the living character of the polymerization is lost, and new peaks can be observed in the ^{1}H-NMR spectrum that we have not yet been able to assign.

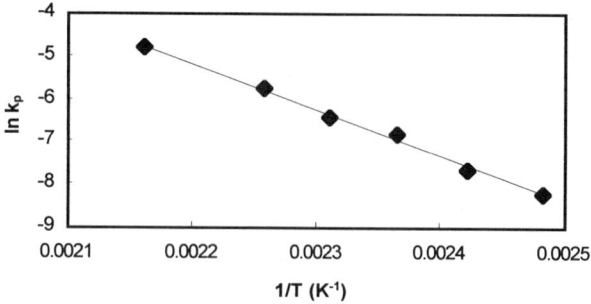

*Figure 3. Arrhenius plot for **1e** polymerization (22.2 mmol **1e**, 0.052 mmol PhSNa, 1.6 ml DMSO)*

The counterion also plays an important role in the rate and selectivity of the polymerization, as shown in Table I. Substituting the sodium cation by a potassium mostly double the reaction rate at 140 °C. Complexing the potassium ion by a crown ether or a cryptand increases even further the polymerization rate by approximately one order of magnitude. However, the livingness of the polymerization is partly lost under these last conditions, as demonstrated by the appearance of a bimodal weight distribution. These experiments suggest that, as is often the case in anionic polymerizations, several propagating species (free-ion and ion pairs, possibly some aggregates) participate in the polymerization, the relative contribution of each being dependent on the nature of the counterion and physical conditions (solvent and temperature).

Table I. Influence of the Counterion on the of Polymerization Diisopropyl Cyclopropane-1,1-Dicarboxylate 1e

Initiator	Time (h)	Conversion (%)	Yield (%)
PhSNa	5	49	52
PhSK	5	86	83
PhSK + 18-crown-6	0.67	53	50
	1.33	73	70
PhSK + Kryptofix 222	0.67	100	96

GENERAL CONDITIONS: T=140°C, 22.2 mmol **1e** in 1.6 mL DMSO, 2.3 mol-% initiator

The results reported here correlate nicely with Harrelson and Arnett's kinetic study on the alkylation of dimethyl ethylmalonate carbanion in DMSO at 25°C.[45-47] This work demonstrated that the counterion (Li, Na, K) of the ethylmalonate carbanion has an influence on the second-order rate constant of about the same order of magnitude as in our study. It was proven that under these conditions, the only reactive species is the free ion and that strong ion-pairing by lithium and sodium ions transform most of the active species into dormant ones. A linear correlation between degree of dissociation of ion-pairs and rate constant was observed.

Results obtained for cyclopropane-1,1-dicarboxylates **1** other than **1e**, in particular with respect to the design of experimental conditions allowing a living polymerization, depend mostly on the solubility of the final polymers. As will be discussed in further details below, polymers **2** are highly crystalline and often have limited solubility in their own monomers or in solvents compatible with anionic polymerizations. As a result, **2** often precipitated out of solutions during the polymerization. If the work-up is carried out under non-acidic conditions, the growing malonate carbanion can still be found present as a carbanionic end-group on the polymer chain (by IR for example). However, the location of the reactive end in the crystal makes it inaccessible to monomer molecules and prevents any further growth. The problem is particularly sensitive with the methyl ester **1a**. The insolubility of the corresponding polymer **2a** is so high that only oligomers of low molecular weight have been obtained so far. On the other hand, *n*-hexyl and *i*-propyl esters (**1d** and **1e**) yield polymers that are soluble in many common solvents and in their own monomer, and living polymerization conditions are easy to find. Ethyl and *t*-butyl esters (**1b** and **1f**) display an intermediate behavior. The polymerizability of **1b** has been extensively studied and a full account has been published recently.[16]

1g

Monomer **1g** was synthesized from cyclopropane-1,1-dicarboxylic acid and isopropenyl acetate.[48] Using our traditional direct alkylation approach for the synthesis of cyclopropane-1,1-dicarboxylate monomers, starting in this case from Meldrum's acid (2,2-dimethyl 1,3-dioxolane-4,6-dione) and 1,2-dibromoethane, did not work here (yield below 2%). The inertness probably arises from the high stability of the corresponding malonic carbanions (pK_a(Meldrum's acid) = 7.3, to be compared to a 'normal' value of about 16 for a non-cyclic malonate).[49] This higher stability results in a higher tendency to ring-open.[37] It also translates into a much higher reactivity in polymerization: both initiation and propagation can take place at room temperature.[50] Unfortunately, the final polymer is highly insoluble and no solvent has been found yet that might allow a spectroscopic analysis in solution. Even oligomers of very low molecular weight appear to be completely insoluble. A polymerization experiment was carried out with a very large amount of thiophenolate ([monomer]$_0$:[initiator]$_0$ = 5) in order to stop the degree of polymerization at five or below. The final product again was insoluble in all tested solvents.

Polymerization of Cyclopropanedinitrile and Cyanocyclopropanecarboxylates

Cyclopropane-1,1-dicarbonitrile **6** and a series of alkyl 1-cyanocyclopropane carboxylates **4** were synthesized from the corresponding malononitrile or cyanoacetates, using the alkylation procedure with 1,2-dibromoethane discussed above. Yields ranged from low (22 % for **6**) to moderate (31-69 % for **4**), while the analytical purity of all samples was consistently high and suitable for further use in polymerization experiments.

$$\underset{4}{\overset{}{\text{cyclopropane with CN and COOR}}} \longrightarrow \underset{5}{\overset{}{\text{polymer with NC, COOR}}}_n \qquad \text{(eq.4)}$$

R = Et (**a**), *i*-Pr (**b**), *n*-Bu (**c**), n-octyl (**d**)

Both monomers polymerize more rapidly than cyclopropane-1,1-dicarboxylates **1**, providing the ring-opened structure only.[51] Monomer **6** is the more reactive: it can be polymerized slowly at room temperature with various initiators, including thiophenolate salts, methylmalonate carbanion, primary and tertiary amines. At 60°C, polymerization is complete in less than 20 minutes (PhSNa, 3.6 mol-%). The polymer is insoluble in all typical solvents, a property that was also observed for its lower homolog, poly(vinylidine cyanide). Its solid-state ^{13}C-NMR and IR spectra are compatible with the expected ring-opened structure, with no visible contaminating functionalities.

$$\underset{6}{\overset{}{\text{cyclopropane with two CN}}} \longrightarrow \underset{7}{\overset{}{\text{polymer with NC, CN}}}_n \qquad \text{(eq. 5)}$$

Not surprisingly, monomers **4** exhibit an intermediate behavior, more reactive than **1** but less than **6**. At 60°C, a temperature at which no polymerization of **1** has ever been observed, polymerization rates were about one order of magnitude lower than for **6**. The final polymers are generally soluble in polar solvents (DMF for example), but their solubity is not influenced by the nature of the alkyl substituent on the ester group.

Synthesis of Poly(trimethylene-1,1-dicarboxylate) Salts

Hydrolysis of poly(di-*n*-propyl cyclopropane-1,1-dicarboxylate) **2c** provides a polyelectrolyte **8** containing two carboxylate anions on every third carbon.[52] The reaction is surprisingly clean, with no decarboxylation taking place as shown by ^{1}H-NMR and solid-state ^{13}C-NMR. In the solid state, loss of the first CO_2 from the hydrated polymer (1.15 molecule of water per carboxylate function) takes place progressively starting at about 390°C. In aqueous solution, progressive precipitation results from heating at 100°C, furnishing a polymer insoluble in water and organic solvents whose structure has yet to be determined.

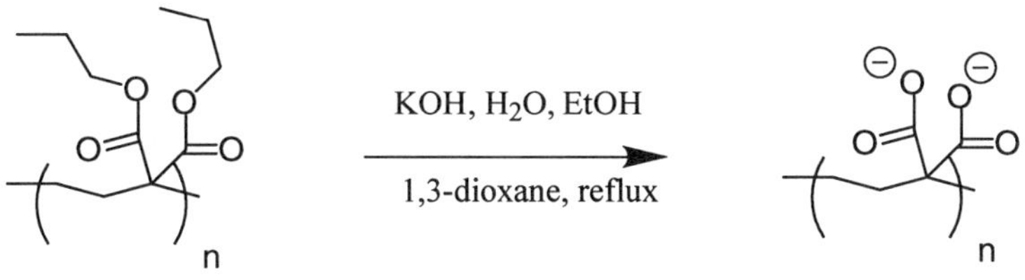

$$\text{2c} \xrightarrow{\text{KOH, H}_2\text{O, EtOH}}_{\text{1,3-dioxane, reflux}} \text{8} \quad \text{(eq. 6)}$$

Recently, it was reported that polymer **9**, which contains a malonic unit in the backbone, is also difficult to decarboxylate upon hydrolysis.[53,54] Some organic literature mentions the difficulty of decarboxylating quaternary malonates, probably due to steric hindrance.[55,56] It seems reasonable to expect that a similar reasoning applies to polymer **8** and explains its unexpected thermal stability.

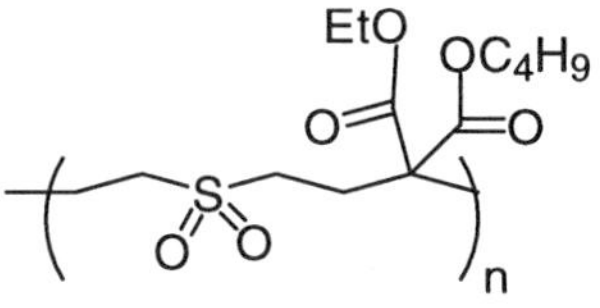

9

Polymer **8** is not very soluble in water, requiring 50 wt-% KOH aqueous solutions to fully solubilize (solubility > 1 g.L^{-1}). The malonic acid substructure can be expected to provide an excellent site for metal complexation, due to its chelate structure. Experiments are currently carried out to check out this possibility.

Ring-Opening Polymerization of Activated Cyclobutanes

All attempts to polymerize diethyl cyclobutane-1,1-dicarboxylate **10**, under the same conditions as its cyclopropyl equivalent, failed completely.[57] At low temperatures, no reaction occured. At higher temperatures (140°C and up), the sodium thiophenolate was consumed by a Krapcho reaction with the ester,[58] providing ethyl cyclobutanecarboxylate and ethyl phenyl thioether as the only products of the reaction (eq 7).

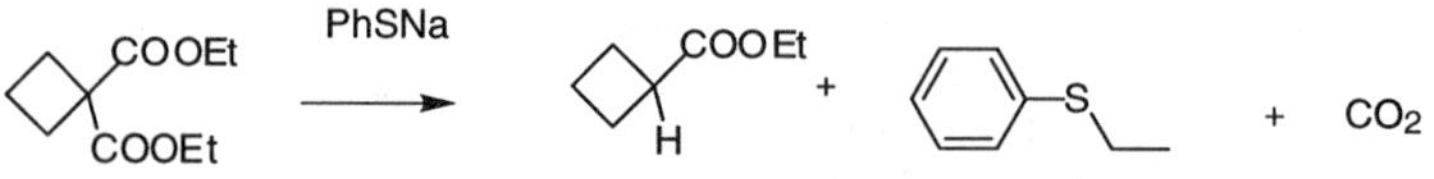

[Structure of compound 10: cyclobutane with two COOEt groups] + PhSNa → [cyclobutane with COOEt and H] + [PhS-CH2CH2- fragment] + CO_2 (eq.7)

10

1,1-Dicyanocyclobutane **11** can be expected to be more reactive than **10** due to the higher stability of the formed dicyano-substituted carbanion. Several experiments showed that ring-opening by sodium thiophenolate does occur, providing the (3-phenylthiopropyl)malononitrile carbanion, but no further reaction takes place (eq. 8).

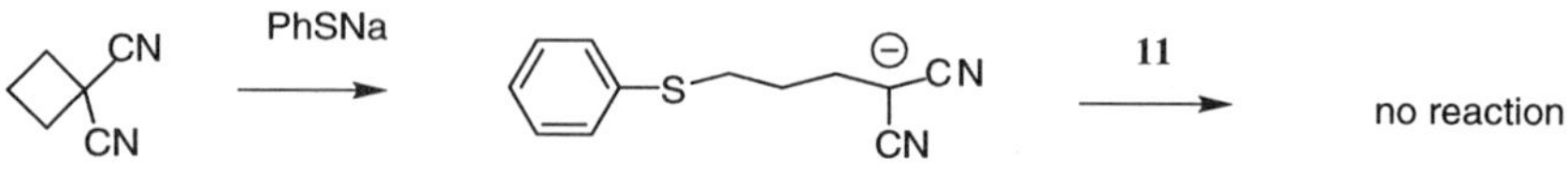

[Structure of 11: cyclobutane with two CN groups] + PhSNa → [PhS-CH2CH2CH2-C(CN)2 anion] $\xrightarrow{11}$ no reaction

11

(eq. 8)

These results contrast with the reported ring-opening polymerization of cyclobutanes **12** containing an additional ether group on the vicinal position.[59-67]

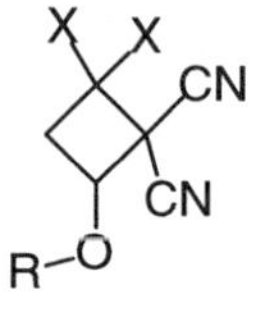

[Structure of 12: cyclobutane with two X groups, two CN groups, and O-R substituent]   **12**

(X = H or CN)

Structural and Physical Properties of Substituted Poly(trimethylenes)

Influence of Architecture on Local Conformation

Traditional carbon-chain polymers of the vinyl and vinylidene types have their substituents located on every second carbon along the backbone. Polymer chemistry students are very familiar with this substitution pattern, but they do not always realize how close side-substituents actually are in such systems. A space filling model of a poly(vinylidene cyanide) segment in an all-trans conformation (left side of Figure 4) can help in visualizing that carbon and nitrogen atoms of the cyanide groups are actually at Van der Waals distances. As a result of these close distances between substituents, strong 1,3 non-bonded interactions force the backbone to adopt conformations minimizing repulsions between side-groups. This is true in the crystal state where helical structures are adopted by tactic vinyl polymers like polypropylene or polystyrene, while unsubstituted linear polyethylene prefers to adopt an all-trans conformation. This is true in solution too, where helicity can be forced by bulky substituents. Minimizing the interactions by shifting to other conformations is of course possible, and is generally easy when only one substituent is placed on every repeating unit. However, when two geminal substituents are positioned on every repeating unit, some strong 1,3 repulsive interaction will be maintained in whatever conformation is finally reached.

Side substituents do not affect only conformations. They also induce lower thermal stabilities for some polymers or the appearance of low ceiling temperatures for some monomers, like 1,1-diphenylethylene, and some α-substituted styrenes and acrylates.[19] It must also be realized that the cyano substituents used in Figure 4 are small, whereas most typical side substituents used on functionalized polymers are large. In most cases, much stronger repulsive interactions have to be expected, including non steric interactions like dipole-dipole or Coulombic interactions.

 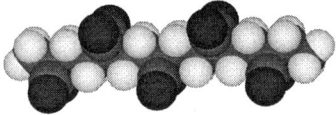

Figure 4. Space filling models for segments of poly(vinylidene cyanide) $(CH_2\text{-}C(CN)_2)_n$ and poly(cyclopropane-1,1-dicarbonitrile) $(CH_2\text{-}CH_2\text{-}C(CN)_2)_n$

Placing two substituents on every third atom maintains a large density of side substituents along the backbone, while minimizing repulsive interactions between next neighbors. A more appropriate arrangement of substituents, where the same cyano substituents are now placed on every third carbon, can be visualized on the space filling model depicted on the right side of Figure 4.

The new substitution pattern influences directly thermal properties. As a result of the new lineup, better thermal stabilities can be observed as demonstrated by TGA. Figure 5 displays the thermal behavior of polymers $((CH_2)_i(COOEt)_2)_n$ with i = 1 and 2.[68] The additional methylene unit in the internal spacer provides for an additional 100 K in thermal stability.

The new architecture has also a dramatic effect on the local conformation of the main chain. Two quaternary centers are now separated by a CH_2CH_2 spacer, making a gauche conformation difficult and a full rotation along the central C-C bond almost impossible, as illustrated by the conformational analysis of Figure 6. Note that this rotational hindrance is not equivalent to a complete freezing of the backbone, as would be the case if a rigid unit like a C=C double bond or an aromatic unit was introduced. The range of available dihedral angles is still large, as indicated by the large valley on the left side of the conformational curve. What is interesting, however, is the possible control over the stiffness of the backbone that can be obtained by fine-tuning the nature of the side substituents rather than the nature of the backbone.

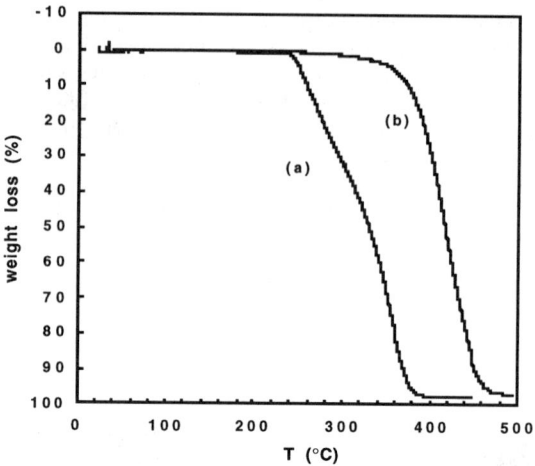

Figure 5. Thermogravimetric analysis (nitrogen, 10 K.min^{-1}) of $(CH_2C(COOEt)_2)_n$ (curve a) and $(CH_2CH_2(COOEt)_2)_n$ (curve b)

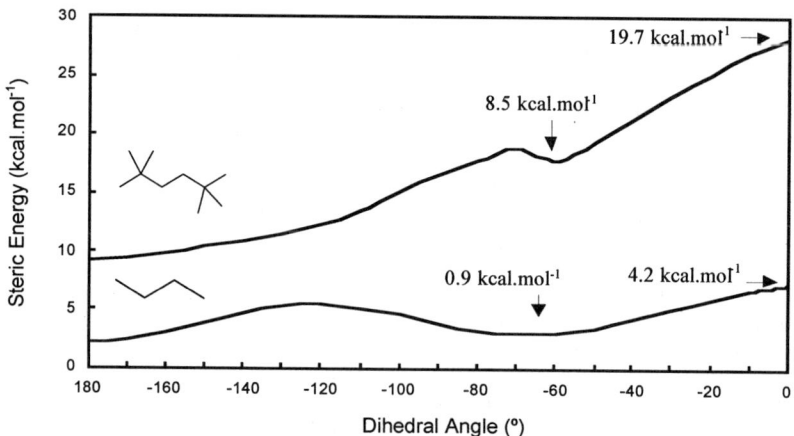

Figure 6. Theoretical conformational analysis of n-butane and 2,2,5,5-tetramethylhexane (molecular mechanics calculation using the MM2 force field)

Experimentally, an increased stiffness is observed indeed in polytrimethylene polymers. A preliminary investigation on the conformation of polymer **2c** $(CH_2CH_2C(COOi-Pr)_2)_n$ was made using Raman spectroscopy and showed no evidence of gauche conformations.[69] The general appearance of the spectra in the 1,000-1,250 cm^{-1} region supports the existence of a rigid backbone, both in the solid state and in solution. Preliminary X-ray results obtained on semicrystalline samples support this view.

The local stiffness has, as expected, a direct influence on ultimate thermal properties. It increases for example, the glass transition temperature of poly(isobutylene) by at least 60 °C when an additional methylene unit is introduced into the backbone, providing poly(l,1-dimethytrimethylene) $(CH_2CH_2C(CH_3)_2)$.[70]

Conclusions

Our study on activated small cycloalkanes and their polymerization has demonstrated that cyclopropanes geminally substituted by two esters, two nitriles, or a combination of an ester and a nitrile, can be polymerized very easily under anionic conditions. The presence of a third activating substituents, while increasing the reactivity, is not a prerequisite for the ring-opening polymerization to occur. On the contrary, cyclobutane monomers bearing the same substituents are much less reactive. Under the best experimental conditions, they can lead only to very low molecular weight oligomers.

The reactivity of cyclopropane monomers depends primarily on the ability of the electron-withdrawing groups to stabilize the propagating carbanion. Diester-substituted cyclopropanes require temperatures higher than 80°C and can be polymerized at temperatures as high as 190°C without the appearance of noticeable side reactions. 1,1-Dicyanocyclopropane, on the contrary, can be polymerized at room temperature. 1-Cyanocyclopropanecarboxylates display an intermediate polymerizability. Counterions also play a role in the polymerizability, free carbanions obtained by complexing the alkali countercation with cryptands providing the most reactive system. The stability of the growing chain makes living polymerizations possible, as demonstrated by a detailed kinetic and mechanistic study on diisopropyl 1,1-cyclopropanedicarboxylate **1c**. This livingness should open access to various controlled macromolecular architectures, including block, star, and branched copolymers. The structure of the end-groups can be controlled by the nature of the initiator and end-capping agents. If necessary, two identical end-groups can be incorporated by end-capping the propagating carbanion with half an equivalent of 1,2-dibromoethane. The large distance between next neighbors along the backbone allows very bulky side groups to be added on the polymer chain. Side-groups incompatible with the polymerization reaction can be added via post-polymerization reactions on the ester and cyano groups. Based on this strategy, a polyelectrolyte containing two carboxylate salt groups on every third carbon has been synthesized by hydrolysing the ester groups on the malonate substructure.

Preliminary studies on the conformational preferences in this new substitution architecture have provided evidence, both at the theoretical and experimental levels, that the all-trans conformation is more favored than for traditional carbon-chain polymers, and that full rotation along the CH_2-CH_2 bond is very difficult. This local preference does not necessarily make the polymer chain a rigid rod, but it allows an all-trans conformation to be easily reached, should repulsive interactions between side groups provide the driving force for the stretching. In the solid state, preliminary evidence supports the formation of an all-trans conformation for the backbone. The substitution pattern also has a direct influence on thermal properties. Poly(diethyl cyclopropane-l,l-dicarboxylate) was found to be about 100 K more thermally stable than its shorter homologue, poly(diethyl vinylidenedicarboxylate). A comparison between poly(isobutylene) and poly(l,1-dimethyltrimethylene) showed that the latter has a glass transition temperature at least 60 °C higher than the former. All poly(trimethylene)s investigated in this study are semicrystalline polymers and show very high tendencies to crystallize.

Acknowledgments

I am deeply indebted to all undergraduate and graduate students who worked on this project in the past few years, both in Belgium and in the US. Their names can be found in the reference list. I am also indebted to my colleagues, Ted Atkins (Bristol), Klaus Schmidt-Rohr, Shaw Ling Hsu and Igor Kaltashov (Amherst) for their collaboration on part of this work. This study was made possible by the financial support of the *Ministère de l'Aménagement du Territoire, de la Recherche, des Technologies et des Relations Extérieures de la Région Wallonne* (Belgium), the *Fonds National de la Recherche Scientifique* (Belgium), the *Fonds de Développement Scientifique de l'Université catholique de Louvain* (Belgium), the *3M Young Faculty Award* program, and the Materials Research Science and Engineering Center at the University of Massachusetts (NSF DMR-9400488 and DMR-9809365).

References

1. de Ilarduya, A. M.; Aleman, C.; Garcia-Alvarez, M.; Lopez-Carrasquero, F.; Munoz-Guerra, S. *Macromolecules* **1999**, *32*, 3257-3263.
2. Jha, S. K.; Cheon, K. S.; Green, M. M.; Selinger, J. V. *J. Am. Chem. Soc.* **1999**, *121*, 1665-1673.
3. Lim, A. R.; Stewart, J. R.; Novak, B. M. *Solid State Commun.* **1999**, *110*, 23-28.
4. Mayer, S.; Maxein, G.; Zentel, R. *Macromolecules* **1998**, *31*, 8522-8525.
5. Murakawa, K.; Furusho, Y.; Takata, T. *Chem. Lett.* **1999**, 93-94.
6. Nakako, H.; Nomura, R.; Tabata, M.; Masuda, T. *Macromolecules* **1999**, *32*, 2861-2864.
7. Nakano, T.; Satoh, Y.; Okamoto, Y. *React. Funct. Polym.* **1999**, *40*, 135-141.
8. Nuckolls, C.; Katz, T. J.; Katz, G.; Collings, P. J.; Castellanos, L. *J. Am. Chem. Soc.* **1999**, *121*, 79-88.
9. Sakajiri, K.; Satoh, K.; Kawauchi, S.; Watanabe, J. *J. Mol. Struct.* **1999**, *476*, 1-8.
10. Maeda, K.; Okamoto, Y. *Macromolecules* **1999**, *32*, 974-980.
11. Aoki, T.; Kobayashi, Y.; Kaneko, T.; Oikawa, E.; Yamamura, Y.; Fujita, Y.; Teraguchi, M.; Nomura, R.; Masuda, T. *Macromolecules* **1999**, *32*, 79-85.

12. Nieh, M. P.; Goodwin, A. A.; Stewart, J. R.; Novak, B. M.; Hoagland, D. A. *Macromolecules* **1998**, *31*, 3151-3154.
13. Schlitzer, D. S.; Novak, B. M. *J. Am. Chem. Soc.* **1998**, *120*, 2196-2197.
14. Penelle, J.; Clarebout, G.; Balikdjian, I. *Polym. Bull.* **1994**, *32*, 395-401.
15. Penelle, J.; Hérion, H.; Sorée, A.; Gorissen, P. *Polym. Prepr. (Am. Chem. Soc., Div. Polym. Chem.)* **1996**, *37(1)*, 208.
16. Penelle, J.; Herion, H.; Xie, T.; Gorissen, P. *Macromol. Chem. Phys.* **1998**, *199*, 1329-1336.
17. Xie, T.; Penelle, J. *Polym. Prepr. (Am. Chem. Soc., Div. Polym. Chem.)* **1998**, *39(2)*, 502.
18. Dainton, F. S.; Ivin, K. J. *Q. Rev., Chem. Soc.* **1958**, *12*, 61.
19. Sawada, H. *Thermodynamics of Polymerization*; Marcel Dekker: New York, 1976.
20. Devlin, T. R. L.; Small, P. A.; Dainton, F. S. *Trans. Faraday Soc.* **1955**, *51*, 1710.
21. Mbala, N. L. Mémoire de Licence, Université Catholique de Louvain, Belgium, 1998.
22. Hall, H. K. J.; Snow, L. G. In *Ring-Opening Polymerization*; Ivin, K. J., Saegusa, T., Eds.; Elsevier: London, 1984; Vol. 1, p 83.
23. Harris, L.; Ashdown, A. A.; Armstrong, R. T. *J. Am. Chem. Soc.* **1936**, *58*, 852.
24. Gunning, H. E.; Steacie, E. W. R. *J. Chem. Phys.* **1949**, *17*, 351.
25. Scott, R. J.; Gunning, H. E. *J. Chem. Phys.* **1952**, *56*, 156.
26. Ivin, K. J. *J. Chem. Soc.* **1956**, 2241.
27. Bordwell, F. G. *Acc. Chem. Res.* **1988**, *21*, 456.
28. Alder, R. W.; Anderson, K. R.; Benjes, P. A.; Butts, C. P.; Koutentis, P. A.; Orpen, A. G. *Chem. Commun.* **1998**, 309-310.
29. Cho, I.; Ahn, K.-D. *J. Polym. Sci., Polym. Chem. Ed.* **1979**, *17*, 3183.
30. Cho, I.; Kim, J.-B. *J. Polym. Sci., Polym. Chem. Ed.* **1980**, *18*, 3053.
31. Kim, J. B.; Cho, I. H. *Tetrahedron* **1997**, *53*, 15157-15166.
32. Olmstead, W. N.; Margolin, Z.; Bordwell, F. G. *J. Org. Chem.* **1980**, *45*, 3295.
33. Donnelly, E. F.; Johnston, D. S.; Pepper, D. C.; Dunn, D. J. *J. Polym. Sci., Polym. Lett. Ed.* **1977**, *15*, 399.
34. Danishefsky, S. *Acc. Chem. Res.* **1978**, *12*, 66.
35. Bone, W. A.; Perkin, W. H. *J. Chem. Soc.* **1895**, *67*, 108.
36. Danishefsky, S.; Rovnyak, G. *J. Org. Chem.* **1975**, *40*, 114.
37. Danishefsky, S.; Singh, R. K. *J. Org. Chem.* **1975**, *40*, 3807.
38. Singh, R. K.; Danishefsky, S. *J. Org. Chem.* **1975**, *40*, 2969.
39. Danishefsky, S. *J. Am. Chem. Soc.* **1979**, *12*, 66.
40. Kierstead, R. W.; Linstead, R. P.; Weedon, B. C. L. *J. Chem. Soc.* **1952**, 3616.
41. Dox, A. W.; Yoder, L. *J. Am. Chem. Soc.* **1921**, *43*, 2097.
42. Stewart, J. M.; Westberg, H. H. *J. Org. Chem.* **1965**, *30*, 1951.
43. Harsanyi, K.; Nemeth, S.; Hegedus, B.; Beszedics, G.; Mondok, J. In *Hung. Teljes* HU, 1989.
44. Zefirov, N. S.; Kuznetsova, T. S.; Kozhushkov, S. I.; Surmina, L. S.; Rashchupkina, Z. A. *Zh. Org. Khim. (Engl. Transl.)* **1983**, *19*, 474.
45. Arnett, E. M.; Maroldo, S. G.; Schilling, S. L.; Harrelson, J. A. *J. Am. Chem. Soc.* **1984**, *106*, 6759.
46. Harrelson, J. A. Ph.D. Dissertation, Duke University, 1986.

47. Arnett, E. M.; Harrelson, J. A. J. *Gazz. Chim. Ital.* **1987**, *117*, 237.
48. Singh, R. K.; Danishefsky, S. *Organic Synthesis* **1990**, *VII*, 401.
49. Zhang, X.-M.; Bordwell, F. G. *J. Org. Chem.* **1994**, *59*, 6456.
50. Herremans, V. Rapport de Stage IPL, Université Catholique de Louvain, 1993.
51. Kagumba, L.; Sorée, A.; Hérion, H.; Penelle, J. .
52. Xie, T.; Penelle, J., unpublished results.
53. Wudl, F.; Weitz, A.; Schmidt-Winkel, P. *Proc. Am. Chem. Soc., Div. Polym. Mater.: Sci. Eng.* **1999**, *80*, 7.
54. Schmidt-Winkel, P.; Wudl, F. *Macromolecules* **1998**, *31*, 2911.
55. Taber, D. F.; Amedio, J. C. J.; Fulino, F. *J. Org. Chem.* **1989**, *54*, 3474.
56. Brown, R. T.; Jones, M. F. *J. Chem. Res.* **1984**, 332.
57. Kagumba, L.; Penelle, J., unpublished results.
58. Krapcho, A. *Synthesis* **1982**, 893.
59. Yokozawa, T.; Fujino, T. *Makromol. Chem., Rapid Commun.* **1993**, *14*, 779-783.
60. Yokozawa, T.; Fukata, H. *J. Polym. Sci., Part A: Polym. Chem.* **1995**, *33*, 1203-1208.
61. Yokozawa, T.; Tsuruta, E. *Macromolecules* **1996**, *29*, 8053-8056.
62. Yokozawa, T.; Mori, T.; Hida, K. *Macromol. Chem. Phys.* **1996**, *197*, 2261-2271.
63. Yokozawa, T.; Suzuki, T. *Macromolecules* **1996**, *29*, 22-25.
64. Yokozawa, T.; Tagami, M.; Takehana, T.; Suzuki, T. *Tetrahedron* **1997**, *53*, 15603-15616.
65. Yokozawa, T.; Toyoizumi, T.; Hayashi, K. *Kobunshi Ronbunshu* **1997**, *54*, 716-722.
66. Yokozawa, T.; Wakabayashi, Y.; Kimura, T. *J. Polym. Sci., Part A: Polym. Chem.* **1997**, *35*, 1563-1570.
67. Yokozawa, T.; Fujino, T.; Morita, I.; Toyoizumi, T. *Macromol. Chem. Phys.* **1998**, *199*, 997-1002.
68. Lebegue, C. Rapport de Stage IPL, Université Catholique de Louvain, 1996.
69. Hsu, S. L., personal communication.
70. *Polymer Handbook*; Brandrup, J., Immergut, E. H., Grulke, E. A., Eds.; Wiley: New York, 1999, p VI 206.

Chapter 5

Nucleophilic Polymerization of Methyl Methacrylate Activated by a Brønsted Acid

Masatoshi Miyamoto, Hirotomo Katano, Chika Kanda, and Yoshiharu Kimura

Department of Polymer Science and Engineering, Kyoto Institute of Technology, Matsugasaki, Sakyo Kyoto 606–8585, Japan

A novel polymerization reaction of MMA comprising the combination of a Brønsted acid and a weak nucleophile was discovered, where the acid acts as the activator for the monomer and the nucleophile functions as the initiator. The polymerization of MMA proceeded smoothly in the presence of, for example, methanesulfonic acid (MsOH) and triphenylphosphine at 40 °C for 2 h to give syndiotactic-rich PMMA in moderate yield. The polymerization preferred non-basic solvents, such as chloroform. The control of the molecular weight was not attained by the change of feed ratio since this polymerization is substantially not living, but is terminated unimolecularly by the proton migration, *i.e.*, the tautomerization. The combinations of Et_2S with MsOH and with trifluoromethanesulfonic acid were also found to be effective for the polymerization of MMA.

Introduction

Many additional polymerization reactions for vinyl and ring-opening monomers can be classified into two categories, although such categorization is not widely accepted. In the first category of polymerization, to which almost all of existing polymerization reactions are classified, the propagation proceeds between a reactive propagating center and a monomer (Scheme 1). The polymerization reactions of this category are further classified into the well-known sub-categories of radical, anionic, cationic, and coordinative polymerizations on the basis of the type of propagating species, which is in accordance with the nature of the respective initiator.

© 2000 American Chemical Society

$$I + M \longrightarrow P_1^*$$

$$\sim\!\!\sim P_n^* + M \longrightarrow \sim\!\!\sim P_{n+1}^*$$

Scheme 1. Concept of conventional polymerization (CP)

The other category is the so-called "polymerization via activated monomer mechanism (PAMM)". Although the polymerization reactions of this category are far less common, they are sometimes observed in ring-opening polymerization.[1-3] In this second category, the monomer reacts with an activator to form an activated monomeric species, which is the real monomeric species and reacts readily with the propagating species (Scheme 2). The propagating species in this category of polymerization is rather stable in comparison with that in the conventional polymerization. An important point in this type of polymerization is that an activated unit is formed at the penultimate position by the propagation, which reacts with the inactivated monomer to regenerate the activated monomer. The most well-known example of this type of polymerization is undoubtedly the polymerization of ε-caprolactam, where a strong base is used as the activator.[3]

$$A + M \longrightarrow M^*$$
(activator)

$$M + M^* \longrightarrow P_1$$

$$\sim\!\!\sim P + M^* \longrightarrow \sim\!\!\sim M^*P$$

$$\sim\!\!\sim M^*P + M \longrightarrow \sim\!\!\sim MP + M^*$$

Scheme 2. Concept of polymerization via activated monomer mechanism (PAMM)

PAMM substantially requires an initiator, whether it is added to the system intentionally or not. In some cases, the inactivated monomer itself acts as the initiator. In other cases, an initiator is added to the system since the inactivated monomer is not enough strong to initiate the polymerization with an appropriate rate. For example, *N*-acyllactam is often added to the polymerization system of ε-caprolactam for this purpose. This kind of polymerization using an additional initiator is also classified into the category of PAMM since the "initiator" is not indispensable, and therefore, the "initiator" is often called promoter or catalyst.

PAMM is also possible for vinyl monomers.[4,5] One example is the polymerization of 2-isopropenyl-2-oxazoline using trimethylsilyl trifluoromethanesulfonate (TMSOTf) as the activator (Scheme 3).[5]

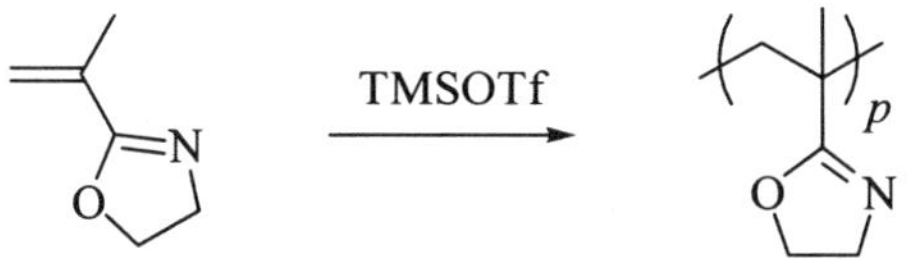

Scheme 3. Polymerization of 2-isopropenyl-2-oxazoline as an example of PAMM

As this example clearly shows, one of the significant characteristics of PAMM is that the property of the propagating species, nucleophilic (anionic) or electrophilic (cationic), does not agree with the nature of the activator: the propagation in the above example is the nucleophilic attack of the propagating species to the activated monomer, although the activator is undoubtedly electrophilic.

In the anionic polymerization of MMA initiated with aluminum porphylines, the addition of a bulky Lewis acid has been reported to accelerate the polymerization drastically.[6] Although this polymerization is examined in the presence of both the initiator and the activator, it is classified into conventional polymerization (CP) since the activator is not indispensable. Similar activator-assisted polymerization has been well known in radical polymerization: the presence of Lewis or Brønsted acid accelerates the rate of radical polymerization of acrylic monomers.[7-9]

Contrary to the description indicated above, an additional category of polymerization is possible, where both initiator and activator are necessary to the progress of the polymerization (Scheme 4).[10,11] Although it resembles the activator-assisted conventional polymerization as well as the PAMM with a promoter, this type of polymerization should be classified into the third category having a mixed character of PAMM and CP. Since PAMM so far known has ionic character, this hybrid-type polymerization (HTP) is also limited to ionic-type polymerization.

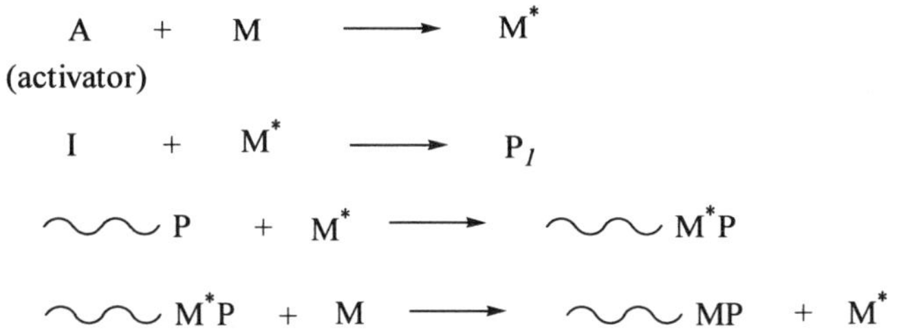

$$A + M \longrightarrow M^*$$
(activator)

$$I + M^* \longrightarrow P_1$$

$$\sim\!\!\sim P + M^* \longrightarrow \sim\!\!\sim M^*P$$

$$\sim\!\!\sim M^*P + M \longrightarrow \sim\!\!\sim MP + M^*$$

Scheme 4. Concept of Hybrid-type polymerization (HTP)

Recently, we reported the polymerization of MMA using a bulky Lewis acid and an enamine (Scheme 5).[10] This is one of rare examples of HTP, where both species are indispensable for the polymerization. The most important point in HTP is to avoid the direct acid-base reaction of the initiator with the activator. It is prohibited by steric reason in this case.

Another possible solution to avoid the direct reaction between the initiator and the activator is to make the reaction reversible. If a Brønsted acid is available as the activator, the acid-base reaction will be reversible. In radical polymerization of ethyl vinyl sulfoxide, phenol coordinates to the carbonyl oxygen of the monomer and alters its electronic property. Namely, Brønsted acid can acts as the activator.[12] However, no study concerning the influence of Brønsted acid on anionic-type polymerization of acrylic monomer has been examined so far. Supposedly, it is because the protonation of nucleophilic propagating end is expected to terminate the propagation. However, a recently study reported by Takeishi et al. has shown that the polymerization of MMA by benzenesulfinic acid as the nucleophilic initiator is not disturbed by the addition of a Brønsted acid.[13]

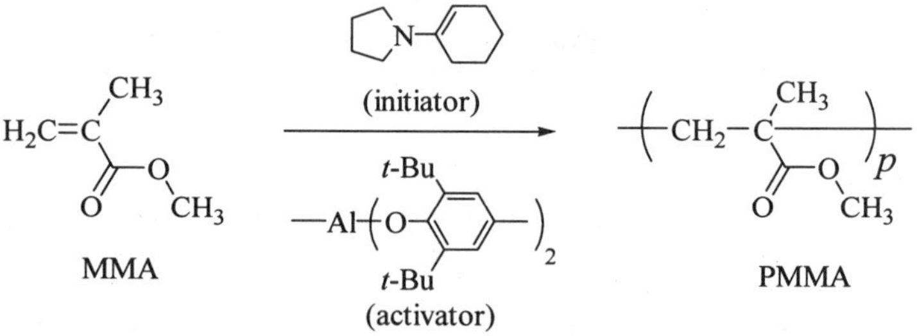

Scheme 5. *Polymerization of MMA by the combination of an enamine and a bulky Lewis acid as an example of HTP.*

In the present report, we show the availability of Brønsted acid as the activator for the novel HTP of MMA.[14] The total concept of the present polymerization is illustrated in Scheme 6. In this expected mechanism, the acid added to the system protonates MMA according to an acid-base equilibrium. The protonated monomer (H-MMA) is electrophilic enough to be attacked by the initiator, which itself has no ability to initiate the inactivated MMA. The propagating species is to have a ketene hemiacetal structure, which is suspected to tautomerize to a keto-form. However, a recent study done by Novak et al. has revealed that the tautomerization of vinyl alcohol to acetaldehyde is slow enough to be capable of radical polymerization.[15] The propagation is the reaction between the nonionic propagating species having a nucleophilic character and H-MMA. The propagating chain will be terminated unimolecularly by the proton migration to form an stable ester, *i.e.*, the tautomerization to a keto-form. Thus, this polymerization is substantially not living, but it provides a new concept of polymerization. Our examination revealed that the combination of triphenylphosphine (as the initiator) with methanesulfonic acid (MsOH, as the activator) was found to construct an effective, novel initiating system for MMA. This type of polymerization using the combination of mild activator and mild initiator is an attractive alternative to replace the conventional radical polymerization, which requires explosive initiators.

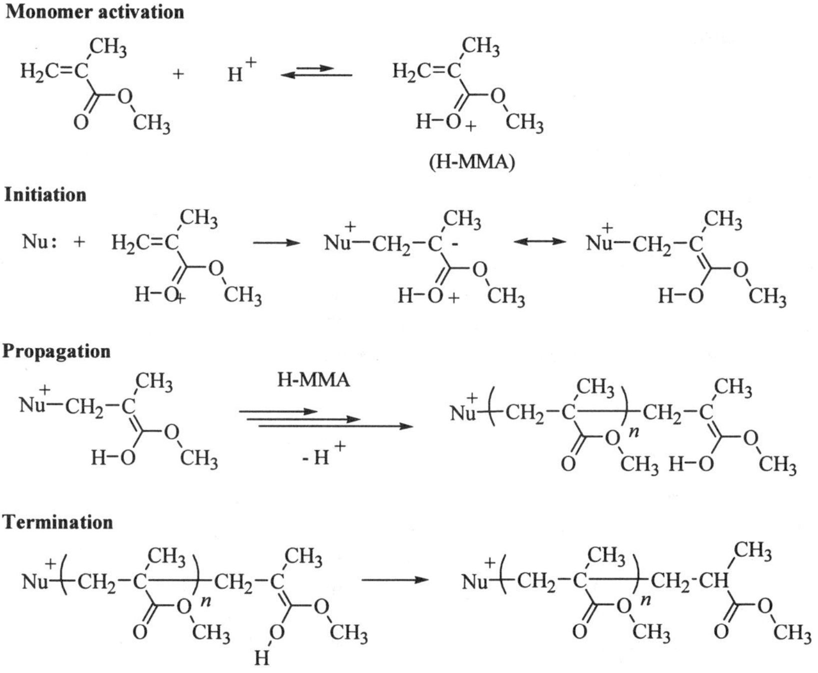

Monomer activation

[Scheme: H₂C=C(CH₃)C(=O)OCH₃ + H⁺ ⇌ H-MMA (protonated MMA)]

(H-MMA)

Initiation

[Scheme: Nu: + H-MMA → Nu-CH₂-C⁺(CH₃)(C(=O)OCH₃) resonance structures]

Propagation

[Scheme: Nu-CH₂-C⁺... + H-MMA, -H⁺ → Nu-(CH₂-C(CH₃)(CO₂CH₃))ₙ-CH₂-C⁺(CH₃)(CO₂CH₃)]

Termination

[Scheme showing termination product]

Scheme 6. Expected mechanism for the Brønsted acid activated polymerization of MMA initiated by a weak nucleophile

Experimental

General

MMA was dried over CaH_2 and distilled under nitrogen before use. Methanesulfonic (MsOH) and trifluoromethanesulfonic acids (TfOH) were distilled under nitrogen. Triphenylphosphine was recystallized from methanol twice. Diethyl sulfide was dried over sodium and distilled under nitrogen. 4-Methoxyphenol and methyl trimethylsilyl dimethylketene acetal (MTDA) were used as received. The solvents were dried according to the conventional methods and distilled before use. ^{1}H NMR spectra were recorded on a Varian Gemini-200 or a Bruker ARX-500 NMR spectrometer, and ^{31}P NMR spectra were recorded on the latter. GPC analysis was performed on a Shimadzu LC-10A system with combined columns of Tosoh TSK gel G4000H8 and G2500H8 in chloroform at 35 °C.

Polymerization

Typical procedure was as follows. In a test tube equipped with a stirrer chip and a three-way stopcock was placed 79 mg (0.30 mmol) of triphenylphosphine and 1.9 mg (15 µmol) of 4-methoxyphenol (as radical inhibitor). After drying in vacuum for 30 min, 0.5 mL of chloroform, 0.32 mL (0.60 g, 3.0 mmol) of MMA, and 0.20 mL (0.29g, 3.0 mmol) of MsOH were introduced to the tube under nitrogen in turn. The mixture was kept at 40 °C for 2h. Then, the contents were diluted with an equi-volume amount of dichloromethane and washed with water three times. The organic layer was separated and added dropwise to hexane to precipitate the polymer, which was purified further by reprecipitation from dichloromethane to hexane, dried in vacuum, and weighed. The yield was 0.29 g (48% yield).

Results and Discussion

The proper selection of the combination of activator and initiator is, of course, the very core of this novel polymerization and probably the most difficult problem to solve. The basicity of the monomer, the acidity of Brønsted acid, and the basicity and the nucleophilicity of initiator are the factors that should be taken into account for the selection. Since the ester group is very weak base (pK_a for alkyl acetate is ca. −6.5 in H_2O at 25 °C),[16] a very strong acid should be used to protonate and activate it. The initiator should have a sufficient nucleophilicity to attack the activated monomer (H-MMA), but at the same time, it should have a low basicity to prevent inactivation by the acid. The conventional initiators applied to the anionic polymerization of MMA, alkyllithiums and the Grignard reagents, undoubtedly cannot be used for this purpose. We started our investigation with the combination from triphenylphosphine ($pK_a = 2.7$) as the initiator and methanesulfonic acid (MsOH, $pK_a = -1.2$) as the activator. Fortunately, this combination caused the polymerization of MMA under a mild condition, although neither of them can induce the polymerization of MMA alone.

Solvent Effect

The results of the polymerization of MMA carried out in various solvents in the presence of 0.1 equivalent of PPh_3 and an equimolar amount of MsOH are shown in Table I, where the pK_a values as well as the donor numbers (DN) of the solvents are indicated as the index for the solvent's basicity. The polymerization proceeded in several solvents including methanol, although the polymer yield generally tended to decrease along with an increase of the solvent's basicity, and no polymerization proceeded in DMSO, the most basic solvent among those we examined. Namely, the present polymerization prefers non-basic solvents, since, presumably, basic solvents interfere the activation of MMA. No clear relationship was found between the solvent's basicity and the molecular weight as well as the molecular weight distribution. In the following experiments chloroform was chosen as the solvent.

The molecular weight of the product was determined from GPC measurement with PMMA standards. The values were much higher than that expected from the feed ratio, 1,000. It is a consequence of the low initiating efficiency of PPh_3. Although we selected PPh_3 as the initiator because of its relatively low basicity combined with its high nucleophilicity, it is still highly basic in comparison with MMA. Therefore, almost all PPh_3 suffers protonation in the presence of a large excess amount of the acid, which greatly reduces the concentration of the unprotonated PPh_3 available for the initiation.

Table I. Solvent Effect on the Brønsted-Acid Activated Polymerization of MMA[a]

solvent			PMMA		
	pK_a[b]	DN[c]	yield, %	M_n, 10^4[d]	M_w/M_n[d]
$CHCl_3$	--	0	49	3.1	3.2
C_6H_6	--	0.1	40	2.7	1.3
MeCN	-10.1	14.1	14	6.0	1.5
None	-6.5[e]	16.5[f]	26	4.9	1.5
MeOH	-2.2	19	36	9.1	3.2
THF	-2.0	20	13	5.9	2.9
DMSO	-1.5	29.8	0	--	--

[a] MMA; 3 mmol, MsOH; 3 mmol, PPh_3; 0.3 mmol, in 0.5 mL of the solvent with 0.015 mmol of p-$MeOC_6H_4OH$, at 40 °C for 24 h.
[b] Data in H_2O at 25 °C from ref. 17 unless otherwise noted.
[c] Donor number from ref. 18.
[d] Determined from GPC with PMMA standards.
[e] Value for alkyl acetate from ref. 16.
[f] Value for methyl acetate.

The experiments summarized in Table I were examined in the presence of 0.5 mol% of 4-methoxyphenol to exclude the possibility of the concurrent occurrence of radical polymerization. The presence of this inhibitor as well as other inhibitors, t-butylcatechol and benzoquinone ($\leq$ 2mol% to MMA), did not influence the polymer yield as well as the molecular weight, confirming that polymerization proceeded via a non-radical mechanism.

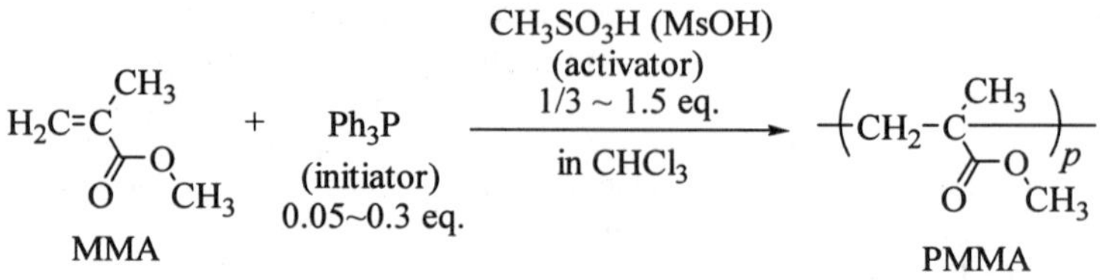

$$H_2C=C(CH_3)(C(O)OCH_3) + Ph_3P \xrightarrow[\text{in CHCl}_3]{\text{CH}_3\text{SO}_3\text{H (MsOH)} \text{ (activator)} \text{ 1/3 ~ 1.5 eq.}} \text{--}(CH_2-C(CH_3)(C(O)OCH_3))_p\text{--}$$

MMA + Ph₃P (initiator) 0.05~0.3 eq. → PMMA

Scheme 7. *Hybrid-type polymerization of MMA*

Effect of the Polymerization Temperature

The results of the polymerization examined in CHCl₃ while varying the polymerization temperature are shown in Table II. The molecular weight of the product increases with a decrease of the polymerization temperature, although it is accompanied by a decrease in yield.

Table II. Effect of Temperature on the Polymerization[a]

Temp., °C	$[PPh_3]_0/[MMA]_0$	yield, %	PMMA M_n, $10^{4[b]}$	M_w/M_n [b]
0	0.1	2.3	--	--
10	0.1	7.3	--	--
20	0.1	16	1.4	1.8
30	0.1	20	0.27	2.2
40	0.1	48	0.72	2.0
50	0.1	32	0.41	1.2
0	0.2	4.7	--	--
10	0.2	26	8.5	1.1
20	0.2	30	2.5	1.5
30	0.2	60	1.1	1.5
40	0.2	71	1.0	2.1
50	0.2	19	0.46	1.1

[a] MMA; 3 mmol, MsOH; 3 mmol, p-MeOC₆H₄OH; 0.015 mmol, in CHCl₃ (0.5 mL) for 2 h.
[b] Determined from GPC with PMMA standards.

Probably, termination on the basis of unimolecular proton transfer becomes less significant at low temperature. The data shown in Table II also give another important piece of information: the usage of 0.2 eq. of the initiator generally gives PMMA in higher yield, while the molecular weight is not so much affected by the feed ratio of the initiator to the activator. The increase in the polymer yield is readily understood since it is a consequence of the increase in the concentration of the unprotonated initiator. The insensitivity of the molecular weight on the feed ratio indicates that it is determined not by the ratio of [H-MMA], which is a function of [MMA] and [MsOH] to [PPh$_3$], but by the relative ratio of the rate of propagation to that of the termination as in the case of conventional radical polymerization.

Figure 1 shows the 500 MHz ^{1}H NMR spectrum of the polymer, whose M_n was determined as 10,300 from the GPC analysis with PMMA calibration. The aromatic signals appears at δ 7.66-7.95 are ascribed not to the contamination by PPh$_3$ (δ 7.29-7.36), but to quaternary phosphonium group, since the chemical shift coincides with that for the phenyl signals of the terminal model compound, ethyltriphenylphosphonium chloride, which appeared at δ 7.72-7.87. The integration of the signals observed in Figure 1 indicated that the polymer had 1.05 PPh$_3^+$ unit per molecule. Thus, chain transfer is considered to be negligible in the present system. The ^{31}P NMR analysis also showed the produced PMMA has a phosphonium end. The quantitative introduction of PPh$_3^+$ unit were also observed in a polymer prepared without solvent, which indicates that the polymerization mechanism in bulk is the same as that in solution. The ^{1}H NMR spectroscopy indicated the produced polymer was syndiotactic-rich (S:H:I = 65:32:3) as in the case of the anionic polymerization in THF or the radical polymerization. It shows no special interaction among the penultimate unit, the propagating center, and the approaching monomer.

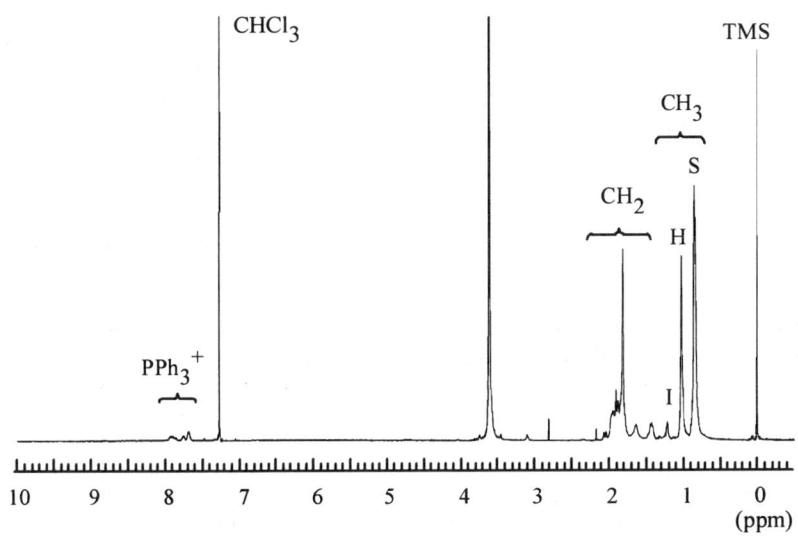

Figure 1. 500 MHz 1H NMR spectrum of the produced PMMA (M_n = 10,300)

Effect of Feed Ratios of Initiator and Activator to the Monomer

The control of the molecular weight is one of the most important features in chain polymerization. Although we did not expect a living nature for this polymerization, the evidence concerning the lack of chain transfer prompted us to examine the effect of the feed ratios of initiator and activator to MMA. The data are summarized in Table III. Increasing the amount of initiator from 0.05 eq. to 0.2 eq., while keeping the amount of the activator constant (1 eq.) caused the polymer yield to increase from 12 to 71%. However, a further increase of the amount of initiator to 0.3 eq. caused a decrease in the polymer yield. Thus, the effect of the feed ratio of initiator on the polymerization was found to be not so simple. Probably, a large amount of PPh_3 diminishes the concentration of MsOH available for the activation of MMA, which results in the decrease in the concentration of the actual monomeric species, H-MMA.

Table III. Effect of the Concentrations of Initiator and Activator on the Polymerization[a]

$[PPh_3]_0$ / $[MMA]_0$	$[MsOH]_0$ / $[MMA]_0$	Yield, %	PMMA M_n, $10^{4[b]}$	M_w/M_n[b]
0.05	1.0	12	0.53	2.0
0.1	1.0	48	0.72	2.0
0.2	1.0	71	1.0	2.1
0.3	1.0	23	0.8	1.3
0.1	0.33	22	2.0	2.2
0.1	0.5	58	0.76	1.9
0.1	1.0	48	0.72	2.0
0.1	1.5	17	4.9	1.5

[a] MMA; 3 mmol, p-MeOC$_6$H$_4$OH; 0.015 mmol, in CHCl$_3$ (0.5 mL) at 40 °C for 2 h.
[b] Determined from GPC with PMMA standards.

The effect of the ratio of the activator to the monomer in feed is not simple, either. The increase in yield when increasing the amount of the activator from 0.33 eq. to 0.5 eq. while keeping the amount of the initiator constant can be attributed to the increase of H-MMA, while the decrease in yield with the further increase of activator from 0.5 to 1.5 can be attributed to the decrease of the concentration of unprotonated PPh_3. The influence of the feed ratios on the molecular weight of the resulting PMMA was ambiguous, and no clear relationship was found. In conclusion, control of the molecular weight could not be attained by a change in the feed ratio.

Appropriate Brønsted Acid as the Activator

Thus, we successfully developed a novel polymerization system according to the new concept of HTP. The combination of PPh$_3$ and MsOH, however, has a number of serious defects. In addition to the lack of control of molecular weight, it requires a very large amount of the acid. Moreover, the polymer yield is not so high in any case: the maximum has been 71%. The relatively low polymer yield is considered to be the inevitable defect of this type of polymerization arising from the difference of the basicities between the monomer and PMMA. PMMA of secondary carboxy ester structure ensures a higher basicity than that of MMA of isopropenyl substituent. Therefore, the protonation of the monomer is disturbed by the PMMA produced by the polymerization. In the other case of the HTP of MMA by the combination of enamine and the Lewis acid (Scheme 5), on the other hand, the Lewis acid coordinates preferably to the less sterically crowded MMA than PMMA, and the quantitative production of the polymer is achieved.[10] The other two defects in the present polymerization, however, may be avoidable by choosing a more proper combination of initiator and activator.

Reduction of the amount of the activator was hoped to be attainable by the use of a stronger acid. The choice of CF$_3$SO$_3$H (pK_a = -14.5) instead of MsOH, however, failed in polymerization (Table IV). It is explained that TfOH is so strong that it protonate PPh$_3$ to inactivate it completely. The selection of a weaker acid was also not effective: the combination of PPh$_3$ with CCl$_2$HCO$_2$H (pK_a = 1.3) failed in polymerization since it is too weak to activate the monomer. Trifluoroacetic acid was effective as the activator, although the polymer yield was low. The usage of TfOH as the activator, however, became possible in the co-existence of diethyl ether, where the actual acidic species was diethyloxonium salt (pK_a = -2.4), a milder acid than TfOH.

Table IV. Effect of Brønsted acid on the polymerization of MMA initiated with PPh$_3$[a]

Acid	pK_a[b]	PMMA yield, %	M_n, 10^{4c}	M_w/M_n[c]
TfOH	-14.5	0	--	--
TfOH /Et$_2$O[d]	-2.4[e]	7.9	2.3	1.7
MsOH	-1.2	71	0.78	1.9
CF$_3$CO$_2$H	0.2	10	7.8	1.3
CCl$_2$HCO$_2$H	1.3	0	--	--

[a] MMA; 3 mmol, Acid; 3 mmol, PPh$_3$; 0.3 mmol, in 0.5mL of CHCl$_3$ with 0.015 mmol of p-MeOC$_6$H$_4$OH at 40 °C for 2 h.
[b] Values in H$_2$O at 25 °C from ref. 19.
[c] Determined from GPC with PMMA standards.
[d] In the presence of 0.1 mL of diethyl ether
[e] Value for diethyl ether from ref. 17.

The polymer yield was, however, lower than the run using MsOH. Obviously, a proper range of acid strength exists for the successful polymerization, which will depend on both initiator and monomer.

Diethyl Sulfide Initiating Systems

To use a stronger acid as the activator for the present polymerization, an initiator of lower basicity must be combined with it. The initiator would be preferably more nucleophilic than PPh$_3$, although it will be difficult to attain since the two characteristics conflict with each other. At present, no satisfactory combination has yet been found, although some combinations except for PPh$_3$/MsOH were found to induce the polymerization.

Diethyl sulfide is less basic than PPh$_3$: although a pK_a for Et$_2$S is not available, that for Me$_2$S is reported as –5.3.[17] The competitive protonation of initiator is, therefore, expected to be less significant with the use of Et$_2$S. Polymerization results using Et$_2$S as the initiator combined with MsOH or TfOH as the activator are summarized in Table V. Unfortunately, Et$_2$S is less nucleophilic than PPh$_3$.[20] The combination of 0.1 eq. of Et$_2$S with 1.0 eq. of MsOH gave no polymeric product after 2 h, supposedly, because of the low initiating ability of Et$_2$S. By prolonging the polymerization for 48 h, PMMA was produced by this combination, but the yield was poor, 2%. The use of a less basic initiator was considered to secure the use of a larger amount of the activator in comparison with the runs with Ph$_3$P. Contrary to our expectation, the highest polymer yield was attained in the run with 0.5 eq. of MsOH. No reasonable explanation is given to this problem at present.

The choice of the less basic initiator is expected to make the use of the stronger activator possible. Actually, TfOH was available as the activator, although the polymer yield was generally lower than the corresponding run using MsOH (Table V) due to the competing deactivation of the initiator. However, the most effective concentration of the activator for the polymerization could be reduced from 0.5 eq. for the MsOH activating system to 0.17 eq. by using TfOH, confirming the higher efficiency of TfOH as the activator. But, the required feed ratio of the initiator to the activator was stricter than the case with MsOH.

The polymer yield in the TfOH/Et$_2$S system could not be improved by the elongation of the polymerization time to 96 h. However, the bulk polymerization gave PMMA in 69% yield after 24 h, although the yield after 2 h was only 10%.

The above results clearly indicate that the proper combination of initiator, activator, and monomer is important, although the considerably low basicity of MMA makes the selection very difficult. Another combination able to catalyze polymerization is methyl trimethylsilyl dimethylketene acetal (MTDA, the representative initiator for GTP) with TfOH. Although the polymerization barely produced the polymer at 40 °C, it gave PMMA in moderate yield at 50 °C as shown in Table VII.

Table V. Polymerization of MMA with Et$_2$S as the Initiator[a]

$\frac{[Et_2S]_0}{[MMA]_0}$	Acid	$\frac{[Acid]_0}{[MMA]_0}$	PMMA yield, %	$M_n, 10^{4}$ [b]	M_w/M_n [b]
0.1	MsOH	0.10	0.5	--	--
0.1	MsOH	0.17	13	11	2.7
0.1	MsOH	0.33	36	7.1	1.3
0.1	MsOH	0.50	42	9.0	1.7
0.1	MsOH	1.0	1.9	--	--
0.2	MsOH	0.10	0.8	--	--
0.2	MsOH	0.17	12	3.4	1.5
0.2	MsOH	0.33	16	5.8	1.4
0.2	MsOH	0.50	33	7.4	1.3
0.2	MsOH	1.0	29	6.7	1.6
0.1	TfOH	0.10	2	--	--
0.1	TfOH	0.17	2.2	--	--
0.1	TfOH	0.33	0	--	--
0.2	TfOH	0.10	0	--	--
0.2	TfOH	0.17	28	2.2	2.5
0.2	TfOH	0.33	0	--	--
0.3	TfOH	0.10	8.4	3.2	1.9
0.3	TfOH	0.17	23	1.3	1.8
0.3	TfOH	0.33	11	2.3	1.7
0.3	TfOH	0.5	0	--	--

[a] MMA; 3 mmol, in 0.5mL of CHCl$_3$ containing 0.015 mmol of p-CH$_3$OC$_6$H$_4$OH at 40 °C for 24h.
[b] Determined from GPC with PMMA standards

Table VI. Time Dependence of the Polymerization of MMA with Et$_2$S and TfOH[a]

Solvent	Time, h	PMMA Yield(%)	M_n, 10^{4}[b]	M_w/M_n[b]
CHCl$_3$	2	3.0	3.5	3.1
CHCl$_3$	24	28	2.2	2.5
CHCl$_3$	96	30	2.8	9.9
Bulk	2	10	4.7	3.7
Bulk	24	69	4.4	10

[a] MMA; 3 mmol, TfOH; 0.5 mmol, Et$_2$S; 0.6 mmol, p-CH$_3$OC$_6$H$_4$OH; 0.015 mmol, at 40 °C.
[b] Determined from GPC with PMMA standards.

Table VII. Bulk Polymerization of MMA with MTDA and TfOH[a]

$[MTDA]_0 / [MMA]_0$	Temp, °C	PMMA yield, %	M_n, 10^{4}[b]	M_w/M_n[b]
0.1	40	0.5	--	--
0.1	50	71	2.2	2.5
0.2	40	0.2	--	--
0.2	50	41	4.7	3.7

[a] MMA; 3 mmol, TfOH; 3 mmol, p-CH$_3$OC$_6$H$_4$OH; 0.015 mmol, at 40 °C for 24 h.
[b] Determined from GPC with PMMA standards

Although the defects for the present Brønsted assisted nucleophilic polymerization have not yet been solved, and the polymerization conditions have not yet been optimized, we advocated the new concept of polymerization and succeeded to prove it. Moreover, this polymerization is applicable to other acrylic monomers. The combination of MsOH with PPh$_3$ was also effective to the polymerization of n-, sec-, and isobutyl methacrylates, methyl acrylate, and, very interestingly, methacrylic acid, but ineffective for 2-hydroxyethyl methacrylate.[21] The experiments are in progress, and the results will be reported elsewhere.

References

1) *Anionic Ring-Opening Polymerization of N-Carboxyanhydrides;* Kricheldorf, H. R.; Eastmont, G. C.; Ledwith, A.; Russo, S.; Sigwalt. P. Eds.; in *Comprehensive Polymer Science*, Pergamon Press, London, 1989, Vol. 3, Chapt. 36.

2) Biedron, T.; Szymanski, R.; Kubisa, P.; Penczek, S. *Makromol. Chem., Macromol. Symp.*, **1990**, *32*, 155.
3) Szwarc, M. *Pure Appl. Chem.*, **1966**, *12*, 127.
4) Otsu, T. *Makromol. Chem., Macromol. Symp.*, **1987**, *10/11*, 253.
5) Miyamoto, M.; Lange, P.; Kanetaka,. S. Saegusa, T. *Polymer Bull.*, **1995**, *34*, 249.
6) Sugimoto, H.; Kuroki, M.; Watanabe, T.; Kawamura, C.; Aida, T.; Inoue S. *Macromolecules*, **1993**, *26*, 3403.
7) Zubov, V. P.; Valuev, L. I.; Kabanov, V. A.; Kargin, V. A. *J. Polym. Sci., A-1*, **1971**, *9*, 833.
8) Rogueda, C.; Tardi, M.; Polton, A.; Sigwalt P. *Eur. Polym. J.*, **1989**, *25*, 885.
9) Srivastava, N.; Srivastava, A. K.; Rai, J. S. P. *Acta Polymerica*, **1989**, *40*, 411.
10) Miyamoto, M.; Kanetaka, S.; *J. Polym. Sci., Polym. Chem. Ed.*, **1999**, *37*, 3671.
11) Kitayama, T.; T. Hirano, T.; Hatada, K. *Polymer J.*, **1996**, *28*, 61.
12) Inoue, H.; Umeda, I.; Otsu, T. *Makromol. Chem.*, **1971**, *147*, 271.
13) Watanabe, T.; Sakai, K.; Sato, R.; Takeishi, M. *Polym. J.*, **1997**, *29*, 6.
14) Miyamoto, M.; Katano, H.; Kanda, C.; Kimura, Y. *Proceedings of ACS Polym. Mat. Sci. Eng.*, **1999**, *80*, 477.
15) Novak, B. M.; Cederstav, A. K. *Polymer Preprints*, **1995**, *36*, 548.
16) *Advanced Organic Chemistry, 4th Ed.*; March, J. John Wiley & Sons, New York, 1992, p. 250.
17) *Organic Solvent, 4th Ed.*; Riddick, J. H.; Bunger, W. B.; Sakano, T. K. John Wiley & Sons, New York, 1986, Chapt. 3.
18) *Donor-Acceptor Approach to Molecular Interaction*; Gutmann, V.; Plenum Press, New York, 1978.
19) Organic Chemistry, 5th Ed.; Pine, S. H. McGraw-Hill, New York, 1987, Chapt. 4.
20) The nucleophilic constant for Et_2S is 5.3, while that for PPh_3 is 7.0.
21) Miyamoto, M.; Maeda, D.; Kimura, Y. unpublished results.

Polymer Modification Using Transition Metals

Given that it is not always possible (or economical) to directly prepare every useful type of functionalized macromolecule, the post-polymerization modification of polymers is an important synthetic strategy. Methodologies ranging from the hydrolysis of poly(vinyl acetate) to give poly(vinyl alcohol) to the surface modification of polyolefins are used widely in industry to create materials with improved chemical and mechanical behavior. Polyolefins, which lack the functionality needed to give certain desirable properties such as adhesion and dyeability, are an especially important target for post-polymerization modification techniques.

Despite the widespread advantages that organometallic catalysts have produced for the development of new polymer syntheses, relatively little work has been carried out in the area of metal-mediated catalytic polymer modification. Once again, the variety of organometallic reaction mechanisms offers many possibilities for the transformation of endgroup, sidechain, and even main chain functionalities into new structures.

In this section, three papers discussing the catalytic modification of polymers are presented. Jun et al. present a brief introduction to known techniques for polybutadiene modification, and report rhodium-mediated methods for hydroacylation and simultaneous hydroacylation / hydrogenation of polybutadiene. The hydrogenation of polybutadiene / polystyrene block copolymers with metallocenes is discussed by Tsiang et al. Finally, a paper by Britcher and Matisons features a nicely developed heterogeneous catalytic polymer modification technique: the hydrosilation of polysiloxanes with platinum complexes supported on reusable glass fibers.

Chapter 6

Modification of Polybutadiene by Transition Metal Catalysts: Hydroacylation of Polybutadiene

Chul-Ho Jun, Hyuk Lee, Jun-Bae Hong, and Dae-Yon Lee

Department of Chemistry, Yonsei University, 120–749, Seoul, Korea

The vinyl groups in polybutadiene are hydroacylated in the presence of a transition metal complex to give acyl-functionalized polybutadiene. Aldimine, aldehyde, and primary alcohol can be used for hydroacylation as substrates. It is noteworthy that the hydrogenation of double bonds as well as hydroacylation are achieved when primary alcohol is used. A variety of functional groups, including acylferrocenyl, benzoyl and acylheteroaromatic, are incorporated into the polymer backbone. The modified polymers are characterized by ^{1}H NMR, ^{13}C NMR, and IR spectroscopy. The brief overviews on the other modification methods are also presented.

The modification of polymers by transition metal catalyst has received much attention as a new method for obtaining functionalized polymers with controlled molecular weights and well-defined microstructures (1, 2). This type of approach has been realized to be more effective than conventional methods. Modified polymers usually possess enhanced physical or chemical properties compared with their precursors. Moreover, the incorporated functional groups can permit further modification of polymers. Polybutadiene has been preferred as a starting material, since it is available at diverse ranges of molecular weight and well-defined microstructures. Numerous examples of the modification of polybutadiene through metal-mediated reactions have been reported, such as hydrogenation, hydroformylation, hydrocarboxylation, and hydrosilylation. Using these processes, desirable functionalized polymers that are difficult to obtain by polymerization of monomers, can be synthesized.

In this chapter, we will discuss the new methods of modifying polybutadiene by hydroacylation, following the brief overviews on the other modification methods previously reported.

© 2000 American Chemical Society

Transition Metal Catalyzed Functionalization of Polybutadiene

There are three isomeric olefin units in polybutadiene that can serve as reactive sites in the course of reaction: 1,2-unit (vinyl group), *cis*-1,4-unit, and *trans*-1,4-unit (internal olefin). All of these olefin units can participate during the reactions. However, regioselectivity have been observed in some cases.

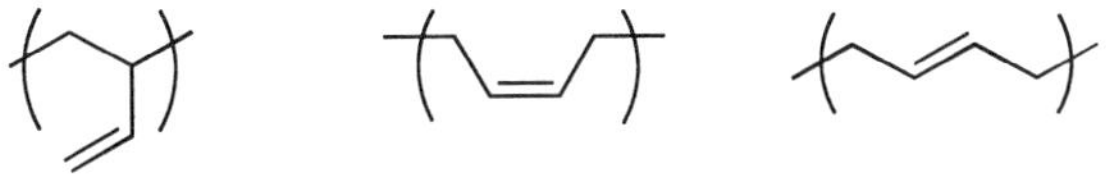

1,2-unit (vinyl group) *cis / trans*-1,4-unit (internal olefin)

One of the widely used methods for modification of polybutadiene is hydrogenation using a homogeneous or heterogeneous catalyst system (2, 3).

Palladium deposited on $CaCO_3$ or $BaSO_4$ is commonly used as a heterogeneous catalyst (4-8) while a rhodium-based catalyst, especially $(PPh_3)_3RhCl$ (Wilkinson's complex), is preferred in the homogeneous system (9-15). Although heterogeneous catalysts are quite reactive and are widely used, most recent research has employed homogeneous catalysis since it proceeds under very mild conditions and shows enhanced selectivity compared with the heterogeneous system. Saturation of residual double bonds results in high stability of the modified polymer toward oxygen, radiation, and thermal exposure (2, 16).

Hydroformylation is another useful method for the modification of polybutadiene, in which the resulting hydroformylated polymer can be easily reduced to polyalcohol (3, 9, 17-21).

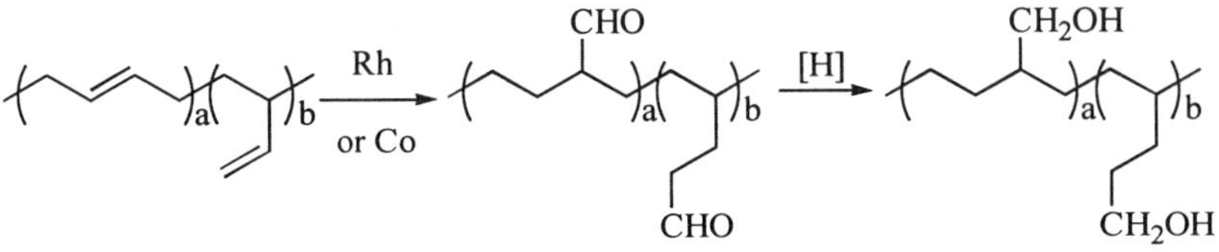

Although both rhodium and cobalt-based catalysts have been known to be active for the hydroformylation of olefin, the cobalt-based catalyst was proven not to be so efficient in reaction with polymer because of the side reaction resulting in gelation of the product (1). With the rhodium-based catalyst, on the other hand, the gelation is no longer a serious problem, and higher efficiency and selectivity than with the cobalt-based catalyst are observed (17-21). The conversion rate usually ranges from 20 to 30 %, but the rate increases up to ca. 80 % when a large excess of phosphine is added (18). To reduce hydroformylated polymer to polyalcohol, a ruthenium catalyst as well as a traditional reducing reagent such as $NaBH_4$ was adopted (3, 9).

Hydrocarboxylation of polybutadiene affords carboxylated polymer that has potential for film and surface coating (22, 23). Palladium catalysts are preferred, and it is noteworthy that regioselectivity can be controlled by changing the catalyst system. When $[PdCl_2(PPh_3)_2]/SnCl_2$ is used, only the vinyl group in polybutadiene is hydrocarboxylated, while both the vinyl group and the internal olefin participated in hydrocarboxyation in the presence of $[PdCl_2(PPh_3)_2]/PPh_3$ or $PdCl_2/CuCl_2/O_2$ (22).

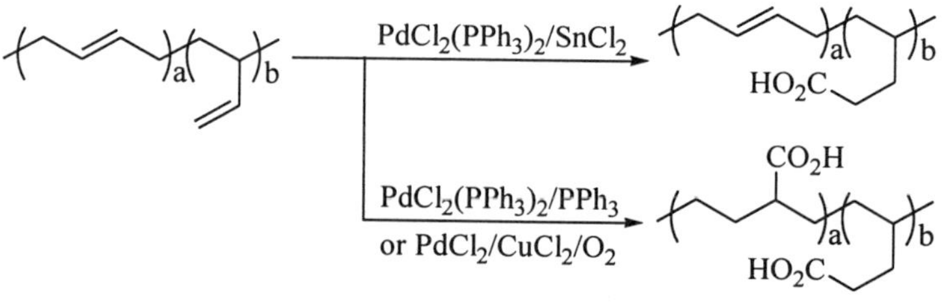

Silane-modified polymers, which can be obtained by hydrosilylation, are known for the enhanced properties of adhesion to fillers and thermal resistance (3, 24-28). In addition to the altering properties, the hydrosilylation of a polymer can offer reactive sites for further modification (26-28). Platinum-based catalysts are frequently used for hydrosilylation. In the case of rhodium-based catalysis, high regioselectivity on the vinyl group over the internal olefin in polybutadiene was observed, and both anti-Markovnikov and Markovnikov addition took place depending on the structure of polymer and silane substrate (24, 25).

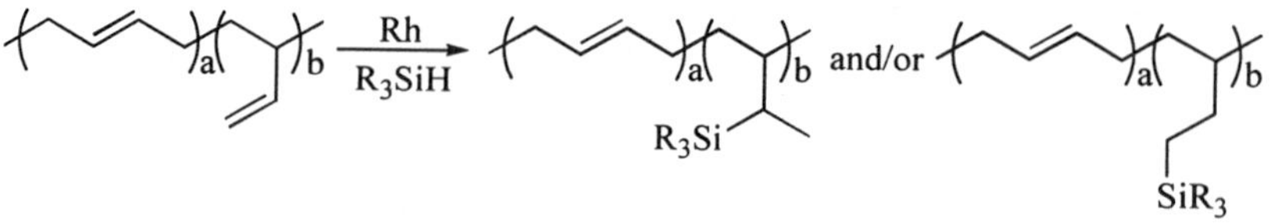

Chelation-Assisted Hydroacylation

Hydroacylation is a direct method to synthesize a ketone from olefin and aldehyde. While intramolecular hydroacylation to produce cyclic ketones has been well developed (29, 30), intermolecular hydroacylation has suffered from decarbonylation (31) occurring in the acylmetal hydride intermediate. There have been many efforts to inhibit decarbonylation by stabilizing the intermediate: pressurizing carbon monoxide (32) or ethylene (33). However, these methods also have limitations such as high-pressure conditions and low selectivity. Another promising method for this purpose is to use cyclometalation that renders to stabilize the intermediate by chelation-assistance (34, 35).

Suggs developed hydroiminoacylation using an aldimine prepared from aldehyde and 2-amino-3-picoline (36, 37). In this reaction, aldimine reacted with 1-alkene in the presence of Wilkinson's complex to give ketimine, which was hydrolyzed to yield ketone.

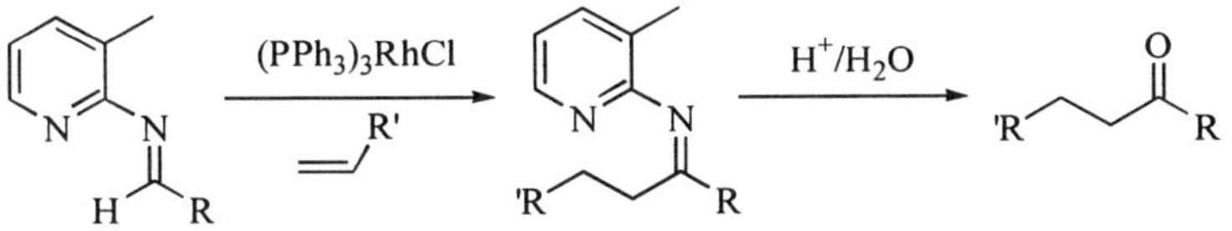

The mechanism of hydroiminoacylation is depicted in Figure 1.

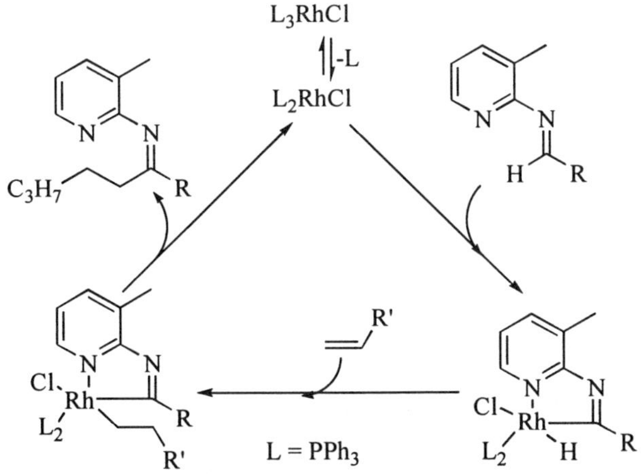

Figure 1. Mechanism of Hydroiminoacylation

The first step is C-H bond activation of aldimine by a Rh(I) catalyst to yield an iminoacylrhodium(III) hydride complex. The hydride insertion into 1-alkene generates iminoacylrhodium(III) alkyl and subsequent reductive elimination produces ketimine with regeneration of the catalyst. Decarbonylation is avoided because, instead of acylmetal hydride, iminoacylrhodium hydride is involved as an intermediate. Although various aldehydes can be transformed to ketone by hydroiminoacylation, two additional steps are required: preparation of aldimine and hydrolysis of the resulting ketimine.

This type of reaction further developed into imine-mediated intermolecular hydroacylation, which adopted a cocatalyst system of organic and organometallic species, 2-amino-3-picoline and Wilkinson's complex (38).

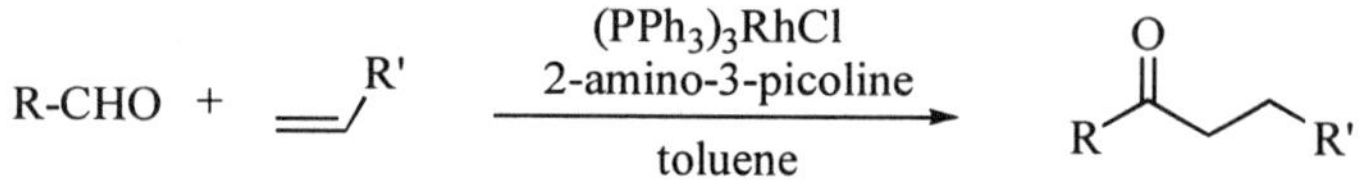

$$\text{R-CHO} + \underset{}{=\!\!\!/^{R'}} \xrightarrow[\text{toluene}]{\substack{(PPh_3)_3RhCl \\ \text{2-amino-3-picoline}}} \underset{R}{\overset{O}{\|}}\!\!\!\diagdown\!\!\diagup^{R'}$$

The most plausible mechanism of the reaction is shown in Figure 2. It is believed that aldimine is generated *in situ* by condensation of aldehyde and 2-amino-3-picoline with the generation of H_2O, then subsequent hydroiminoacylation proceeds to give ketimine. The resulting ketimine is hydrolyzed by previously formed H_2O to give ketone with regeneration of 2-amino-3-picoline.

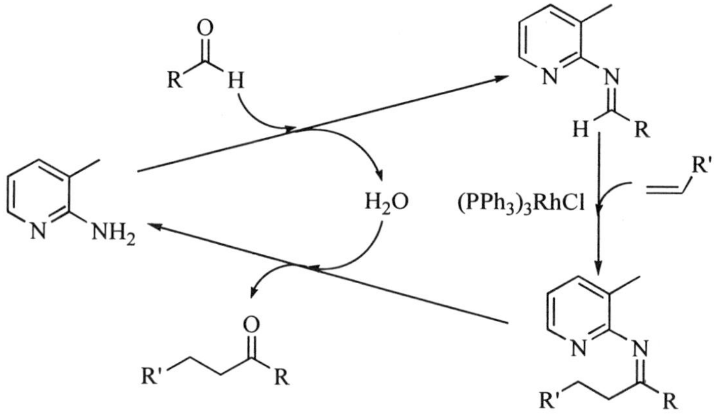

Figure 2. Mechanism of Imine-Mediated Hydroacylation

Chelation-assisted hydroacylation has been extended further to the use of primary alcohol instead of aldehyde as a substrate. Primary alcohol reacts with 1-alkene to give a mixture of ketone and alkane (39).

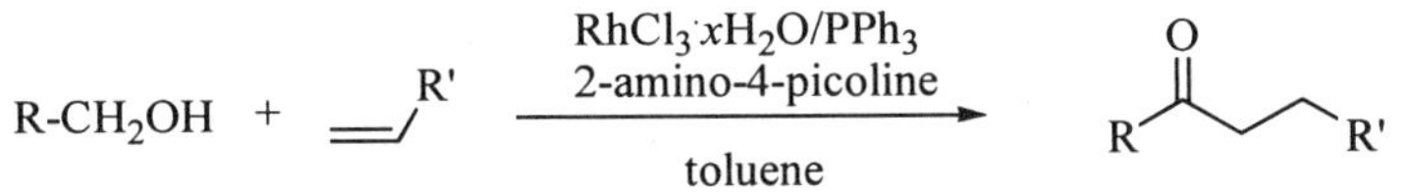

$$\text{R-CH}_2\text{OH} + \underset{}{=\!\!/}^{\text{R'}} \xrightarrow[\text{toluene}]{\substack{\text{RhCl}_3\cdot x\text{H}_2\text{O/PPh}_3 \\ \text{2-amino-4-picoline}}} \underset{\text{R}}{\overset{\text{O}}{\bigsqcup}}\!\!\diagdown\!\!\text{R'}$$

The reaction consists of two consecutive reactions: oxidation and hydroacylation. The first step is the oxidation of primary alcohol to aldehyde during which time 1-alkene serves as a hydrogen acceptor and is reduced to alkane (40). Once aldehyde is formed, it reacts with 1-alkene to give ketone by chelation-assisted hydroacylation. Since 1-alkene is consumed in both steps of oxidation and hydroacylation, at least two equivalents of 1-alkene are required. The catalytic system is optimized as 2-amino-4-picoline and $\text{RhCl}_3 \cdot x\text{H}_2\text{O}$ and PPh_3, which might be freshly transformed into $(\text{PPh}_3)_3\text{RhCl}$.

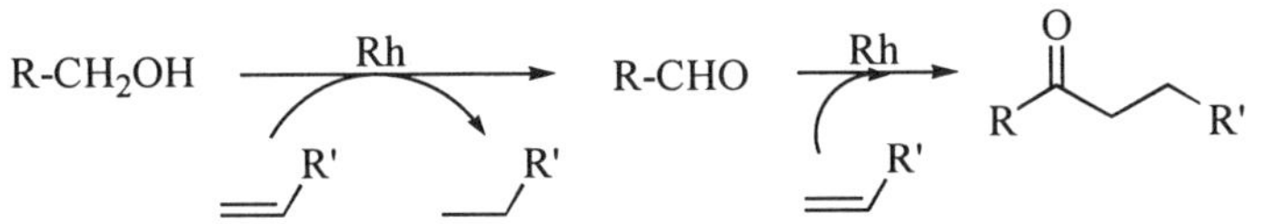

$$\text{R-CH}_2\text{OH} \xrightarrow{\text{Rh}} \text{R-CHO} \xrightarrow{\text{Rh}} \text{R}\!\!-\!\!\overset{\text{O}}{\underset{}{\bigsqcup}}\!\!-\!\!\text{R'}$$

The vinyl groups in polybutadiene as well as 1-alkenes have been assumed to be the reactive sites for chelation-assisted hydroacylation, by which various acyl groups could be incorporated into the polybutadiene backbone. Furthermore, selective modification on the vinyl group in polybutadiene was expected since the internal olefin is less reactive than the terminal olefin in hydroacylation. In the following sections, some of these results will be presented.

Hydroacylation of Polybutadiene

In this work, various acyl-functionalized polybutadienes were synthesized by hydroacylation methods described in the previous section. Two types of polybutadiene were chosen as substrates: one contains 27 % of the vinyl group and 73 % of the internal olefin (**1a**); the other contains 45 % of the vinyl group (**1b**). Produced polymers were characterized by IR, ^1H NMR, and ^{13}C NMR spectroscopy.

Hydroiminoacylation of Polybutadiene

Transition metal-containing polymers are of interest because they have potential as new types of catalyst and useful materials (41, 42). Polymers especially bearing a ferrocenyl group are known as additives for solid propellants (43). This type of polymer has been commonly synthesized through polymerization of monomers bearing ferrocenyl group moieties (44-46).

Hydroiminoacylation is another method for incorporating the ferrocenyl group directly into polybutadiene (47).

The substrate for hydroiminoacylation is ferrocenecarboxaldimine, prepared by condensation of 2-amino-3-picoline and ferrocenecarboxaldehyde. The hydroiminoacylation of PTPB (phenyl terminated polybutadiene) **1a** with ferrocene-carboxaldimine was carried out in the presence of Wilkinson's complex to give ketimine-incorporated PTPB. After acid-catalyzed hydrolysis of the resulting ketimine-incorporated polymer, the acylferrocene-impregnated PTPB **2** was isolated (Scheme 1). The conversion rate was 74 %, which is calculated by measuring the integration of peaks in the ^{1}H NMR spectra. The partially acylferrocenyl group-impregnated PTPB can be rehydroacylated under identical reaction conditions to give the completely hydroacylated PTPB **3**.

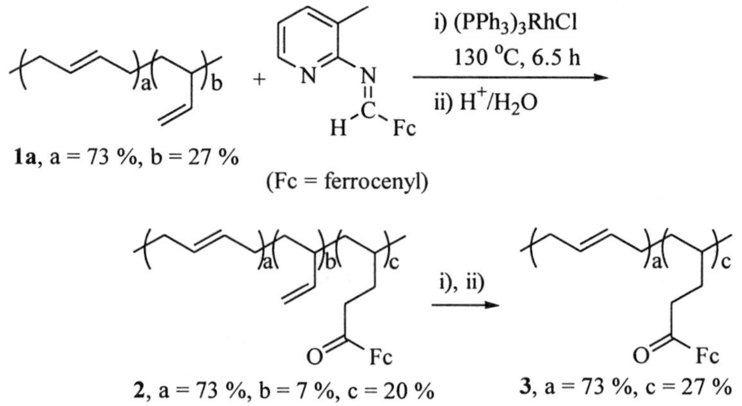

Scheme 1

The IR spectra of PTPB **1a** and hydroacylated PTPB are shown in Figure 3. There are new peaks at 1680 cm^{-1} due to the carbonyl group in the spectra of hydroacylated PTPB. The characteristic band of the vinyl group at 910 cm^{-1} is dramatically decreased in **2** and finally disappears in **3**, while those of *trans*-internal olefin at 964 cm^{-1} and *cis*-internal olefin at 725 cm^{-1} remain unchanged. In the ^{1}H NMR spectra of **2** and **3**, there appear new peaks for CH$_2$ α to the carbonyl group at 2.7 ppm (Figure 4). The degree of hydroacylation is calculated by measuring the ratio of the vinyl group and the internal olefin in the ^{1}H NMR spectra. The peaks of vinyl CH$_2$ are in the range of 4.9-5.0 ppm, and those of the internal olefinic –CH=CH– and the vinyl –CH= are in 5.3-5.6 ppm. According to the ^{1}H NMR spectra, conversion rate for **2** is measured to be 74 % and that for **3** is 100%. If hydroacylation of internal olefin occurs, the peak for CH α to the carbonyl group should have appeared. However only peak for CH$_2$ α to the carbonyl group is observed. These results from IR and NMR spectra show that hydroacylation exclusively occurs on the vinyl group in PTPB. The ^{13}C NMR spectra also showed the characteristic peaks of **3**. While the peaks of carbons in the vinyl group at 142.6 and 114.3 ppm in **1a** have completely disappeared, new characteristic peaks of the acylferrocenyl group have appeared at 79.0 ppm (C-1

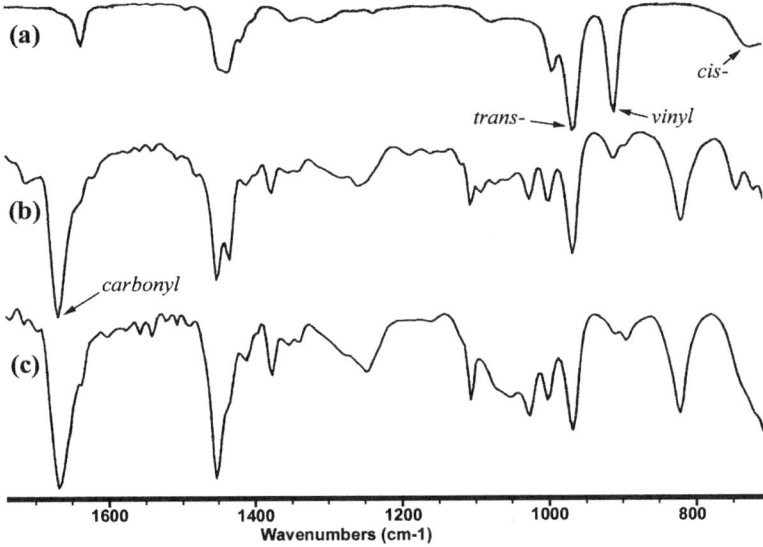

*Figure 3. IR Spectra of PTPB: (a) **1a**, (b) **2**, and (b) **3***

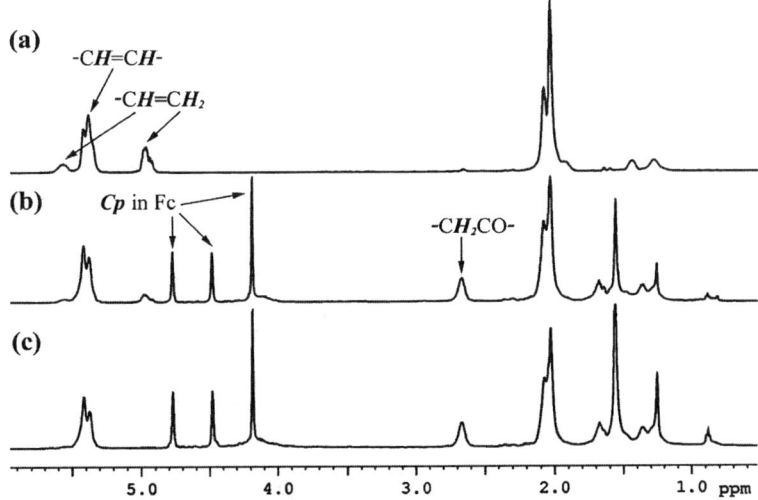

*Figure 4. ^{1}H NMR Spectra of PTPB: (a) **1a**, (b) **2**, and (c) **3**. (Reproduced with permission from reference 47. Copyright 1993 Elsevier Science.)*

in substituted Cp group), 72.0 ppm (C-2, 5 in substituted Cp group), 69.6 ppm (unsubstituted Cp group), 69.3 ppm (C-3, 4 in substituted Cp group), and 204.6 ppm for the carbonyl group.

Procedure of Hydroiminoacylation of PTPB with Ferrocenecarboxaldimine

A screw-capped pressure vial was charged with 37.2 mg (0.0402 mmol) of Wilkinson's complex dissolved in 3 mL of toluene. To the solution was added 122 mg (0.402 mmol) of ferrocenecarboxaldimine (47) and 84.3 mg of PTPB **1a**, and the mixture was heated at 130 °C for 6.5 h. The reaction mixture was hydrolyzed with 10 mL of 1 N HCl aqueous solution. The product was extracted with 20 mL of chloroform, and purified by column chromatography (n-hexane / ethyl acetate = 5/2) to give 99 mg (67 % yield based upon **1a**) of **2**.

Imine-Mediated Hydroacylation of Polybutadiene

Although the acyl group can be incorporated into PTPB successfully by hydroiminoacylation, the reaction still needs multi steps. With the development of imine-mediated hydroacylation of 1-alkene by cocatalyst system, we applied this system to the modification of PTPB (48).

PTPB **1a** was allowed to react with various aromatic aldehydes under the cocatalyst system of Wilkinson's complex and 2-amino-3-picoline (Scheme 2).

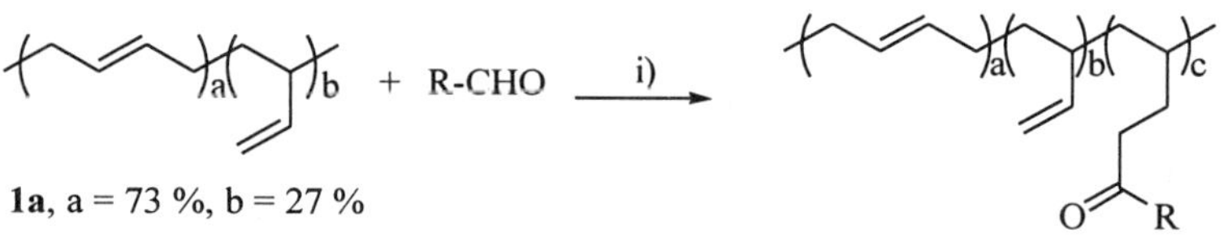

1a, a = 73 %, b = 27 %

i) 10 mol% $(PPh_3)_3RhCl$, 10 mol% PPh_3,
100 mol% 2-amino-3-picoline,
70 mol% H_2O, 130 °C, 24 h

	R =	conversion rate (c/b+c)(%)
4	$Me_2N-C_6H_4-$	60
5	$CH_3-C_6H_4-$	42
6	Ph-	32
7	$F-C_6H_4-$	23
8	$CF_3-C_6H_4-$	8
9	Fc-	21

Scheme 2

PPh_3 and H_2O were added to improve the efficiency of the catalyst system. PPh_3 is supposed to enhance the activity of the Rh(I) complex by replacing the oxidized triphenylphosphine, and additional H_2O might constrain an intermediate ketimine to hydrolyze to give ketone. The reactions of various aldehydes including ferrocenecarboxaldehyde with PTPB were examined. For ferrocenecarboxaldehyde, only 21% of vinyl groups were hydroacylated. For aldehydes bearing an electron-donating substituent such as *N,N*-dimethylamino group, a high conversion rate was observed compared with ones bearing an electron-withdrawing substituent such as trifluoromethyl group.

General Procedure of Hydroacylation of PTPB with aldehyde

A screw-capped vial was charged with 37 mg (0.04 mmol) of Wilkinson's complex dissolved in 1 mL toluene. To the solution were added 100 mg of PTPB **1a**, 0.4 mmol of aldehyde, 43.3 mg (0.4 mmol) of 2-amino-3-picoline, 10.5 mg (0.4 mmol) of PPh_3 and 0.28 mmol of H_2O, then the mixture was heated at 130 °C for 24 h. After evaporating solvent, the mixture was purified by column chromatography (*n*-hexane/ethylacetate = 2/5) to give the corresponding hydroacylated polymers (**4-9**).

Simultaneous Hydrogenation and Hydroacylation of Polybutadiene

When primary alcohol is used for the chelation-assisted hydroacylation instead of aldimine or aldehyde as a substrate, hydrogenation as well as hydroacylation is observed. It is quite interesting that PTPB acts as both a hydrogen acceptor and a hydroacylation substrate.

PTPB **1a** reacts with various aromatic primary alcohols under the cocatalyst system (10 mol% of $RhCl_3 \cdot xH_2O$, 20 mol% of PPh_3, and 100 mol% of 2-amino-4-picoline) at 150°C for 24 h (Scheme 3). After the reaction, the produced polymers were purified by column chromatography, then characterized by IR, ^1H NMR and ^{13}C NMR spectroscopy (49). The results are summarized in Table 1.

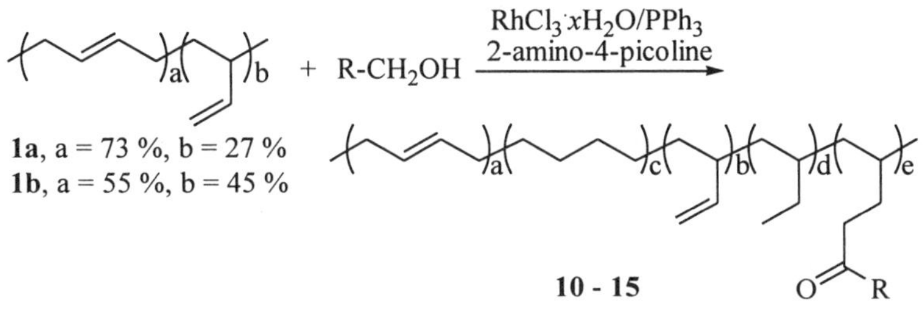

Scheme 3

In IR Spectrum of **10a**, the peak of the vinyl group at 910 cm^{-1} disappeared completely, and new peak of the carbonyl group appeared at 1687 cm^{-1} (Figure 5). The ^{13}C NMR spectrum provided the proof of the hydroacylation with the peak for CH_2 α to the carbonyl group at 36.0 ppm and of carbonyl carbon at 200.5 ppm from benzoylmethylene group (Figure 6). The CH_3 peak of the ethyl group generated from the hydrogenation of the vinyl group also appeared at 10.9 ppm with disappearance of the signals of vinyl carbons at 142.6 and 114.3 ppm. The ^1H NMR spectrum of **10a** is shown in Figure 7. It presents crucial information about the amounts of hydrogenation (*c, d*) as well as those of hydroacylation (*e*) of the vinyl groups by measuring the integration of the peaks at 0.83 ppm (CH_3 of the ethyl group) and at 2.93ppm (CH_2 α to the carbonyl group). For **10a**, 37 % of vinyl groups and 12 % of internal olefins were hydrogenated and 63 % of vinyl groups were hydroacylated. That is, all the vinyl groups are either hydrogenated or hydroacylated. The internal olefin only serves as a hydrogen acceptor, but the vinyl group participates in both hydrogenation and hydroacylation. For hydrogenation, the vinyl group is slightly more reactive than the internal olefin (8 % and 10 % for the internal olefin and vinyl group, respectively) The different reactivity between two types of double bonds can be explained in terms of steric influence. Because the vinyl group is less sterically hindered than the internal olefin, the catalyst can access vinyl group more easily.

When PTPB 1b was used instead of 1a under identical conditions, 51 % of the vinyl groups and only 7 % of the internal olefins were hydrogenated, while 49 % of the vinyl group was hydroacylated (10b). In this case, the amount of hydrogenation exceeds that of hydroacylation, which implies hydroacylation is more susceptible to steric hindrance than is hydrogenation. The reason is probably that the bulky intermediate, iminoacylrhodium(III) hydride, is involved in the course of hydroacylation, while the less hindered rhodium(III) hydride is involved in hydrogenation. Various primary alcohols were examined for this reaction to give the acyl-functionalized PTPB (Table 1).

Table 1. Hydroacylation of PTPB 1 with Various Primary Alcohols

Entry	Product	R	Proportion[a] (%)					Yield [b] (%)
			a	b	c	d	e	
1	10a	Phenyl	65	0	8	10	17	75
2	10b		51	0	4	23	22	79
3	11a	4-Methoxy phenyl	68	0	5	10	17	68
4	11b		45	0	10	22	23	70
5	12a	2-Naphthyl	64	2	9	11	14	79
6	12b		42	4	13	23	18	75
7	13a	4-Biphenyl	59	1	14	12	14	80
8	13b		42	4	13	24	17	84
9	14a	3-Pyridyl	72	5	1	9	13	72
10	14b		42	7	13	27	11	80
11	15a	2-Thiophenyl	63	4	10	10	13	73
12	15b		48	5	7	19	21	75

[a] Calculated by measuring the integration of ^{1}H NMR spectra (a = 1,4-unit, b = hydrogenated 1,4-unit, c = 1,2-unit, d = hydrogenated 1,2-unit, e = acylated 1,2-unit, total = 100 %) [b] Based on PTPB.(Reproduced with permission from reference 49. Copyright 1998 Elsevier Science.)

Procedure of Hydroacylation of 1a with Benzyl Alcohol

In a 25-mL stainless autoclave were placed 737 mg (3.5 mmol) of 1a, 40 mg (0.18 mmol) of RhCl$_3$·xH$_2$O (purchased from Pressure Chemical Co. and used without further purification), 94 mg (0.36 mmol) of PPh$_3$, 382 mg (3.5

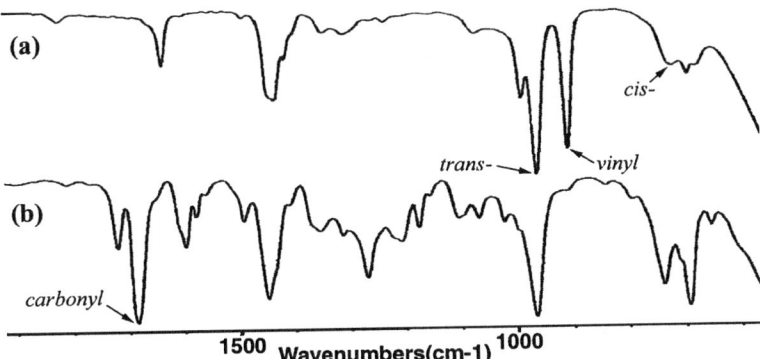

*Figure 5. IR Spectra of PTPB: (a) **1a** and (b) **10a***

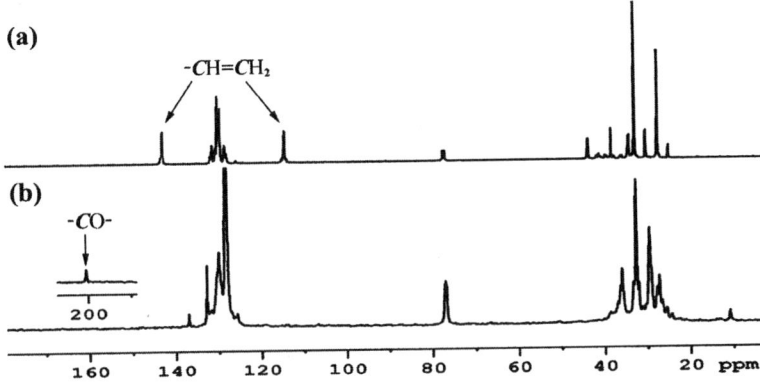

*Figure 6. ^{13}C NMR Spectra of PTPB: (a) **1a** and (b) **10a***

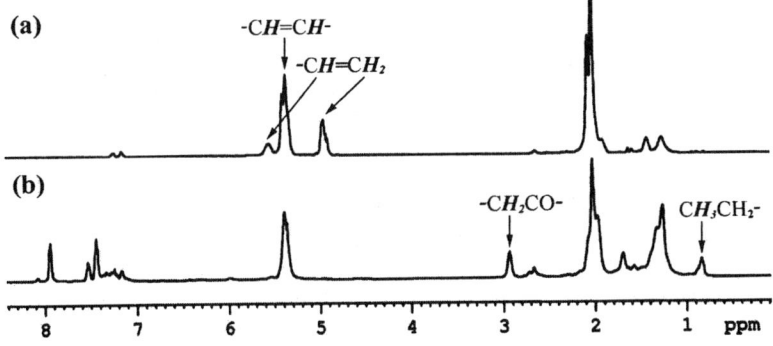

*Figure 7. ^{1}H NMR Spectra of PTPB: (a) **1a** and (b) **10a***

mmol) of 2-amino-4-picoline, 1923 mg (17.8 mmol) of benzyl alcohol, and 3692 mg (40.1 mmol) of toluene. After heating at 150 °C for 24 h, the mixture was purified by column chromatography (*n*-hexane/ethyl acetate = 5/2) to give 746 mg (75 % yield based upon **1a**) of **10a**. For the other primary alcohols , the reactions were conducted in 1 mL screwed-capped pressure vial (49).

Conclusion

We have presented transition metal catalyzed modification of polybutadiene to afford the acyl-functionalized polymers by means of hydroacylation. When aldimine or aldehyde was used as a substrate, hydroacylation occurred exclusively on the vinyl group. The hydroacylation of polybutadiene with primary alcohol accompanied simultaneous hydrogenation of the vinyl group and the internal olefin. These types of hydroacylation are the new methods of incorporating functional groups or new pendant groups into polybutadiene. By using these methods, new functionalized polymer with various properties could be synthesized.

Literature Cited

1. McGrath, M. P.; Sall, E. D.; Tremont, S. J. *Chem. Rev.* **1995**, *95*, 381-398.
2. McManus, N. T.; Rempel, G. L. *Rev. Macromol. Chem. Phys.* **1995**, *C35*, 239-285.
3. Mohammadi, N. A.; Rempel, G. L. In *Chemical Reactions on Polymers;* Benham, J. L., Kinstle, J. F., Eds.; ACS Symposium Series 364; American Chemical Society: Washington, DC, 1988; pp 393-408.
4. Rachapudy, H.; Smith, G. G.; Raju, V. R.; Graessley, W. W. *J. Polym. Sci.: Polym. Phys. Ed.* **1979**, *17*, 1211-1222.
5. Zhongde, X.; Hadjichristidis, N.; Carella, J. M.; Fetters, L. J. *Macromolecules* **1983**, *16*, 925-929.
6. Carella, J. M.; Graessley, W. W.; Fetters, L. J. *Macromolecules* **1984**, *17*, 2775-2786.
7. Rosedale, J. H.; Bates, F. S. *J. Am. Chem. Soc.* **1988**, *110*, 3542-3545.
8. Gehlsen, M. D.; Bates, F. S. *Macromolecules* **1993**, *26*, 4122-4127.
9. Mohammadi, N. A.; Ling, S. S. M.; Rempel, G. L. *Polym. Prepr.* **1986**, *27*, 95-96.
10. Doi, Y.; Yano, A.; Soga, K.; Burfield, D. R. *Macromolecules* **1986**, *19*, 2409-2412.
11. Gilliom, L. R. *Macromolecules* **1989**, *22*, 662-665.
12. Mohammadi, N. A.; Rempel, G. L. *J. Mol. Catal.* **1989**, *50*, 259-275.
13. Guo, X.; Rempel, G. L. *J. Mol. Catal.* **1990**, *63*, 279-298.
14. Mohammadi, N. A.; Rempel, G. L. *Macromolecules* **1987**, *20*, 2362-2368.
15. Bhattacharjee, S.; Bhowmick, A. K.; Avasthi. B. N. *Ind. Eng. Chem. Res.* **1991**, *30*, 1086-1092.
16. Schultz, D. N. In *Handbook of Elastomers;* Bhowmick, A. K., Stephens, H. L., Eds.; Marcel Dekker: New York, 1988; pp 75-100.

17. Azuma, C.; Mitsuboshi, T.; Sanui, K.; Ogata, N. *J. Polym. Sci.: Polym. Chem. Ed.* **1980**, *18*, 781-797.
18. Tremont, S. J.; Remsen, E. E.; Mills, P. L. *Macromolecules* **1990**, *23*, 1984-1993.
19. Mills, P. L.; Tremont, S. J.; Remsen, E. E. *Ind. Eng. Chem. Res.* **1990**, *29*, 1443-1454.
20. Scott, P. J.; Rempel, G. L. *Macromolecules* **1992**, *25*, 2811-2819.
21. McGrath, M. P.; Sall, E. D.; Foster, D.; Trement, S. J.; Sendijarevic, A.; Sendijarevic, V.; Primer, D.; Jiang, J.; Iyer, K.; Klempner, D.; Frisch, K. C. *J. Appl. Polym. Sci.* **1995**, *56*, 533-543.
22. Narayanan, P.; Clubley, B. G.; Cole-Hamilton, D. J. *J. Chem. Soc., Chem. Commun.* **1991**, 1628-1630.
23. Ajjou, A. N.; Alper, H. *Macromolecules* **1996**, *29*, 1784-1788.
24. Guo, X.; Farwaha, R.; Rempel, G. L. *Macromolecules* **1990**, *23*, 5047-5054.
25. Guo, X.; Farwaha, R.; Rempel, G. L. *Macromolecules* **1992**, *25*, 883-886.
26. Iraqi, A.; Watkinson, M.; Crayston, J. A.; Cole-Hamilton, D. J. *J. Chem. Soc., Chem. Commun.* **1991**, 1767-1769.
27. Hempenius, M. A.; Michelberger, W.; Moller, M. *Macromolecules* **1997**, *30*, 5602-5605.
28. Xenidou, M.; Hadjichristidis, N. *Macromolecules* **1998**, *31*, 5690-5694.
29. Barnhart, R. W.; Wang, X.; Noheda, P.; Bergens, S. H.; Whelan, J.; Bosnich, B. *J. Am. Chem. Soc.* **1994**, *116*, 1821-1830.
30. Bosnich, B. *Acc. Chem. Res.* **1998**, *31*, 667-674.
31. Colquhoun, H. M.; Thompson, D. J.; Twigg, M. V. *Carbonylation: Direct Synthesis of Carbonyl Compound*; Plenum: New York, 1992; pp205-225.
32. Kondo, T.; Akazome, M.; Tsuji, Y.; Watanabe, Y. *J. Org. Chem.* **1990**, *55*, 1286-1291.
33. Marder, T. B.; Roe, D. C.; Milstein, D. *Organometallics* **1988**, *7*, 1451-1453.
34. Bruce, M. I. *Angew. Chem., Int. Ed. Engl.* **1977**, *16*, 73-86.
35. Jun, C.-H.; Hong, J.-B.; Lee, D.-Y. *Synlett* **1999**, 1-12.
36. Suggs, J. W. *J. Am. Chem. Soc.* **1979**, *101*, 489.
37. Suggs, J. W. U.S. Patent 4,241,206, Dec 23, 1980.
38. Jun, C.-H.; Lee, H.; Hong, J.-B. *J. Org. Chem.* **1997**, *62*, 1200-1201.
39. Jun, C.-H.; Huh, C.-W.; Na, S.-J. *Angew. Chem., Int. Ed.* **1998**, *37*, 145-147.
40. Zassinovich, G.; Mestroni, G. *Chem. Rev.* **1992**, *92*, 1051-1069.
41. Hartley, F. R. *Supported Metal Complexes*; D. Reidel Publishing Co.: Dordrecht, Netherlands, 1985.
42. Shirai, H.; Hojo, N. In *Functional Monomers and Polymers*; Takemoto, K., Inaki, Y., Ottenbrite, R. M. Eds.; Marcel Dekker: New York, 1987; pp49-148.
43. Reed, S. F., Jr. U.S. Patent 3,813,304, May 28, 1974.
44. Landuyt, D. C.; Reed, S. F., Jr. *J. Polym. Sci.: Part A-1* **1971**, *9*, 523-529.
45. Reed, S. F., Jr. *J. Polym. Sci.: Polym. Chem. Ed.* **1981**, *19*, 1867-1869.
46. Manners, I. *Adv. Organomet. Chem.* **1995**, *37*, 131-168.
47. Jun, C.-H.; Kang, J.-B.; Kim, J.-Y. *J. Organomet. Chem.* **1993**, *458*, 193-198.
48. Kim, J.-H.; Jun, C.-H. *Bull. Korean Chem. Soc.* **1999**, *20*, 27-29.
49. Jun, C.-H.; Hwang, D.-C. *Polymer* **1998**, *39*, 7143-7147.

Chapter 7

Metallocene Catalysts Used for the Hydrogenation of Polystyrene-*b*-Polybutadiene-*b*-Polystyrene Block Copolymers

Raymond Chien-chao Tsiang, Wen-shen Yang, and Ming-der Tsai

Department of Chemical Engineering, National Chung Cheng University, Chiayi, Taiwan, Republic of China

Catalytic effectiveness of three metallocenes (Cp_2Co, Cp_2Ni, and Cp_2TiCl_2), in the presence of various reducing agents (*n*-butyllithium, phenyllithium, and triethylaluminum), have been studied for hydrogenating polystyrene-*b*-polybutadiene-*b*-polystyrene (SBS) thermoplastic elastomers. The Cp_2Co / *n*-butyllithium system was found the most active. A *n*-butyllithium to cobaltocene molar ratio of 1~2, a 15~25 kg/cm^2 H_2 pressure, and a minimal 80°C temperature were necessary in order to achieve a high extent of hydrogenation. UV spectroscopy of the catalyst system showed the formation of CpLi indicating that some Cp ligands in the original Cp_2Co were displaced by *n*-butyllithium.

Introduction

Metallocene catalysts have been of great interest to the polyolefin industry for stereospecific polymerizations such as synthesis of syndiotactic polystyrene (*1-3*), copolymerization of ethene, propene and other olefinic monomers (*4-6*), polymerization of diolefin monomers (*7-9*), or copolymerization of styrene with diolefin (*10,11*). Recently, their catalytic effectiveness for hydrogenating conjugated diolefin polymers have been studied. Y. Kishimoto and H. Morita found that bis-(cyclopentadienyl)titanium (IV) compounds, in the presence of an alkyl lithium compound, were effective (*12*), and later a bis-(cyclopentadienyl)titanium (III) compound was also proved effective by L. Chamberlain, et. al. (*13*) The alkyl lithium was later found unnecessary when Y. Kishimoto and T. Masubuchi applied a specific titanium(IV) compound, Ti(η^5-C_5H_5)$_2R_2$, where R were any tri-substituted phenyl groups (*14*), to their hydrogenation reaction. F. Parellada, et. al. provided a hydrogenation process using Ti(η^5-C$_5$H$_5$)$_2$(PhOR')$_2$ where R' is an alkyl group (*15*). S. Hahn and D. Wilson used certain monocyclopentadienyl metal compound containing Ti, Zr, or Hf for hydrogenating ethylenic unsaturations (*16*). Metallocene catalysts based on Group IV metals were heavily studied in those works.

This paper compares three metallocene catalysts (Cp$_2$Co, Cp$_2$Ni, and Cp$_2$TiCl$_2$) comprising various central metals regarding their effectiveness for hydrogenating the conjugated dienes in a polystyrene-*b*-polybutadiene-*b*-polystyrene (SBS) thermoplastic elastomer. The hydrogenation of the SBS block copolymer would eliminate its unsaturated double bonds and improve its long term heat, weather and UV stability (*17,18*). All three metallocenes were tested in the presence of various reducing agents (*n*-butyllithium, phenyllithium, and triethylaluminum) in order to achieve a higher productivity. Using the cobaltocene / *n*-butyllithium catalyst system as an example, the optimal reaction conditions were then studied. Furthermore, limited studies concerning a fundamental understanding of the catalytic species during a hydrogenation reaction have been made.

Experimental

Materials

A linear SBS block copolymer was obtained from Taiwan Synthetic Rubber Corp (TSRC). This copolymer was made via an anionic synthesis with a weight average molecular weight of 120,000, a polydispersity of 1.1, and a polystyrene content of 30 wt%. Its polybutadiene block contains a total double bonds of 13.0 mmole/g with a microstructure of 43.8% 1,2-unit, 31.4% *trans* 1,4-unit, and 24.8% *cis* 1,4-unit.

Cobaltocene (Cp$_2$Co), nickelocene (Cp$_2$Ni), and titanocene (Cp$_2$TiCl$_2$) were purchased from Aldrich.

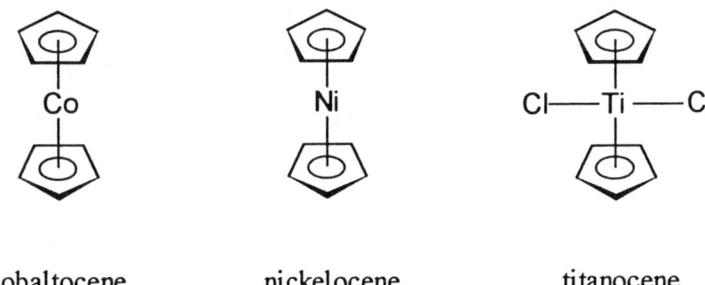

n-Butyllithium was purchased as a 15 wt% solution in hexane from Merck. Phenyllithium as a 17 wt% solution in cyclohexane and triethylaluminum as a 15% solution in hexane were purchased from TCI (Japan). Cyclohexane solvent was obtained from TSRC and was dried with activated alumina (from Alcoa).

Hydrogenation Procedures

1×10^{-3} mole of metallocene was dissolved into 100 ml cyclohexane under an air-free environment to make the catalyst system required for later use. An appropriate amount of reducing agent solution, calculated based on a specified reducing agent / metallocene mole ratio, was then added and the reaction mixture was stirred for 5 hrs at which time the mixture looked significantly thickened.

The hydrogenation reaction was carried out in a 2-liter mechanically stirred autoclave (made by PPI, rated to 6000 psi) at 80℃. 330 g of the SBS copolymer solution previously prepared at a concentration of 7.3 wt% SBS in cyclohexane solvent was first charged into the autoclave, followed by 30 ml of the catalyst solution. The hydrogen pressure was then kept constant at a specified setting with continual gas addition, and the reaction mixture was saturated with the hydrogen using a gas dispersing impeller. Samples of the reaction mixture were taken at fixed time intervals, and were repeatedly washed with 4% sulfuric acid solution via rigorous shaking to remove the residual catalyst. The hydrogenated SBS copolymer was then precipitated in isopropanol and dried at 40°C in a vacuum oven. No chain scission or crosslinking has occurred during the reaction under such hydrogenation conditions (*19*).

Sample Analyses

The microstructures of the polymers were determined by ^{13}C-NMR spectra using Bruker AMX400 100.61 MHz spectrometer at 25°C in $CDCl_3$ following a well established method (*20*). The disappearance of the double bonds was observed using a Shimadzu FTIR-8101M instrument with a liquid N_2 cooled MCT detector. The spectral resolution was 2 cm^{-1} and the samples were prepared as cast films on KBr plates for IR scans. While the degree of hydrogenation for 1,2-unit and trans-1,4 unit could be determined directly from the FTIR spectrum, the degree of hydrogenation for cis-1,4 unit could not. This was because of a significant overlapping of characteristic peaks for cis-1,4 unit (at 724 cm^{-1}) and polystyrene (at 699 cm^{-1}). In order to quantitatively determine the degree of hydrogenation for cis-1,4 unit, an additional ^{1}H-NMR analysis was needed. Although the degrees of hydrogenation for cis-1,4 and trans-1,4 polybutadiene units were difficult to separate in a ^{1}H-NMR spectrum, the degree of hydrogenation for the total 1,4-polybutadiene units could be determined. Since the fractions of cis-1,4 and trans-1,4 polybutadiene units for the copolymer feed had been previously determined, the degree of hydrogenation of cis-1,4 polybutadiene could be back calculated from the following equation:

$$X_{1,4} = F_{cis\text{-}1,4} \cdot X_{cis\text{-}1,4} + F_{trans\text{-}1,4} \cdot X_{trans\text{-}1,4}$$

Therefore, such a combined ^{1}H-NMR and FTIR analysis enabled us to determine the extent of hydrogenation for each type of isomeric unit of polybutadiene.

Results and Discussion

Comparison among Various Metallocenes

Three metallocenes (Cp_2Co, Cp_2Ni, and Cp_2TiCl_2), combined with n-butyllithium at a 2:1 ratio of n-butyllithium to metallocene, were used to hydrogenate the SBS block copolymer under experimental conditions of 80°C, 25 kg/cm^2 pressure, 7.3 wt% polymer concentration. The extents of hydrogenation for 1,2-unit and *trans* 1,4-unit were directly monitored from the FTIR spectrum, and that for *cis* 1,4-unit was determined from a combination of FTIR and ^{1}H-NMR analyses due to a significant overlapping of characteristic peaks for *cis* 1,4-unit (at 724 cm^{-1}) and polystyrene (at 699 cm^{-1}).

The effectiveness of Cp_2Co / n-butyllithium, Cp_2Ni / n-butyllithium, and Cp_2TiCl_2 / n-butyllithium for hydrogenating various isomeric units are shown in Figure 1. Regardless of the metallocene used, the 1,2-unit was easier to hydrogenate than 1,4-unit. The higher reactivity of 1,2-unit might arise from the less steric hindrance and hence the higher accessibility of the double bond. In addition, the *cis*-isomer of the 1,4-units was hydrogenated faster than the *trans*-isomer, which could be ascribed to the higher stability of the *trans* configuration. While the Cp_2Ni / n-butyllithium catalyst system was unable to completely hydrogenate the 1,4-units, the Cp_2Co / n-butyllithium and the Cp_2TiCl_2 / n-butyllithium systems produced satisfactory results. The Cp_2Co / n-butyllithium system, in particular, provides a complete hydrogenation in 5 hrs.

Comparison among Various Reducing Agents

Three different reducing agents (n-butyllithium, phenyllithium, and triethylaluminum), combined with cobaltocene have been evaluated. The effectiveness of Cp_2Co / n-butyllithium, Cp_2Co / phenyllithium, and Cp_2Co / triethylaluminum for hydrogenating various isomeric units are shown in Figure 2. While no hydrogenation was observed for Cp_2Co / triethylaluminum, both Cp_2Co / n-butyllithium and Cp_2Co / phenyllithium catalyst systems have been found to be active. The reason causing slightly lower extent of hydrogenation for the latter is

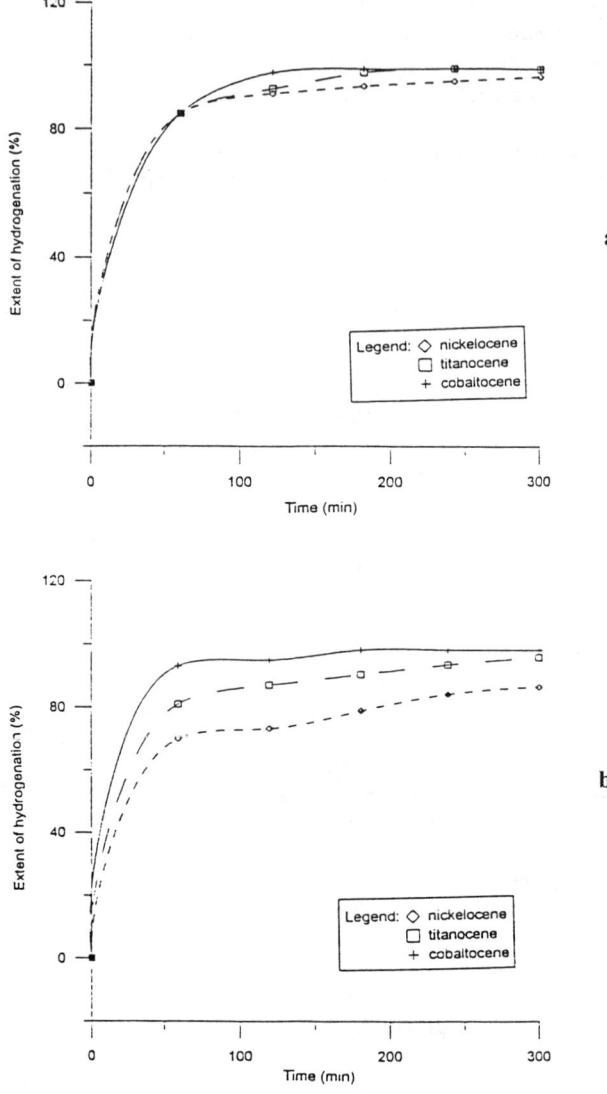

Figure 1 Extents of Hydrogenation for isomeric units of polybutadiene segment using various metallocenes with n-butyllithium under conditions: T=80 °C, P=25 kg/cm^2, [polymer]=0.073g/g solution; n-BuLi / metallocene=2 (a: 1,2 units; b: cis-1,4 units; c: trans-1,4 units)

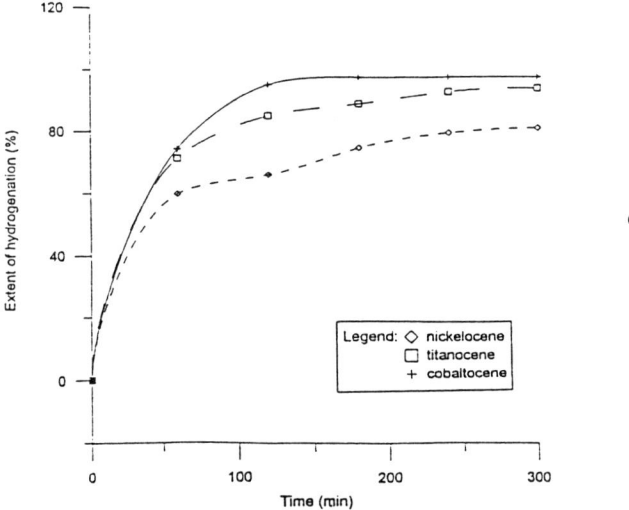

Figure 1. *Continued*

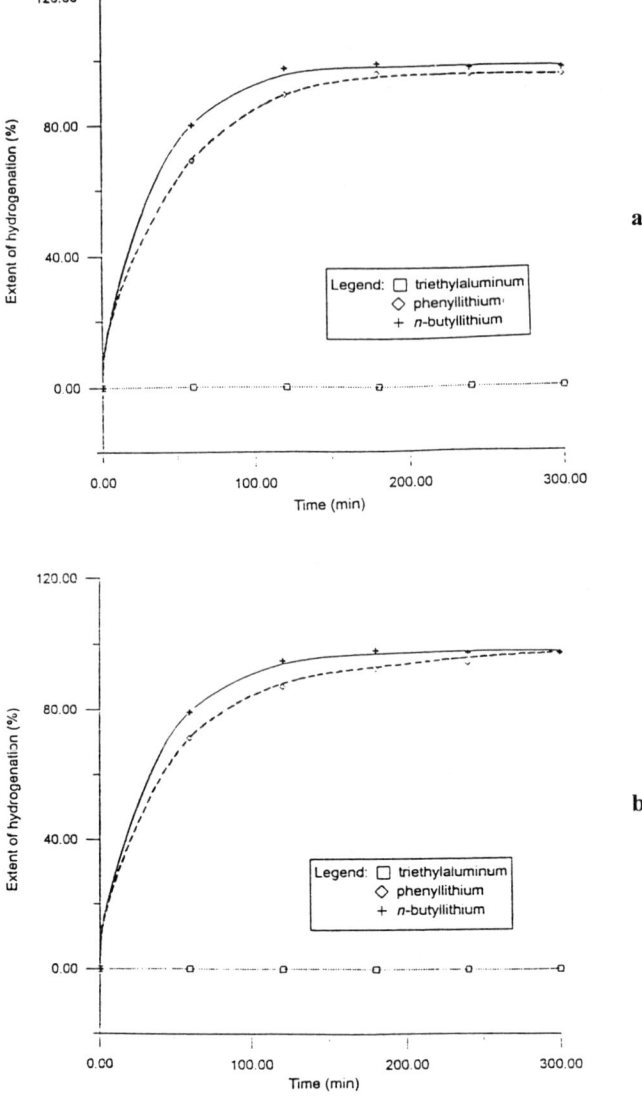

Figure 2 Extents of Hydrogenation for isomeric units of polybutadiene segment using cobaltocene with various reducing agents under conditions: T=80 °C, P=25 kg/cm^2, [polymer]=0.073g/g solution; reducing agent / cobaltocene=2 (a: 1,2 units; b: cis-1,4 units; c: trans-1,4 units) (Reproduced with permission from reference 19. Copyright 1999 John Wiley & Sons, Inc.)

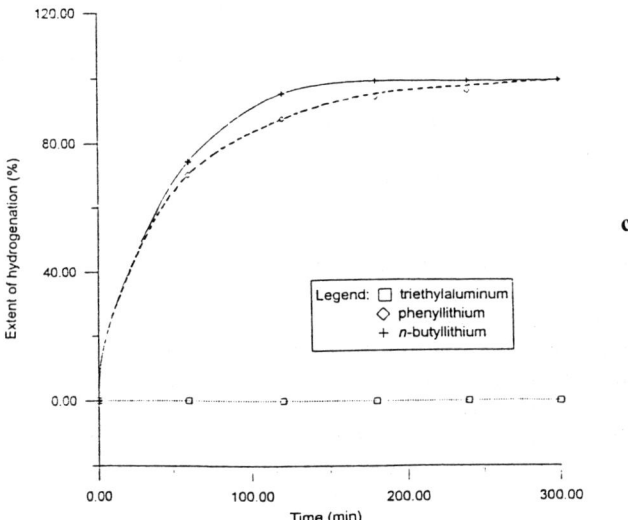

Figure 2. *Continued.*

unknown, but, if such difference is not a run-to-run difference, could likely be ascribed to the formation of different organometallic intermediate and/or the change in the oxidation state of the central metal ion.

Effect of Reaction Temperature

The effect of temperature has been studied for a hydrogenation reaction using the cobaltocene / n-butyllithium catalyst system. The H_2 was maintained at 25 kg/cm^2 and the n-butyllithium to cobaltocene molar ratio was 2. The extents of hydrogenation of 1,2-polybutadiene, cis-1,4 polybutadiene and trans-1,4 polybutadiene are shown in Figure 3. Both the kinetics and extent of hydrogenation increase with an increase in the reaction temperature. To ensure a complete hydrogenation, the reaction temperature ought to be higher than 80°C. The ultimate extents of hydrogenation are in this descending order: 1,2-unit > cis-1,4 unit > trans-1,4 unit. The 1,2-unit is easier to hydrogenate due to the less steric hindrance. Conversely, the higher stability of trans-1,4 unit caused by the structural symmetry makes it the most difficult to hydrogenate.

Effect of the Ratio of n-Butyllithium to Metallocene

The effect of the ratio of reducing agent to metallocene has been studied based on the Cp$_2$Co / n-butyllithium catalyst system. The extents of hydrogenation have been measured for reactions catalyzed at various n-butyllithium :cobaltocene ratios. The results are shown in Table I. Maximal extents of hydrogenation were achieved at a n-butyllithium to cobaltocene ratio of 1~2, regardless of the type of isomeric units. Any excessive amount would probably cause a complexation with the cobaltocene and consequently decreased the activity of the catalyst.

Effect of H_2 Pressure on Extent of Hydrogenation

The hydrogenation of SBS has been conducted using Cp$_2$Co / n-butyllithium catalyst system under four different H_2 pressures, 10, 15, 25, and 40 kg/cm^2. Experimental data are shown in Table II. While complete hydrogenation was discerned for a reaction taken place under a constant H_2 pressure of either 15 kg/cm^2 or 25 kg/cm^2, no hydrogenation occurred at 10 kg/cm^2 H_2 pressure. Conversely, 40 kg/cm^2 pressure was too high and the extent of hydrogenation for all three isomeric units decreased. It is speculated that an overpressure of hydrogen caused an undesirable reaction converting the catalytic cobaltocene into the inert cobalt hydride losing all the Cp ligands, thus resulting in a decrease in the extent of hydrogenation.

Mechanistic Studies of the Metallocene / Reducing Agent System

The catalyst system studied here was prepared, by adding n-BuLi to cobaltocene at a 2:1 mole ratio, in cyclohexane solvent as described in the previous section. Using a UV-VIS spectroscopy, the spectra of n-BuLi, cobaltocene and their mixture were collected (21) and compared as shown in Figure 4(a)-4(c). The lack of any peak in between 250 nm and 300 nm in Figure 4(c) manifested the disappearance of n-BuLi and Cp_2Co which would otherwise exhibit characteristic peaks as were shown in Figure 4(a) and 4(b). Conversely, the new peak shown in Figure 4(c) located near 240 nm indicated the generation of a new intermediate species. This location happened to be the characteristic UV-VIS location of CpLi as was observed on a pure CpLi sample in Figure 4(d). As a further verification, ^{1}H-NMR spectra taken of the catalyst system and CpLi were compared. Figure 5(a), taken from the catalyst system, does exhibit characteristic CpLi peaks as are shown in Figure 5(b) for the pure CpLi (The three peaks of equal area for the CpLi are indicative of the protonation, which commonly occurs in $CDCl_3$ solvent during the ^{1}H-NMR analysis. The protonated Cp possesses two types of olefinic sp^2 proton near 6.49 and 6.29 ppm at 2H each, and one type of sp^3 proton near 2.68 ppm at 2H). Consequently, it was believed that the role of n-BuLi in our catalyst system was to cause a displacement of the Cp ligand(s) from cobalt center atom. Although lithiation of the Cp ligand (replace H on a Cp ligand by Li) is also a possibility based on ferrocene studies, but our spectral evidence is consistent with formation of CpLi instead, and may not be unlikely since Co is less thermally stable. However, because our catalyst system was prepared in the absence of hydrogen, it was entirely likely that reaction with hydrogen could change the nature of the catalytic species during the SBS hydrogenation.

The complexity of the coordination catalysis and the lack of reference standards made it extremely difficult to investigate the exact steps of the catalytic mechanism. Nevertheless, since it now seems to be well established that the catalytic mechanism of metallocene dichloride in the presence of methyl aluminoxane (MAO) comprises an alkylation of the metal and a hydrogenolysis of the metal-alkyl bond (22-24), it is likely bis(cyclopentadienyl)metal hydride or (cyclopentadienyl)metal hydride will form in our system. In fact, highly reactive species, such as Cp_2TiH and Cp_2Ti, had been postulated in earlier titanocene studies (25) to exist as reaction intermediates forming a dimeric compound, μ -($\eta^5:\eta^5$-fulvalene)-di- μ -hydrido-bis(cyclopentadienyltitanium) (26,27). Existence of other intermediate species and a possible change in the oxidation state of the metal are also likely. However, the nature of these intermediate species in the hydrogenation reaction is not precisely known, and their roles in the reaction mechanism has yet to be studied.

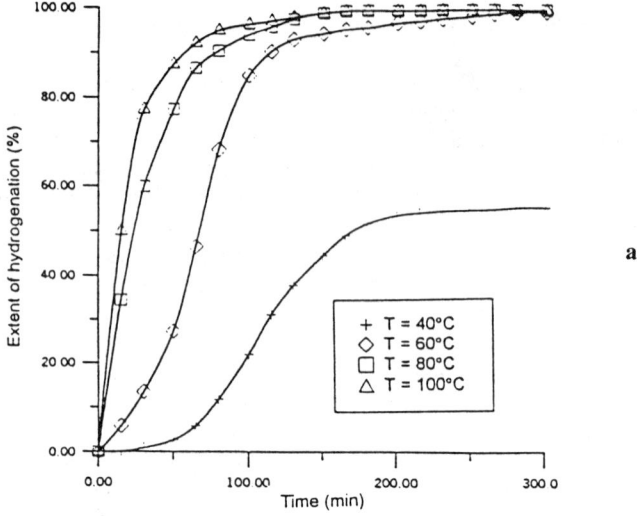

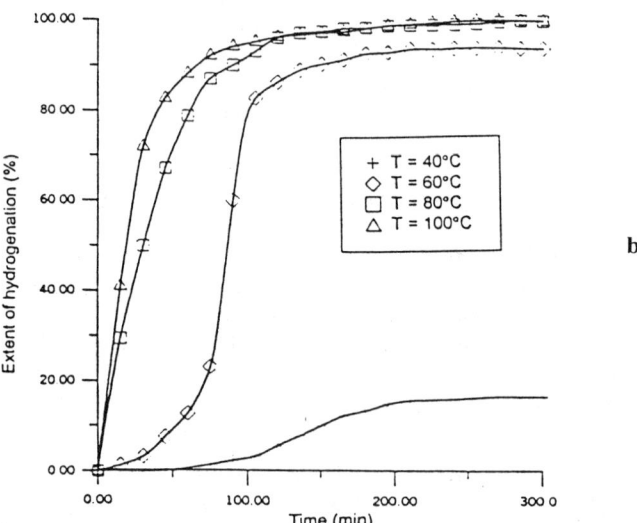

Figure 3 Extents of Hydrogenation for isomeric units of polybutadiene segment at various temperatures under conditions: $P=25$ kg/cm^2, [polymer]=0.073g/g solution; n-BuLi / cobaltocene=2)
(a: 1,2 units; b: cis-1,4 units; c: trans-1,4 units)

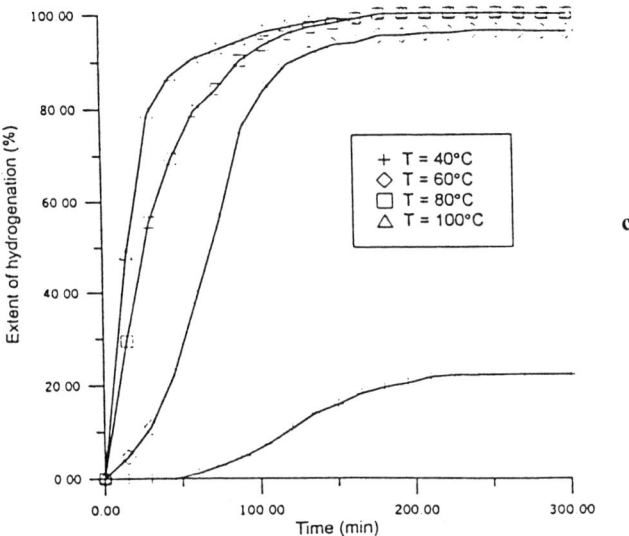

Figure 3. *Continued.*

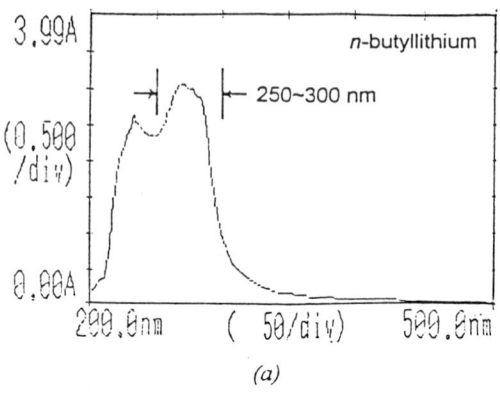

(a)

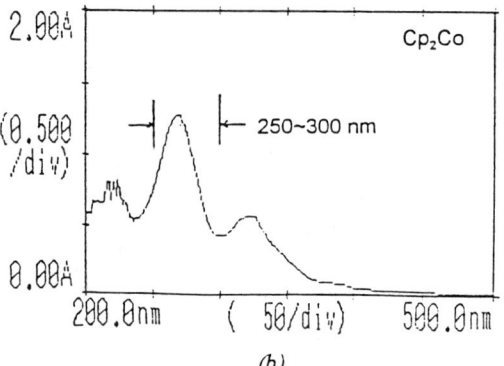

(b)

Figure 4 UV-VIS spectra of (a) n-butyllithium, (b) Cp$_2$Co, (c) Cp$_2$Co plus n-butyllithium, and (d) CpLi (Reproduced with permission from reference 20. Copyright 1999 Elsevier Science)

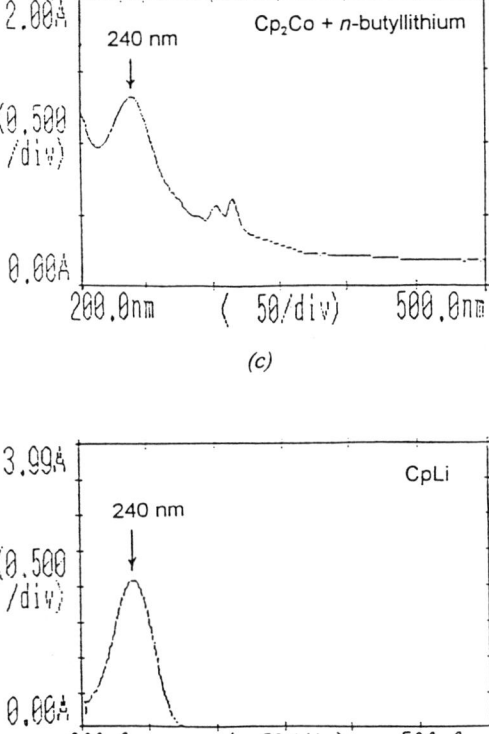

Figure 4. *Continued.*

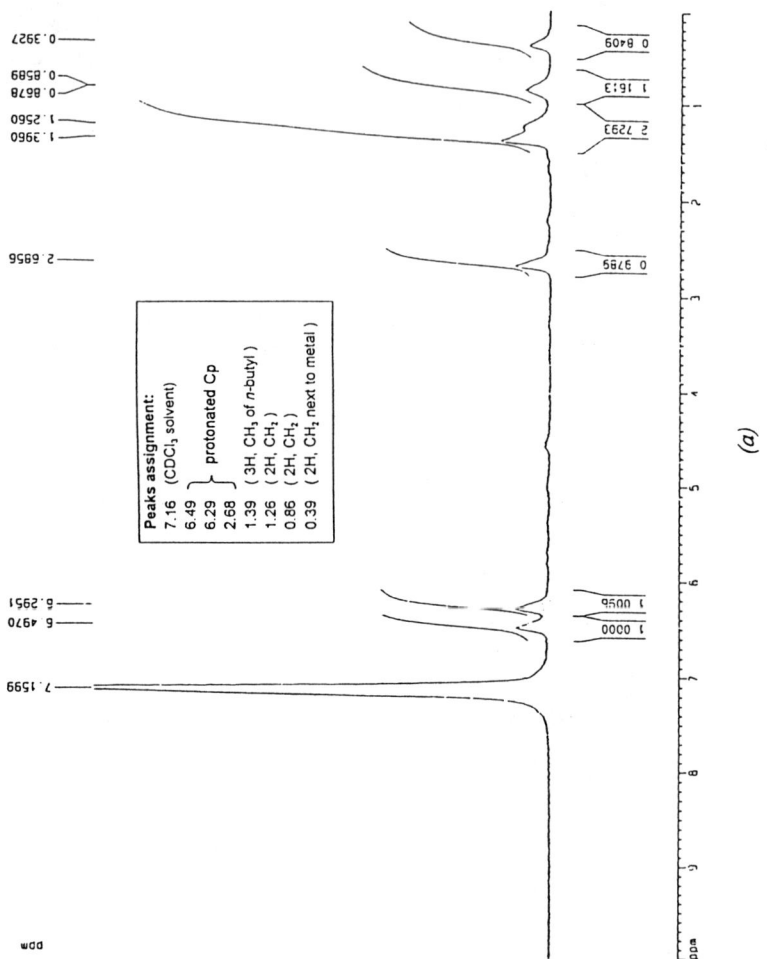

(a)

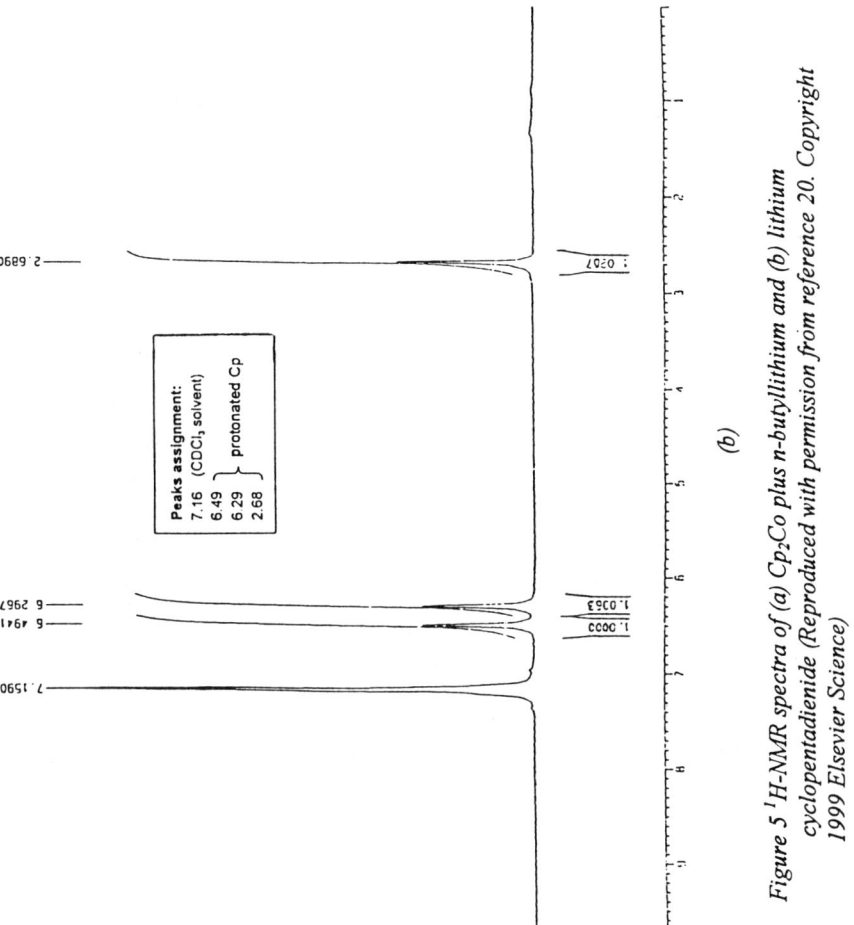

Figure 5 ^{1}H-NMR spectra of (a) Cp$_2$Co plus n-butyllithium and (b) lithium cyclopentadienide (Reproduced with permission from reference 20. Copyright 1999 Elsevier Science)

Table I. Hydrogenation efficiencies at various *n*-butyllithium to Cp$_2$Co ratios

	0	0.5	1	2	4	8	16
1,2-unit	0%	82%	100%	100%	99%	79%	0%
Cis-1,4	0%	44%	99%	100%	94%	42%	0%
Trans-1,4	0%	37%	99%	100%	92%	33%	0%

NOTE: Conditions are T=80°C, P=25 kg/cm^2, [polymer]=0.073 g/(g solution).

Table II. Hydrogenation efficiencies at various H$_2$ pressure

	10 kg/cm^2	15 kg/cm^2	25 kg/cm^2	40 kg/cm^2
1,2-unit	0%	~100%	~100%	99.2%
Cis-1,4	0%	~100%	~100%	79.2%
Trans-1,4	0%	~100%	~100%	63.5%

NOTE: Conditions are T=80°C, *n*-butyllithium/cobaltocene=2, [polymer]=0.073 g/(g solution)

Conclusions

Catalyst systems consisting of a metallocene and a reducing agent at a 1:2 mole ratio has been evaluated for their effectiveness in the hydrogenation of SBS block copolymers. The metallocenes evaluated include Cp_2Co, Cp_2Ni, and Cp_2TiCl_2, and the reducing agents include n-butyllithium, phenyllithium, and triethylaluminum. The Cp_2Co paired with n-butyllithium rendered the highest hydrogenation effectiveness for all types of isomeric butadiene units. A n-butyllithium to cobaltocene molar ratio of 1~2, a H_2 pressure of 15~25 kg/cm^2, and a minimal 80°C reaction temperature were required to achieve high extent of hydrogenation. Based on UV spectroscopies of the catalyst system it has been found that CpLi was formed indicating a displacement of some Cp ligands in the original Cp_2Co by n-butyllithium.

References

1. Ishihara, N.; Kuramoto, M.; Uoi, M. *Macromolecules* **1988**, *21*, 3356-3360.
2. Pellecchia, C.; Longo, P.; Grassi, A.; Zembelli, A. *Makromol. Chem., Rapid Commun.* **1987**, *8*, 277-283.
3. Zambelli, A.; Oliva, L.; Pellecchia, C. *Macromolecules* **1989**, *22*, 2129-2130.
4. Sernetz, F. G.; Mulhaupt, R. *J. Polym. Sci., Part A., Polym. Chem.* **1997**, *35*, 2549-2560.
5. Aaltonen, P.; Fink, G.; Lofgren, B.; Seppala, J. *Macromolecules* **1996**, *29*, 5255-5260.
6. Aaltonen, P.; Lofgren, B. *Macromolecules* **1995**, *28*, 5353-5361.
7. Ricci, G.; Italia, S.; Comitani, C.; Porri, L. *Polymer Communications* **1991**, *32*, 514-517.
8. Oliva, L.; Longo, P.; Grassi, A.; Ammendola, P.; Pellecchia, C. *Makromol. Chem., Rapid Commun.* **1990**, *11*, 519-525.
9. Takeuchi, M.; Shiono, T.; Soga, K. *Polym. International* **1992**, *29*, 209-212.
10. Longo, P.; Proto, A.; Oliva, P.; Sessa, I.; Zambelli, A. *J. Polym. Sci., Part A, Polym. Chem.* **1997**, *35*, 2697-2702.
11. Zambelli, A.; Proto, A.; Longo, P.; Oliva, P. *Macromol. Chem. Phys.* **1994**, *195*, 2623-2631.
12. Kishimoto, Y.; Morita, H. U.S. Pat. 4,501,857, assigned to Asahi Company, Japan, 1985.
13. Chamberlain, L.; Gibler, C. U.S. Pat. 5,141,997, assigned to Shell Oil Company, U.S.A. 1993.
14. Kishimoto, Y.; Masubuchi, T. U.S. Pat. 4,673,714, assigned to Asahi Company, Japan, 1987.
15. Parellada, F.; Maria, D.; Barrio, C.; Juan, A. U.S. Pat. 5583185, assigned to Repsol Quimica, Spain, 1996.
16. Hahn, S.; Wilson, D. U.S. Pat. 5789638, assigned to Dow Chemical Company, U.S.A. 1998
17. Hsieh, H. L.; Quirk, R. P. *Anionic Polymerization, principles and practical applications*, Marcel Dekker, New York, NY, 1996.

18. Hoxmeier, R. J. U. S. Pat. 4,879,349, assigned to Shell Oil Company, U.S.A. 1989.
19. Yang, W.; Hsieh, H. C.; Tsiang, R. C. *J. Appl. Polym. Sci.* **1999**, *72*, 1807-1815.
20. Huang, D. C.; Tsiang, R. C. *J. Appl. Polym. Sci.* **1996**, *61*, 333-342.
21. Tsiang, R. C.; Yang, W.; Tsai, M. *Polymer*, **1999**, *40*, 6351-6360.
22. Chien, J. C. W.; Wang, B. P. *J. Polym. Sci., Polym. Chem. Ed.* **1988**, *26*, 3089-3102.
23. Giannetti, E.; Nicoletti, G. M.; Mazzocchi, R. *J. Polym. Sci., Polym. Chem. Ed.* **1985**, *23*, 2117-2134.
24. Zambelli, A.; Pellecchia, C.; Oliva, L.; Longo, P.; Grassi, A. *Macromol. Chem.* **1991**, *192*, 223-231.
25. Bercaw, J.; Marvich, R.; Bell, L.; Brintzinger, H. *J. Am. Chem. Soc.* **1972**, *94:4*, 1219-1238.
26. Davison, A.; Wreford, S. *J. Am. Chem. Soc.*, **1974**, *96:9*, 3017-3018.
27. Zhang, Y.; Liao, S.; Xu, Y. *J. Organometallic Chem.*, **1990**, *382*, 69-76.

Chapter 8

E-Glass Fiber Supported Hydrosilation Catalysts

L. G. Britcher and J. G. Matisons

Ian Wark Research Institute, University of South Australia,
The Levels, Mawson Lakes, South Australia 5095, Australia

Mercapto-platinum complexes supported on E-glass fibers have shown catalytic activity for hydrosilating a poly(methyl hydrogen)siloxane (d.p.=33) with 1-hexene. The supported complexes were prepared by modifying E-glass fibers with mercaptopropyltrimethoxysilane for the subsequent attachment of platinum. To analyze the complexes formed, Diffuse Reflectance Fourier Transform Infrared (DRIFT) Spectroscopy and X-ray Photoelectron Spectroscopy (XPS) were employed. Two different sources of E-glass fibers proved to have a consistent effect on the amount of platinum and sulfur adsorbed onto the surface. All of the catalysts could be recycled several times, with little loss in activity.

Introduction

To combine the advantages of homogeneous catalysis (specificity and activity) with those of heterogeneous catalysis (reuse and easy removal), "supported catalysts" have been developed and used. Supported catalysts are produced by complexing transition metals to solid inorganic or organic suppports. Table I lists some properties comparing homogeneous catalysts to supported catalysts. The principal advantage of supported catalysts lies in their ready recovery from the reaction mixture, and their subsequent reuse without appreciable loss in activity.

Two general classes of supported complexes have been developed [1]:

1. A metal complex is linked to the support through attachment to one of its ligands.

e.g.

$\}$—(PPh$_2$)Rh(PPh$_3$)$_2$Cl   where $\}$—PPh$_2$ = supported diphenylphosphine

© 2000 American Chemical Society

2. A metal complex reacts with the support and displaces a ligand, by substituting the ligand with a group from the support.

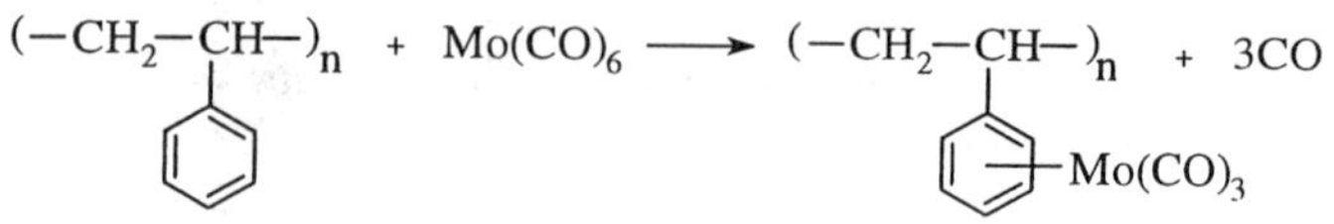

$$(-CH_2-CH-)_n\text{(Ph)} + Mo(CO)_6 \longrightarrow (-CH_2-CH-)_n\text{(Ph-Mo(CO)}_3\text{)} + 3CO$$

Table I. Comparison of Homogeneous and Supported Catalysts.

Property	Homogeneous catalysts	Supported catalysts (or heterogeneous)
recovery of catalyst	not simple	removed by filtration
protection against catalyst poisoning (e.g. oxygen, moisture)	not possible	possible
efficiency (are all of the molecules available for catalysis?)	all molecules are theoretically available	not all atoms are at the surface where catalysis takes place, therefore not all are used.
reproducibility	a definite stoichiometry and structure.	heavily dependent on the structure of the support, which can change.
control	groups on the complexes can be modified to control a reaction	ill-defined surface sites can make systematic design and improvement difficult
activity	usually better than supported complexes	variable (only a few are more active than their homogeneous counterparts)

Inorganic supports such as silica, alumina, glasses, clays, or zeolites have rigid structures, an advantage over polymer supported complexes as uncomplexed ligands do not block coordination sites on metal complexes by their inherent mobility. Thermal and mechanical stability is improved in inorganic oxide supports too. Polymers also swell in solvents making it difficult to use them in columns or confined spaces (2).

Between 1945 and 1947 hydrosilation emerged from about six separate laboratories in the United States (3). The hydrosilation reaction involves the addition of one or more Si-H groups to unsaturated organic reagents (Figure 1). All group eight transition metals display some catalytic activity in hydrosilation, with platinum being the most active, and therefore the most widely used catalyst.

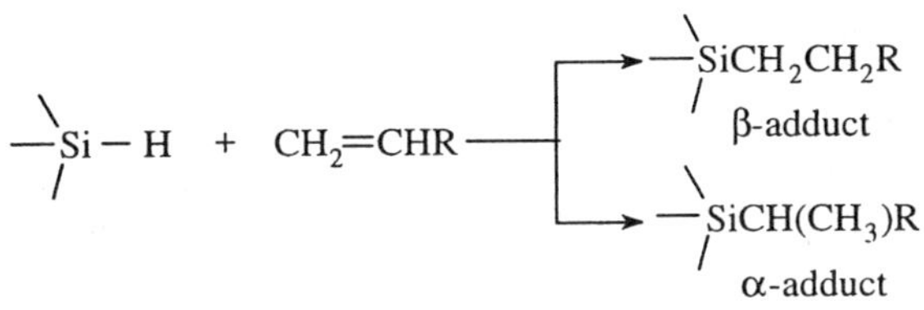

$$\begin{array}{c}\diagdown\\ -Si-H\\ \diagup\end{array} + CH_2=CHR \longrightarrow \begin{cases} \begin{array}{c}\diagdown\\ -SiCH_2CH_2R\\ \diagup\end{array} \quad \beta\text{-adduct}\\[2ex] \begin{array}{c}\diagdown\\ -SiCH(CH_3)R\\ \diagup\end{array} \quad \alpha\text{-adduct}\end{cases}$$

Figure 1. The hydrosilation reaction.

The substituent R (see Figure 1) can be varied rather widely to accommodate various organic groups. Hydrosilation reactions generate two adducts, α and β (major product) along with some minor products from side reactions (4-8). Figure 1 represents the simplest form of hydrosilation; as the silicon hydride can also add across alkynes producing a different range of adducts depending on the catalyst, silane, and alkyne. Alkoxy or aryloxy compounds can also be produced by addition of silicon hydrides to carbonyl compounds in the presence of a catalyst (9).

Supported Hydrosilation Catalysts

Initially, ion exchange resins made of styrene-divinylbenzene were used as supports for platinum complexes (10). The resulting heterogeneous catalysts were effective for the hydrosilation of phenylmethylsilanes to 1-alkenes, 1-alkynes, and ethyl acrylate remaining active even after being used over 20 times. The catalysts were prepared by interaction of functionalized ammonium supports with hexachloroplatinic acid as shown in Figure 2. Polystyrene can be functionalized with diphenylphosphine (11,12), cyano (12), and amine groups (12) to support rhodium and palladium catalysts. Such heterogeneous complexes hydrosilate butadiene, 1-hexene, and 1-heptene. Likewise, polyamides have been successfully used as supports for H_2PtCl_6, $PtCl_2(CH_3CN)_2$, and $[RhCl(CO)_2]_2$ (13). Diphenylphosphino-alkylorganosiloxanes have been immobilized on silica and subsequently complexed with $Rh(PPh_3)_3Cl$, to produce heterogeneous hydrosilation and hydrogenation catalysts *(14)*. Similar insoluble supports prepared by the polycondensation of suitable trialkoxysilyl-substituted organo-sulphides, phosphines, and amines interact with platinum complexes to form hydrosilation catalysts *(15)*.

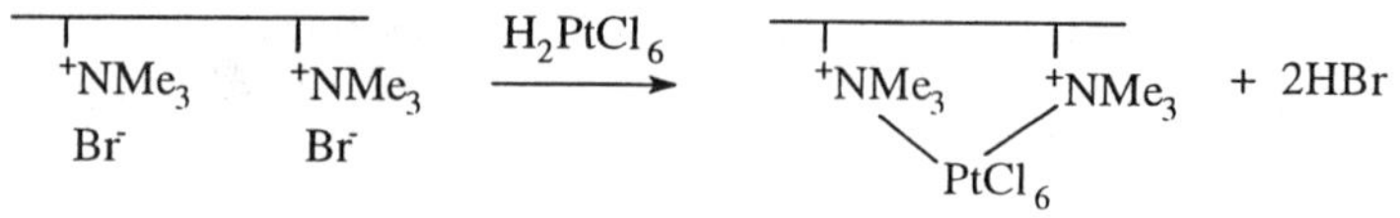

$$\overset{|}{\underset{Br^-}{^+NMe_3}} \quad \overset{|}{\underset{Br^-}{^+NMe_3}} \quad \xrightarrow{H_2PtCl_6} \quad \overset{|}{^+NMe_3} \underset{PtCl_6}{\diagup} \overset{|}{^+NMe_3} \quad + \; 2HBr$$

Figure 2. Immobilization of hexachloroplatinic acid on ion exchange resins.

Capka and Hetflejs *(16)* have used γ-alumina, silica, zeolites such as molecular sieves and glass as supports for group VIII metal hydrosilation catalysts. Phosphinated chrysotile or asbestos has been used to support $Rh(PPh_3)_3Cl$, for the hydrosilation of 1-hexene to $HSi(OEt)_3$ *(17)*. A cyanoethylmercapto silane was synthesized and then immobilized onto silica, followed by addition of potassium chloroplatinate, to give a catalyst which could be reused several times *(18)*. Some common ligands used to attach metal complexes to supports include -SH *(19)*, $-NH_2$ *(20)*, $-PPh_2$ *(21)*, $-NMe_2$ *(16)*, and CN *(16)*. Silica bound crown ethers were used as supports for K_2PtCl_4, for the subsequent hydrosilation of triethoxysilane to 1-decene, 1-dodecene, allyl benzene, and allyl glycidyl ether *(22)*. Photoactive $Pt(C_2O_4)[(OCH_3)_3Si(CH_2)_2P(CH_2CH_3)_2]_2$ was complexed to silica, and subsequently used as a hydrosilation catalyst. Upon irradiation of the photoactive complex CO_2 is released, and a surface attached $PtP(CH_2CH_3)_2$ species forms *(23)*.

The activity of the various catalysts depends to a large extent on the ligands used to attach the metal complex to the support. For instance, several supported rhodium complexes were prepared by reacting chlorobis(ethylene)rhodium (I) with supported alkylene phosphine complexes; where the chain length of the phosphine-alkylene links $[-(CH_2)_nPPh_2]$ could be varied (n= 1-6). The catalytic activity was ten times greater for the complex linked to silica through a short alkylene chain (containing one methylene group), than for the rhodium complexes linked having alkylene chains containing two to six carbon atoms *(24)*. It is interesting that platinum complexes were inactive when bound through pyridyl ligands to silica, but when phosphine ligands were substituted for pyridine, activity was restored. Similarly, rhodium complexes were more active when bound through phosphine, rather than amine ligands, to silica *(16)*.

Glass fibers are widely used as reinforcing agents in polymer composites and treatment with various silane coupling agents can provide a water-resistant bond between the glass fiber interface and the polymer matrix *(25)*. There are several different oxide components in E-glass fibers, therefore the fibers will have different surface properties (e.g. isoelectric point and adsorption sites) when compared to a single oxide support. This work aimed to investigate whether the glass fibers could be used successfully to prepare catalytically active supported complexes. The glass fibers were modified with mercaptopropyltrimethoxysilane then hexachloroplatinic acid to prepare catalytic materials for hydrosilation. Diffuse Infrared Fourier Transform Spectroscopy (DRIFT) Spectroscopy and Xray Photoelectron Spectroscopy (XPS) where used to characterize the supported platinum complexes formed.

Experimental

Untreated E-glass fibers (water washed only) were obtained from ACI Australia and used without further modification. The constituent elemental profile of the glass fibers as determined by X-ray fluorescence measurements carried out by the supplier is as follows: 55.0 % SiO_2; 21.5% CaO; 14.5% Al_2O_3; 6.0% B_2O_3; 0.8% Na_2O; 0.6% MgO; 0.8% Fe_2O_3; 0.3% TiO_2; 0.6% F_2; 0.3% FO_2; 0.1% K_2O.

Similarly, untreated E-glass fibers (water washed only) were also obtained from Ahlstrom corporation, and were used without further modification. The average composition of the E-glass fibers as supplied by the manufacturer is as follows: 55% SiO_2; 22% CaO; 15% Al_2O_3 + Fe_2O_3; 6% B_2O_3; < 1% alkali content; < 1% other impurities.

3-(mercaptopropyl)trimethoxysilane (Hüls, America and Aldrich) and hydrogen hexachloroplatinate hexahydrate (IV) (Sigma, 40 wt% Pt) were used as supplied. Toluene (ACE chemicals) and ethanol (ACE chemicals) were distilled before use.

Preparation of the E-glass Fiber Supported Complexes.
ACI E-glass fibers (2.5 g) were placed in ethanol (50 mL) containing 3-(mercaptopropyl)trimethoxysilane (1 mL) for 16 hours at room temperature under a nitrogen atmosphere. The solution was then decanted off and the fibers dried under vacuum (50 °C, 0.5 kPa) for 4 hours, followed by washing with 4x 60 mL portions of toluene. The fibers were dried under vacuum (4 h, 0.5 kPa, 50 °C), and then placed in a solution of ethanol (100 mL) and $H_2PtCl_6.xH_2O$ (0.2 g) for 24 hours at room temperature under nitrogen. The yellow solution was decanted off and the fibers washed with 5x 60 mL portions of ethanol, and dried under vacuum (4 h, 0.5 kPa, 50 °C). The same method was repeated using Ahlstrom E-glass fibers. XPS and DRIFT spectroscopy were used to analyze the supported complexes.

Hydrosilation of Hexene using the Glass Supported Platinum complexes as Catalysts.
Poly[(methylhydrogen)siloxane] (1.00 g, 0.06 mmol), hexene (0.72 g, 8.45 mmol), toluene (20 mL) and the glass fiber supported complexes prepared above (0.2 g) were stirred together at 70 °C under nitrogen for 16 hours. The solution was decanted off and analyzed by ^{1}H NMR. To remove any physisorbed reactants, the fibers were washed with three 30 mL portions of toluene. Refluxing washed complexed fibers with toluene, then analyzing the toluene for the presence of any reagents,

indicated no platinum or other reagents were released from the glass surface. Each supported complex was reused five times (or less if no activity was detected).

The products were analyzed by ^{1}H NMR to determine the amount of hydrosilation on the siloxane, by using the number of hydrogens from Si<u>H</u> (4.7 ppm) compared to the number of Si<u>CH$_2$</u> hydrogens.

Instrumental Methods

DRIFT Spectroscopy. The treated E-glass Fibers were analyzed with a single-beam Nicolet Magna Model 750 spectrometer in the wavenumber region 4000-650 cm^{-1}, with the use of an MCT-A liquid nitrogen cooled detector. The interferogram was apodized using the boxcar method installed in the FTIR software (OMNIC FTIR software version 2.0). The fibers were mounted parallel to each other in a specially constructed sample holder, which was placed in a Spectra Tech diffuse reflectance apparatus. Each glass fiber sample was scanned 256 times, with the fibers at an angle of 90°, with respect to the direction of the infrared beam, for maximum signal-to-noise ratio. The background spectrum was taken from high purity (IR) grade KBr powder (Merck), placed in a sample cup, and levelled to the top of the cup using a spatula. Signal-to-noise ratio was generally better than 100 to 1, using 256 scans.

XPS. XPS spectra were obtained using a Perkin-Elmer PHI 5600 XPS system with a concentric hemispherical analyzer and Mg Ka X-ray source operating at 300 W. Pass energies of 89 and 18 eV were used for survey and elemental spectra, respectively. Atomic concentrations were obtained using standard sensitivity factors. The angle between the sample surface and the analyzer was fixed at 45. The pressure during analysis ranged from 10^{-8} to 10^{-9} Torr. A five minute Argon ion etch was applied after the intitial XPS analysis; this sputter rate being calibrated for the complete removal of a Ta_2O_5 film on Tantalum (5 nm thick). All binding energies (BEs) were referenced to the C1s neutral carbon peak at 284.6eV, to compensate for the effect of surface charging.

Nuclear Magnetic Resonance (NMR). NMR spectra were obtained with a Varian Gemini Fourier Transform NMR spectrometer (200MHz) and associated software. Both ^{13}C and ^{1}H NMR spectra were obtained on this instrument using CDCl$_3$ (Cambridge Isotope Laboratories) as solvent and internal standard (^{1}H NMR d 7.25 ppm; ^{13}C NMR d 77.0 ppm) unless stated otherwise. The delay between successive pulse sequences was 1-10 seconds. The number of transients for ^{1}H NMR and ^{13}C NMR was generally 32 and 2000 respectively.

Results and Discussion

Sulfur compounds usually retard or prevent hydrosilation occurring *(3,4)*, however in those few instances when a sulfur ligand is present on silica to immobilize platinum complexes, good catalytic activity has resulted *(19, 26-28)*. This work investigates the preparation of immobilized sulfur-platinum complexes using E-glass fibers as the support. The samples were prepared by applying the mercaptosilane to glass fibers in ethanol, removing the fibers from solution, drying in vacuum, and washing with toluene before drying again. Hexachloroplatinic acid (H_2PtCl_6) was then reacted with the treated fibers in ethanol. Two different sources of E-glass fibers were used ACI E-glass fibers (complex **1**) and Ahlstrom E-glass fibers (complex **2**). To analyze the complexes formed on the surface DRIFT spectroscopy and XPS were used.

Analysis of the Supported Complexes

DRIFT spectroscopy. The DRIFT spectra of unsized E-glass fibers for both ACI glass and Ahlstrom glass are shown in Figure 3. There are strong glass absorbances below 1600 cm^{-1}, and broad silanol and adsorbed water absorbances between 3100 to 3700 cm^{-1} (Figure 3). It is difficult to detect adsorbed reagents upon the E-glass surface in these regions. Two broad, strong IR vibrations at approximately 1450 cm^{-1} and 1248 cm^{-1} are attributed to the B-O, Si-O-B, and Si-O-Si absorbances (Figure 3). The first B-O overtone occurs at ~ 2690 cm^{-1}. No C-H IR vibrations are visible between 2700 cm^{-1} to 3000 cm^{-1}.

A DRIFT spectrum was obtained of the E-glass fibers after adsorption of the silane to compare the differences seen when the platinum was adsorbed. The DRIFT spectrum for the mercaptosilane adsorbed onto the ACI E-glass fibers (Figure 4) shows asymmetric and symmetric C-H stretching vibrations between 2850 cm^{-1} and 2964 cm^{-1}. These absorptions differ from the neat silane which has two strong C-H stretching vibrations at 2848 cm^{-1} (assigned to OCH_3) and 2945 cm^{-1} assigned to asymmetric CH_2 stretching. After adsorption, different stretching vibrations are observed on the glass surface similar to those seen when the mercaptosilane is adsorbed onto other mineral surfaces such as kaolin *(29)*. Sulfur atoms or other electronegative atoms, such as oxygen or nitrogen, attached to carbon generally display greater separation of asymmetric and symmetric vibrations *(30)*. The mercaptosilane C-H absorbances could arise from the labeled carbons shown in Figure 4. The α carbons generally show two absorbances in the regions 2840 cm^{-1}-2880 cm^{-1} and 2915 cm^{-1}-2960 cm^{-1}, while the β carbons similarly show two absorbances in the regions 2842 cm^{-1}-2863 cm^{-1} and 2916 cm^{-1}-2936 cm^{-1}. The γ carbon absorbances occur in the region 2810 cm^{-1}-2850 cm^{-1}. For the neat silane the α and β C-H stretching vibrations are not separated. On the glass fibers, new absorbances are seen. The methoxy peak has disappeared and the CH_2 stretching vibrations have shifted.

Using the stretching vibrations in Figure 4 as a guide the following assignments can be made; the symmetric and asymmetric stretches of the β carbons are at 2853 cm^{-1} and 2924 cm^{-1} respectively, while the α carbons have symmetric and asymmetric stretches at 2886 cm^{-1} and 2964 cm^{-1} respectively. The decrease of the γ carbons indicates that silane adsorption to the glass surface has occurred with concomitant loss of methanol. The DRIFT spectrum has provided clear information on the adsorption of mercaptopropyltrimethoxysilane to glass fibers. Mercaptosilanes, when adsorbed onto surfaces, do display different FTIR stretching vibrations which can be reasonably assigned.

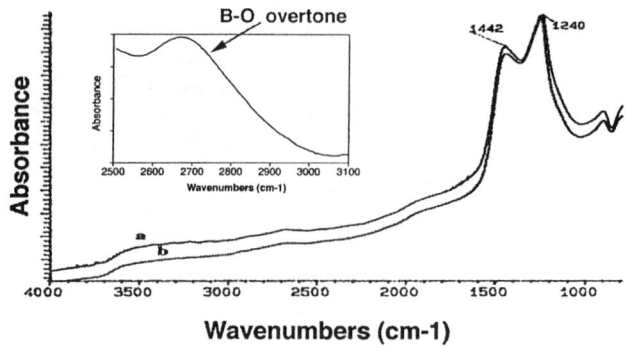

Figure 3. DRIFT Spectra of a) Ahlstrom E-glass Fibers and b) ACI E-glass Fibers.

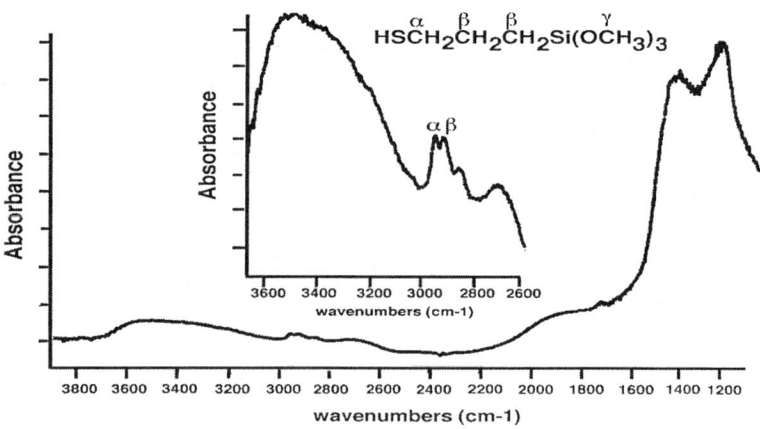

Figure 4. ACI E-glass fibers modified with mercaptopropyltrimethoxysilane

The DRIFT spectrum (Figure 5) of complex **2** shows weak symmetric and asymmetric C-H stretches between 2850 cm^{-1} and 2962 cm^{-1} (complex **1** gave a similar spectrum). The platinum surface complex is unlikely to have any additional absorbances in the C-H stretching vibration region of the FTIR. Compared to Figure 4, the silane adsorbances have decreased as the C-H stretching vibrations are weaker and harder to separate or distinguish. The only distinct peak is at 2964 cm-1 which has been assigned as the asymmetric stretch of the CH_2S group, suggesting that while the mercaptosilane adsorbs to the glass surface, most of the mercapto groups may now be clearly well oriented away from the actual glass surface. The assignments for the other C-H stretching vibrations follow those of the silane only treated sample. The silane treated surface was washed several times before the platinum was attached, and after attachment there was an overall decrease in the C-H stretching vibration intensity. As the platinum complexes to the mercaptosilane its large atomic radius will cause a decrease of all C-H absorbances. In summary then, the DRIFT spectra not only gives information about the silane on the glass surface, but indirectly also shows that the platinum has been attached to the treated glass surface, though not what type of platinum complexes were formed. To get more information on the latter we turn to XPS analysis.

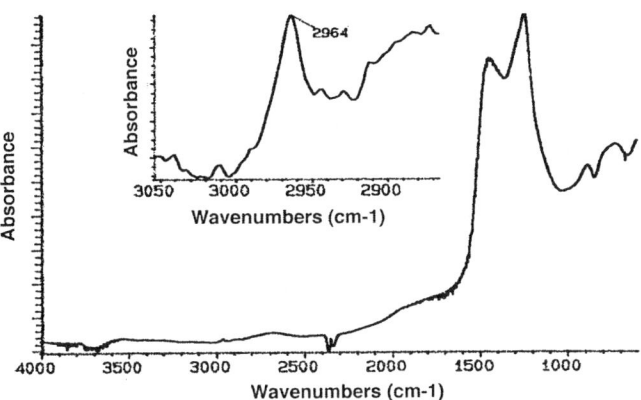

Figure 5. DRIFT spectrum of complex 2.

XPS. Table II shows the atomic concentrations from the XPS data for the untreated E-glass fibers and complexes **1** and **2**. As the silane is added to the glass surface the silicon level is expected to increase, but all or some of the minor glass elements will decrease as they are covered by the silane during the treatment. Therefore, the minor glass element ratios with silicon (i.e. B/Si, Ca/Si, Al/Si, Na/Si) will all decrease. If the silane has not adsorbed uniformly on the glass surface producing a 'patchy' coating then some sites may still be exposed and the ratio of these elements with the silicon may only decrease marginally or remain unchanged. The O/Si, Ca/Si, Al/Si, and B/Si ratios have all decreased compared to clean glass fibers (Table II), indicating the silane is chemisorbed to the surface. Sulfur was detected on the surface as is platinum for complex **2**; though the calcium, aluminium, and boron levels for complex **2** have not decreased as much as for complex **1**.

Although the silane is adsorbed to the surface, aluminium and calcium are still prevalent so the silane layer formed appears to cover the glass surface in patches. After etching the mercaptosilane treated glass surface the concentration of sulfur increases, indicating the silane/siloxane is not positioned on the surface with all of the sulfur groups pointing out well away from the surface, but some of the mercapto groups must be directed or bonded to the surface as represented in Figure 6. In this surface configuration the XPS sulfur signal is attenuated by the carbon content above. The carbon is removed first on etching to reveal more of the surface bonded species. After etching complex **2** the sulfur and platinum concentrations remain the same, again indicating that some of the mercapto groups are near the surface. The mercapto groups proximal to the surface make it more difficult for the platinum to attach through covalent Pt-S bonds. Platinum, a Lewis acid, may also prefer to attach to the Lewis-base, aluminium, calcium, or boron sites resident on the glass. To determine whether the platinum has coordinated with the mercaptosilane on the glass surface, changes in sulfur and platinum binding energies were examined.

Table II. XPS Data for the Supported Complexes

Element	Atomic Concentrations (%)							
	ACI Fibers	Complex 1	ACI Fibers (5nm etch)	Complex 1 (5nm etch)	Ahlstrom Fibers	Complex 2	Ahlstrom Fibers (5 nm etch)	Complex 2 (5nm etch)
C 1s	29.3	45.3	10.7	33.7	31.3	35.7	14.9	9.6
O 1s	50.0	33.3	61.3.	40.0	44.0	38.0	54.9	56.7
Si 2p	12.2	17.6	15.7	20.0	16.8	19.7	21.1	21.9
Al 2s	3.5	1.2	4.0	1.5	3.4	2.6	4.2	4.6
Ca 2p	2.9	0.8	4.2	1.4	1.8	1.5	2.3	3.7
Na 1s	0.9	<0.05	1.6	<0.05	0.2	<0.05	0.2	<0.05
B 1s	0.6	<0.05	1.3	<0.05	2.5	0.8	2.3	2.8
Cl 2p	0.6	0.6	0.7	0.8	0.2	<0.05	0.1	0.8
F 1s	0.4	<0.05	0.4	<0.05	<0.05	<0.05	<0.05	<0.05
S 2p	<0.05	0.5	<0.05	0.8	<0.05	1.6	<0.05	1.6
Pt 4f	<0.05	0.8	<0.05	1.3	<0.05	0.2	<0.05	0.2

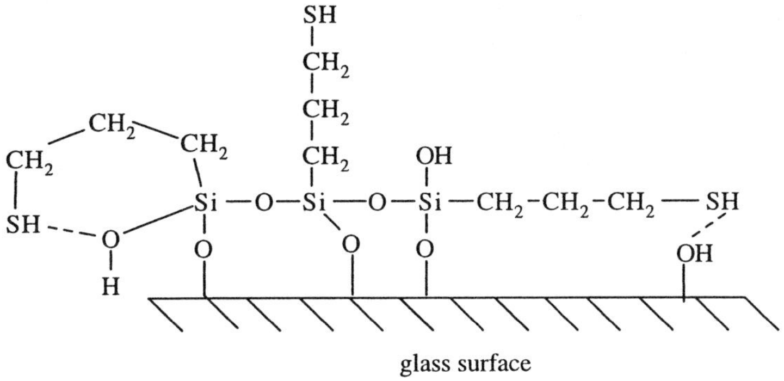

```
            SH
            |
            CH₂
            |
       CH₂  CH₂
  CH₂ /    \CH₂   |
  |    \    /   CH₂   OH
  |     Si—O—Si—O—Si—CH₂-CH₂-CH₂——SH
  SH---O     |      |       |
       |     O      O       O        OH
       H
```
                    glass surface

Figure 6. Diagram of how the mercaptosilane could be adsorbing to the glass.

When mercaptopropyltrimethoxysilane is adsorbed onto glass fibers the sulfur (2p) binding energy changes from 162.9 eV to 162.5 eV on etching. After H_2PtCl_6 is adsorbed onto the treated fibers the sulfur binding energy increases to 163.4 eV for complex **1** and 163.6 eV for complex **2**. This 0.6-1.0 eV binding energy increase, coupled with the platinum (4f) binding energy decrease to 72.3 eV (compared to H_2PtCl_6 at 75.6 eV), are clear indicators that a Pt-S bond is formed.

Jiang *et al.(19,26-28)* prepared a series of sulfur containing silica compounds. To these silica surfaces, platinum complexes were attached (Figure 7) generating catalytic surfaces with various S/Pt ratios. Complex **a** was prepared by:-
1) Refluxing toluene, silica, γ-mercaptopropyltriethoxysilane, ethanol, together with 10% HCl.
2) Filtering the silica and drying in vacuum.
3) Extracting with toluene then drying.
4) The treated silica was refluxed with a solution of $H_2PtCl_6 6H_2O$ in ethanol, and dried *(27)*.

Complex **b** was prepared by the same procedure except the silane used in this instance was 3,3 bis(triethoxysilyl)dipropylthioether. The sulfur (2p) binding energy of both complexes increases by 1.1 eV on reaction with the H_2PtCl_6. Concomitantly the Pt (4f) binding energy decreases by 1.8-2.1 eV (compared to H_2PtCl_6), while the Cl (2p) binding energy remains similar to that of H_2PtCl_6 (198.8 eV).

The S (2p) binding energy difference for the complexes is comparable with the published results *(19,27-28)*, however the Pt (4f) difference is greater (2.9 eV). If the platinum has not attached to the sulfur then other possibilities for what is surrounding the platinum needs to be discussed. A PtO(adsorbed) species has been reported with a Pt 4f binding energy of 72.2 eV *(31)* for Pt/Al_2O_3 when either oxidized in air or reduced in hydrogen at temperatures above 150 °C. The binding energy values of the Pt/Al_2O_3 before being exposed have not been reported, as this is the usual method for preparing dispersed metal catalysts. Consequently there is a possibility that a

PtO(adsorbed) species has formed on our treated glass surfaces, though the conditions used for preparing our Pt complexes were much milder than the conditions used to form PtO(adsorbed) species on the Pt/Al$_2$O$_3$ catalysts. The Pt (4f) binding energy of the complexes is higher than H$_2$PtCl$_6$ adsorbed onto glass fibers. It appears the H$_2$PtCl$_6$ is being reduced but not by the unreacted silanol groups on the glass. Possibly the mercapto groups proximal to the glass surface synergistically assist the reduction of H$_2$PtCl$_6$ on the glass surface. The question still remains as to what groups surround the platinum on the glass surface?

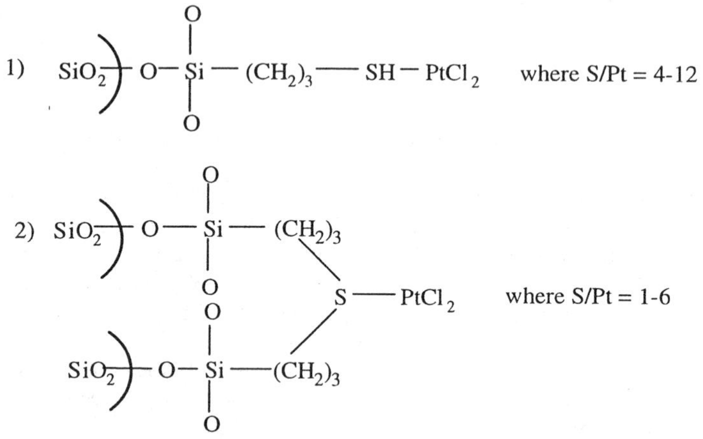

1) SiO$_2$—O—Si(=O)(O)—(CH$_2$)$_3$—SH—PtCl$_2$ where S/Pt = 4-12

2) SiO$_2$—O—Si(=O)—(CH$_2$)$_3$
   SiO$_2$—O—Si(=O)—(CH$_2$)$_3$
   }S—PtCl$_2$ where S/Pt = 1-6

Figure 7. Supported mercaptosiloxane platinum complexes prepared by Jiang (19, 26-28).

The Cl (2p) binding energy (199.3 eV before etching and 199.7 eV after etching) for complex **1** has increased compared to H$_2$PtCl$_6$ (198.8 eV). On clean glass fibers the chlorine (2p) binding energy is 197.7 eV. If the platinum was still coordinated to the chlorine, then the Cl (2p) binding energy should be close to H$_2$PtCl$_6$ (±0.5 eV). Therefore the chlorine is not coordinated to the platinum but adsorbing onto the glass fibers in another form. When glass fibers are treated with SOCl$_2$ the surface Si-OH groups are replaced with surface Si-Cl groups; and the Cl (2p) binding energy is 198.8 eV *(32)*. It is still not clear whether the chlorine is interacting with the free OH groups on the glass surface, or if they remain coordinated with a reduced form of the platinum. Chlorine was not detected at all for complex **2**.

The XPS and DRIFT spectra combined did give information about the extent of surface treatment and metal adsorption for complexes **1** and **2**. The sulfur binding energy indicates that a Pt^{II}-S complex has formed. Complex **1** shows silane hydrolysis with some of the sulfur groups pointing down onto the surface, while DRIFT spectra did show the majority of the sulfur groups were pointing away from the surface. A comparison with samples prepared previously revealed a distinctive platinum binding energy *(19,27-28)*. The platinum complexes were then tested for their hydrosilating activity.

Catalytic Activity

Hydrosilations have traditionally been studied using homogeneous catalysts, and hydrido silanes *(9)* with a number of organic reagents. To evaluate supported, or heterogeneous hydrosilation catalysts a number of different methods are used. The simplest method for evaluation is to measure the yield of the reaction (preferred by a number of workers) *(16, 33-35)*. A number of different silanes and organic reagents *(1)* are used in these evaluations, making it difficult to directly compare one catalyst to another. Other evaluation methods use Gas Liquid Chromatography (GLC) *(36)* and Gas Chromatography (GC) *(21)* to measure conversions and rates. Here, a suitable standard mixture (hydrido-silane and unsaturated organic reagents) is available so the percentage of various products could be measured. Kinetic data has also been collected from such studies *(21, 36, 37)*. Our interest focuses on supported hydrosilation catalysts in preparing functionalized siloxane polymers rather than functionalized silanes. GLC and GC are not viable techniques in rapidly assessing the progress of reactions on such polymers. Therefore, different analytical techniques need to be used for catalyst evaluation. Routinely ^{1}H NMR is used in our laboratories to test the extent of hydrosilation as it is a quick and simple technique. Surface analytical techniques, discussed earlier, have been useful to help define the nature of the heterogeneous catalysts made, and these have affirmed the integrity of the catalysts used in the hydrosilation.

The test reaction used to evaluate the supported complexes prepared previously, uses the hydrosilation of poly(methylhydrogen)siloxane (dp=33) with hexene (Figure 8). The siloxane and hexene were stirred together under nitrogen in toluene and 0.2 g of the supported catalytic complex at 70 °C for 12 hours, at which time the reaction was found to generally reach equilibrium. The hydrosilation reaction was monitored by ^{1}H NMR, Figure 9 shows the ^{1}H NMR of the hydrosilation using supported complex **1**. For this sample, toluene was removed under vacuum; however a clear NMR can still be obtained even if toluene is not removed, making it possible to monitor the reaction in situ. There are several general principle's in monitoring a hydrosilation reaction. Firstly, the decrease of the Si-H resonance at 4.7 ppm, associated with the poly(methylhydrogen)siloxane indicates reaction in the siloxane side chain. Secondly, the disappearance or decrease of the vinyl resonances at 5 ppm and 5.8 ppm, affirms coupling of the hexene to the polymer side chain. Finally, successful hydrosilation shows resonances at 0.5 ppm (CH_2 α to silicon), at 1.1 ppm (CH_2 β to silicon) overlapping with another CH_2 in the alkyl chain (Figure 9). Such resonances were assigned on the basis of two papers where the proton spectrum *(38, 39)* were clearly interpreted. The initial ^{1}H NMR did not show evidence for the formation of the α-adduct. ^{13}C NMR confirmed that only the β-adduct had formed.

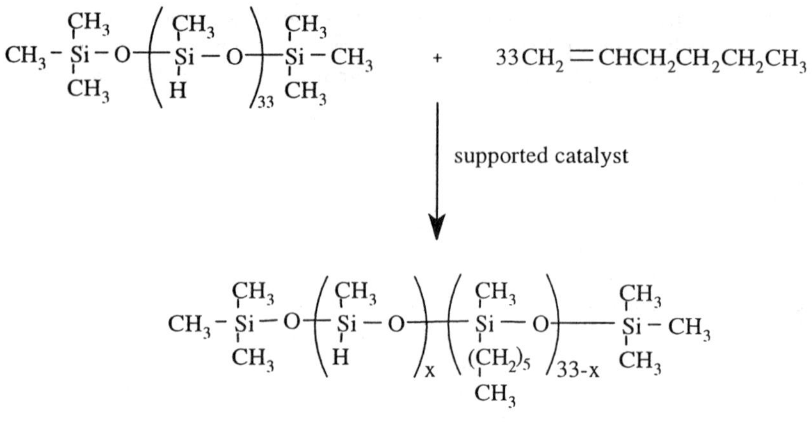

$$\text{CH}_3-\underset{\underset{\text{CH}_3}{|}}{\overset{\overset{\text{CH}_3}{|}}{\text{Si}}}-\text{O}\left(\underset{\underset{\text{H}}{|}}{\overset{\overset{\text{CH}_3}{|}}{\text{Si}}}-\text{O}\right)_{33}\underset{\underset{\text{CH}_3}{|}}{\overset{\overset{\text{CH}_3}{|}}{\text{Si}}}-\text{CH}_3 \quad + \quad 33\,\text{CH}_2\!=\!\text{CHCH}_2\text{CH}_2\text{CH}_2\text{CH}_3$$

$$\Big\downarrow \text{supported catalyst}$$

$$\text{CH}_3-\underset{\underset{\text{CH}_3}{|}}{\overset{\overset{\text{CH}_3}{|}}{\text{Si}}}-\text{O}\left(\underset{\underset{\text{H}}{|}}{\overset{\overset{\text{CH}_3}{|}}{\text{Si}}}-\text{O}\right)_{x}\left(\underset{\underset{\underset{\text{CH}_3}{|}}{(\text{CH}_2)_5}}{\overset{\overset{\text{CH}_3}{|}}{\text{Si}}}-\text{O}\right)_{33-x}\underset{\underset{\text{CH}_3}{|}}{\overset{\overset{\text{CH}_3}{|}}{\text{Si}}}-\text{CH}_3$$

Figure 8. Hydrosilation reaction used to evaluate supported complexes.

To determine the hydrosilation efficiency on the polymer, the frequency of the SiCH$_2$- groups (33-x) that had reacted, were determined by using the peak area integrals from the ^{1}H NMR to solve the equation below. The protons associated with the SiCH$_2$ (33-x) groups will be inversely related to the number of SiH (x) protons that remain unreacted both of which can be determined from their respective ^{1}H NMR integrals.

$$\frac{\int \text{SiCH}_2 \text{ protons from } ^1\text{H NMR}}{\int \text{SiH protons from } ^1\text{H NMR}} = \frac{2(33-x)}{x}$$

Each supported complex was tested for a maximum of 5 catalytic runs. A plot of 33-x, representing the hydrosilation efficiency versus the number of runs, was obtained for each heterogeneous catalyst.

Side Reactions

The ^{1}H NMR before removal of toluene, indicated the presence of an internal double bond at 5.3 ppm, in addition to the usual hydrosilation products with the CH$_2$ next to the internal double bond obscured by the methyl group of toluene also at 2 ppm. The compound containing such an internal double bond, is removed in vacuum together with toluene, so is not present in the ^{1}H NMR spectrum shown in Figure 9. Lewis and Uriarte *(39)* also formed a similar internal olefin, on hydrosilation of, 1-hexene or 1-octene with trimethylsilane. Such olefins are then unreactive to subsequent hydrosilation, so are competing side reactions to hydrosilation.

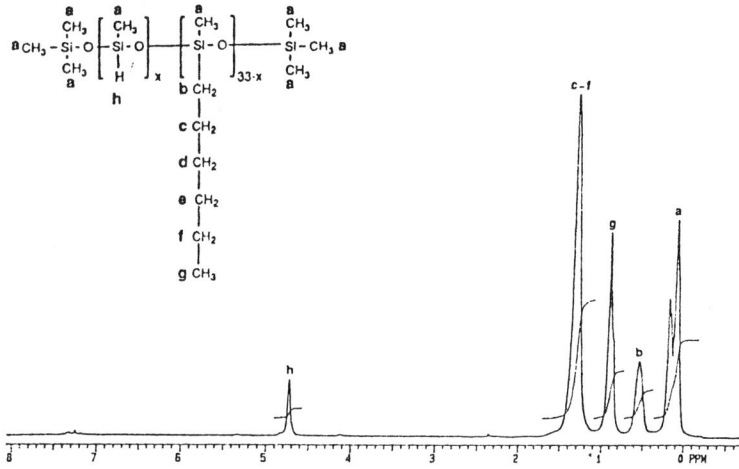

Figure 9. ^1H *NMR of the hydrosilation of hexene with an SiH containing siloxane, using the supported complexes as catalysts.*

Hydrogen usually evolves during the observed induction period, using homogeneous catalysts, leading to the formation of Si-Si bonds. Such side reactions occur during the hydrosilation of vinyltrialkoxysilanes with different pendant hydridosiloxane backbones *(40)*. Though no hydrogen evolution during the course of our reactions is seen, it is still possible the side reaction occurs to a minor extent using supported complexes as catalysts.

Figure 10 compares ACI and Ahlstrom E-glass fiber mercaptoplatinum supported complexes. Complex **2** compared to Complex **1** did not perform as well and lost catalytic activity after only three runs. This is not unexpected, as the S/Pt ratio is higher and there is less surface platinum. Clearly, subtle differences in E-glass fiber composition between the various manufacturers has a noticeable effect on surface composition, and the amount of catalyst bound. Such differences are evident, in catalytic efficiency, rate and turnover numbers. Without systematic variations of glass fiber composition being correlated to surface catalyst content and efficiency, existing differences can only be speculative. However, the sudden loss of activity in complex **1** run 4, is likely to coincide with loss of surface platinum into solution. Apart from this decrease in catalytic activity, all other runs maintained catalytic activity and efficiency at reasonably consistent levels (though complex **1** was better than complex **2**). Interestingly, the catalytic activity of complex **1** increased after run 2 before decreasing after run 4. It is not clear why this happens, but the catalytic species maybe rearranging on the surface, making more surface platinum available for reaction. The results were consistent and reproducible for each type of E-glass fiber used.

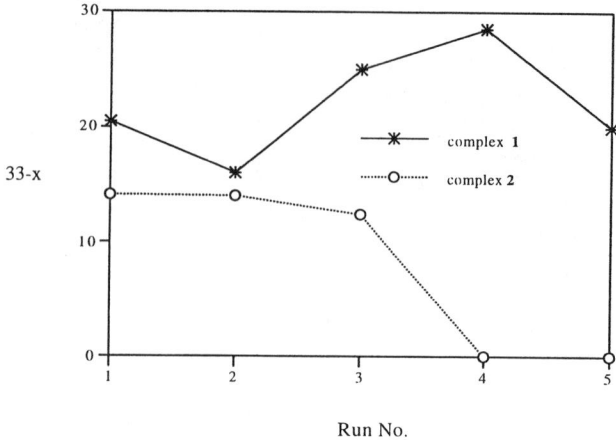

Figure 10. Comparison of activity between ACI and Ahlstrom glass fiber supported mercaptoplatinum complexes.

Catalyst deactivation is usually caused by metal leaching from the surface *(1)*. Ejike *(41)* similarly found that supported platinum complexes lost the greatest amount of adsorbed metal complex during the 3rd catalytic run; however such heterogeneous catalysts still remained active for up to six catalytic cycles in hydrosilating bistrimethylsiloxymethylhydrosilane with 1-decene. A similar catalyst was prepared by Wang and Jiang *(19)* for hydrosilation of triethoxysilane with hexene; and their catalysts remained active over 20 consecutive hydrosilations. Analogous amineplatinum supported complexes lost catalytic activity dramatically after 6 recycles *(19)*. The atomic ratio of S/Pt affects product yield with an increase in the S/Pt ratio causing a decrease in yield *(19)*.

Poly(methylhydrogen)siloxane has 33 pendant Si-H groups to hydrosilate, whereas silane reagents generally have only one Si-H bond. Poly(methylhydrogen)siloxane is also a much larger molecule than a silane, so it is increasingly difficult for organic double bonds to diffuse to the reactive sites on the siloxane as more SiH groups are hydrosilated. Homogenous catalysts show a corresponding decrease in activity when polymeric substrates for hydrosilations are used *(21)*. How the reaction proceeds on the silica or glass surface and the mechanism involved, remains an area for further study. The purpose of these simple catalytic tests was to investigate the effectiveness of our supported complexes for hydrosilations using siloxanes. Even though hydrosilation was incomplete and some pendant Si-H groups remained on the siloxane, these catalysts did show catalytic activity.

Sulfoxide and sulfide complexes of platinum have been previously prepared and are active hydrosilation catalysts *(42)*. Sulfur is generally known to inhibit hydrosilation reactions. The fact that our supported complexes are catalytically active, further confirms that a Pt-S complex forms on the glass surface. Significantly, when H_2PtCl_6 was adsorbed onto the untreated glass surface the heterogeneous catalyst was active for 1 run only, consistent with the complex desorbing from the glass surface *(40)*. Weak binding of H_2PtCl_6 to the untreated glass surface allows desorption or 'catalyst leaching' to occur, so hydrosilation is driven by the desorbed homogeneous catalyst, that cannot be renewed. Glass appears to be a good support for mercaptoplatinum complexes that can effectively hydrosilate siloxanes.

Conclusions

E-glass fibers can be effectively modified with a mercaptosilane, followed by adsorption of the metal complex H_2PtCl_6, to give an active supported catalyst. The binding energy information supplied by XPS confirmed a $Pt^{II}S$ species was formed. Small differences in the composition of the glass fibers used, do affect the amount of platinum adsorbed onto the surface and also the catalytic activity of the supported complexes. E-glass fiber supported mercaptoplatinum complexes are effective as catalysts and therefore the use of glass fibers for supporting metal complexes, does look promising. Greater control of the S/Pt ratio, would make the mercaptoplatinum supported complexes even better, and see these materials as the preferred catalysts for hydrosilation.

References

1. Hartley, F. R. *Supported Metal Complexes*; Reidel: The Netherlands: Utrecht, 1985.
2. Bailey, D. C.; Langer, S. H. *Chem. Rev.* **1981**, *81*, 109.
3. Meals, R. N. *Pure Appl. Chem.* **1966**, *13*, 141.
4. Speier, J. L. *Adv. Organomet. Chem.* **1979**, *17*, 407.
5. Brown-Wesley, K. *Organometallics* **1987**, *6*, 1590.
6. Lapp, A.; Xiongwei, H.; Herz, J. *J. Makromol. Chem.* **1988**, *189*, 1061.
7. Herz, J.; Belkebir-Mrani, A.; Rempp, P. *Eur. Polym. J.* **1973**, *9*, 1165.
8. Lestel, L.; Cheradame, H.; Boileau, S. *Polymer* **1990**, *31*, 1154.
9. Marciniec, B; Gulinski, J.; Urbaniak, W.; Kornetka, Z. W. In *Comprehensive Handbook on Hydrosilylation;* Marciniec, B., Ed.; Pergamon: Oxford, 1992; pp 83-94.
10. Reikhsfel'd, V. O.; Vinogradov, V. N.; Fillipov, N. A. *Zh. Obshch. Khim.* **1978**, *48*, 2383.
11. Capka, M.; Svoboda, P.; Hetflejs, J. *Coll. Czech. Chem. Commun.* **1973**, *38*, 1242.

12. Capka, M.; Svoboda, P.; Cerny, M.; Hetflejs J. *Chem. Ind. (London)* **1972**, 650.
13. Michalska, Z. M.; Ostaszewski, B. *J. Organomet. Chem.* **1986**, *299*, 259.
14. Bartholin, M.; Graillat, C.; Guyot, A.; Coudurirer, G.; Bandiera, J.; Naccache, C. *J. Mol. Catal.* **1977**, *3*, 17.
15. Panster, P.; Kleinschmit, P. Proc. VII Int. Symp. Organosilicon Chem., Kyoto, Japan **1984**, 150.
16. Capka, M.; Hetflejs, J. *Coll. Czech. Commun.* **1974**, *39*, 154.
17. Cai, M; Song, C.; Chen, J. *Yinyong Huaxue* **1997**, *14*, 37 CA *127:248503*.
18. Marciniec, B.; Urbaniak, W. *J. Mol. Catal.* **1983**, *18*, 49.
19. Wang, L.; Jiang, Y. *J. Organomet. Chem.* **1983**, *251*, 39.
20. Ruiz, J.; Bentz, P. O.; Mann, B. E.; Spencer, C. M.; Taylor, B. F.; Maitlis, P. M. *J. Chem. Soc., Dalton Trans.* **1987**, 2709.
21. Marciniec, B.; Urbaniak, W.; Pawlak, P. *J. Mol. Catal.* **1982**, *14*, 323.
22. Chen, Y.; Meng, L.; Li, L., Luo, J. *Chin. J. Polym. Sci.* **1993**, *11*, 22.
23. Prignano, A. L.; Trogler, W. C. *J. Am. Chem. Soc.* **1987**, *109*, 3586.
24. Michalska, Z. M.; Capka, M.; Stoch, J. *J. Mol. Catal.* **1981**, *11*, 323.
25. Plueddemann, E. P. In *Silane Coupling Agents*; Plenum: New York, 1982
26. Hu, C. Y.; Han, X. M., H.; Jiang, Y. Y. *Kexue Tongbao* **1988**, *33*, 843.
27. Hu, C. Y.; Han, X. M.; Jiang, Y. Y., Liu, J. G., Shi, T. Y. *J. Macromol. Sci. Chem.* **1989**, *A26*, 349.
28. Hu, C. Y.; Han, X. M.; Jiang, Y. Y. *J. Mol. Catal.* **1986**, *35*, 329.
29. Porro, T. J.; Pattacini, S. C. *Appl. Spect.* **1990**, *44*, 1170.
30. Lin-Vien, D.; Colthup, N. B.; Fateley, W. G.; Grasselli, J. G. *The Handbook of Infrared and Raman Characteristic Frequencies of Organic Molecules;* Academic: Boston, MA, 1991, p 485.
31. Shyu, J. Z.; Otto, K. *Appl. Surf. Sci.* **1988**, *32*, 246.
32. Britcher, L.G. PhD Thesis, University of South Australia, Mawson Lakes, South Australia, 1997
33. Marciniec, B.; Kornetka, Z.; Urbaniak, W. *J. Mol. Catal.* **1981**, *12*, 221.
34. Williams, R. E. US Patent 4 503 160 1985.
35. Chunye, H.; Xuemeng, H.; Yingyan, J. *Kexue Tongbao* **1988**, *33*, 843.
36. Michalska, Z. M. *J. Mol. Catal.* **1977**, *3*, 125.
37. Kraus, M. *Coll. Czech. Commun.* **1974**, *39*, 1318.
38. Lewis, L. N. *J. Am. Chem. Soc.* **1990**, *12*, 16.
39. Lewis, L. N.; Uriarte, R. J. *Organometallics* **1990**, *9*, 621.
40. Britcher, L .G.; Kehoe, D. C.; Matisons, J. G.; Swincer, A. G. *Macromolecules* **1995**, *28*, 3110.
41. Ejike, E. N. PhD Thesis University of Manchester, West Yorkshire, United Kingdom, 1984.
42. Trofimov, A. E.; Spevak, V. N.; Lobadyuk, V. I.; Skvortsov, N. K.; Reikhsfel'd, V. O. *Zh. Obshch. Khim.* **1989**, *59*, 2048

Olefin / Alkyne Addition Polymerization

Undoubtedly, the most widely used application of transition metal catalysis in polymer synthesis is the addition polymerization of monomers with unsaturated carbon-carbon bonds. While the industrial importance and intensely studied mechanistic aspects of metallocene-catalyzed polyolefin synthesis are well-known, the polymerization of acetylenes with organometallics is also an important technique for the preparation of conjugated macromolecules. The continued development and understanding of olefin / alkyne addition polymerizations is thus an area of intense current research in both academia and industry.

The first paper in this section, by Tang et al., reports on the polymerization of a wide variety of substituted acetylenes with transition metal complexes, yielding polar-functionalized and stereoregular polyacetylene materials. The polycyclotrimerization polymerization of diacetylenes to give hyperbranched poly-(phenylenealkenes) is also discussed. Next, Wu et al. report the use of monocyclopentadienyltitanium complexes to prepare vinyl-type polynorbornenes. The last paper in this section is a thorough mechanistic study by Resconi et al. of important chain transfer and isomerization reactions present during propylene polymerization with a highly stereo- and regiospecific zirconocene initiator.

Chapter 9

New Catalysts for Polymerizations of Substituted Acetylenes

Ben Zhong Tang, Kaitian Xu, Qunhui Sun,
Priscilla P. S. Lee, Han Peng, Fouad Salhi, and Yuping Dong

Department of Chemistry, Hong Kong University of Science and Technology,
Clear Water Bay, Kowloon, Hong Kong, China (e-mail: tangbenz@ust.hk)

A variety of transition metal catalysts have been developed for the polymerizations of different types of substituted acetylenes. [Rh(nbd)Cl]$_2$ (nbd = 2,5-norbornadiene) and WCl$_6$–Ph$_4$Sn/dioxane are effective catalysts for polymerizations of acetylenes containing polar functional groups such as cyano, ether, and ester units. Stereoregular poly(phenylacetylenes) are produced by aqueous polymerizations catalyzed by Rh complexes including [Rh(nbd)Cl]$_2$, [Rh(cod)Cl]$_2$, Rh(cod)(pip)Cl, [Rh(cod)Cl]$_2$(pda), Rh(cod)(bbpmt), Rh(cod)(NH$_3$)Cl, [Rh(cod)(mid)$_2$]PF$_6$, Rh(nbd)(tos)(H$_2$O), and Rh(cod)(tos)(H$_2$O) (cod = 1,5-cyclooctadiene, pip = piperidine, pda = o-phenyldiamine, bbpmt = bis(4-t-butyl)-2-pyridylmethanethiolate, mid = N-methylimidazole, and tos = p-toluenesulfonate). Air- and moisture-stable metal carbonyl complexes W(CO)$_3$(mes) (mes = mesitylene) and Mo(CO)$_4$(nbd) effectively initiate polymerizations of phenylacetylene and 1-chloro-1-octyne without the use of cocatalyst and photoirradiation. A number of transition metal compounds including TaCl$_5$, NbCl$_5$, Mo(CO)$_4$(nbd), [W(CO)$_3$cp]$_2$ (cp = cyclopentadiene), PdCl$_2$–ClSiMe$_3$, and Pd/C–ClSiMe$_3$ catalyze polycyclotrimerizations of terminal and internal diacetylenes (diynes) such as 1,8-nonadiyne, 1,9-decadiyne, 3,9-dodecadiyne, and 1,9-bis(trimethylsilyl)-1,8-nonadiyne, yielding soluble hyperbranched poly(phenylenealkenes) under mild conditions.

Introduction

The investigation of electroactive polymers and other functional materials has become one of the most important areas of research in polymer and materials science during past two decades (*1-3*). Polyacetylenes, the best-known conjugated polymers,

© 2000 American Chemical Society

have attracted much interest among scientists and technologists. The exploration of highly efficient and easily manipulatable catalysts for the polymerizations of acetylenes is of importance in views of both theoretical interest and practical applications.

A number of transition metal compounds based on Mo, W, Nb, and Ta have been reported as effective catalysts for the polymerizations of acetylenes (*2,4,5*). Soluble polyacetylenes with high molecular weights have been obtained in good yields. However, because almost all the catalysts are only effective for nonpolar or low-polarity acetylenes, most polyacetylenes prepared so far bear only alkyl and/or aryl groups. Although introduction of polar functional groups such as cyano (–CN), amino (–NH$_2$), hydroxy (–OH), and carboxy (–CO$_2$H) units into polyacetylenes may facilitate the development of polyacetylenes with novel properties such as liquid crystallinity, chirality, enantiopermselectivity, photoconductivity, and magnetism, functional acetylene monomers have not yet been well polymerized. Because the polar groups are "toxic" to most transition metal catalysts, polymerizations of polar acetylenes often gave oligomers or insoluble products, normally in low yields (*4,6,7*).

Most of the "conventional" catalysts for the polymerizations of acetylenes are air- and moisture-sensitive and must be handled in dry and inert atmosphere. The polymerizations of acetylenes therefore need to use organic solvents as reaction media, most of which are toxic and hazardous. Replacement of the organic solvents by water or aqueous media will undoubtedly contribute to the protection of public health and living environment. In addition to the air- and moisture-sensitivity, most catalysts for acetylene polymerizations need additional Lewis acids as cocatalysts and/or photoirradiation to generate the active species (*4,8*). The polymerizations must be conducted under stringent conditions. Development of air- and moisture-stable catalysts without use of additional cocatalysts and photoirradiation is thus of obvious practical significance (*9*).

Diynes are another type of acetylenes, i.e. diacetylenes. A variety of transition metal catalysts can cause diynes to undergo cyclotrimerization reactions. Although cyclotrimerization of diynes has been extensively investigated (*10*), the previous attention was mainly focused on the low molecular weight compounds, such as cyclodimers and cyclotrimers. So far there have been only a few scattered reports concerning polycyclotrimerizations of diynes (*11-13*). In the attempts to prepare soluble polyphenylenes via cyclotrimerization reaction, Chalk et al. conducted some preliminary investigations, mainly on cocyclotrimerization of diynes with monoynes (*14-16*). No further study was reported. When diynes undergo intermolecular polycyclotrimerization, hyperbranched polyphenylenes will form. Exploration of new catalysts for polycyclotrimerization of diynes will potentially open up a new avenue in the development of hyperbranched polymers with unique molecular architectures and materials properties.

We have conducted a series of investigations to explore new transition metal catalysts for the polymerizations of substituted acetylenes. From these studies, we have successfully developed a variety of novel and effective catalysts for different kinds of acetylene polymerizations. Our catalyst systems include: (1) robust catalysts for the polymerizations of polar acetylenes which has led to the development of a new class of liquid crystalline, light emitting, and photoconductive polyacetylenes, (2) water-soluble catalysts for aqueous polymerizations of arylacetylenes, (3) air-stable and easy-to-handle catalysts for acetylene polymerizations in which no cocatalysts and photoirradiation are required, and (4) new catalysts for the polycyclotrimerizations of diynes which yield soluble hyperbranched poly(phenylenealkenes). This chapter is to summarize our findings in this research area.

Results and Discussion

Robust Catalysts for Polymerizations of Polar Substituted Acetylenes

The polymerization of polar acetylenes such as cyanoacetylenes was quite difficult. Many different kinds of catalysts (e.g. anionic, Ziegler-Natta, metathesis, and photochemical initiation systems) have been attempted, most of which failed to produce soluble high-molecular-weight polymers (*4,6,17*). The polymerization of cyanoacetylene or propiolonitrile (HC≡CCN) catalyzed by $MoCl_5$–Ph_4Sn, for example, gave only 3% insoluble black powdery substance after a very long polymerization time (92 h), while the polymerization of 5-cyano-1-pentyne [HC≡C(CH$_2$)$_3$CN] initiated by $MoCl_5$–Me_4Sn yielded polymers with molecular weights of only a few thousand. Gorman et al. (*18*) have recently succeeded in developing Pd- and Ni-based late-transition-metal catalysts for synthesizing poly(cyanoacetylene)s in good yields, although the polymers are only partially soluble and contain substantial amounts of impurities (or catalyst residues).

We synthesized two types of cyanoacetylene monomers (*19,20*), whose molecular structures are shown in Chart 1. The monomers **1** are phenylacetylene derivatives while the monomers **2** are alkylacetylene derivatives.

Chart 1

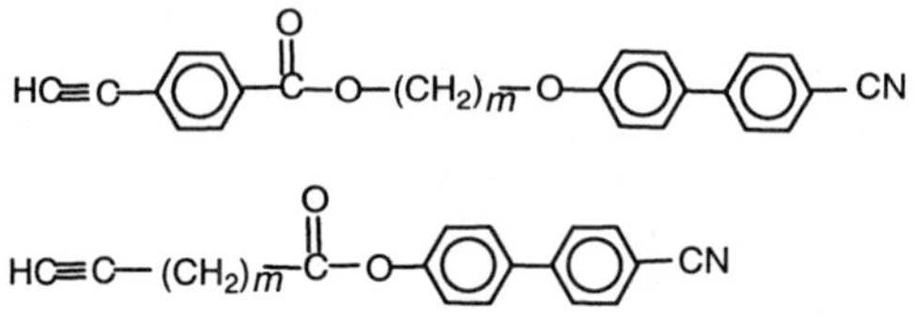

HC≡C—⌬—C(O)—O—(CH$_2$)$_m$—O—⌬—⌬—CN m = 6 **(1a)**
                                           12 **(1b)**

HC≡C—(CH$_2$)$_m$—C(O)—O—⌬—⌬—CN m = 2 **(2a)**
                                       3 **(2b)**
                                       8 **(2c)**

We first tried to polymerize **1**, {4-[({n-[(4'-cyano-4-biphenylyl)oxy]alkyl}oxy)-carbonyl]phenyl}acetylene, using the classical metathesis catalysts (*19*). When **1a** was mixed with WCl_6–Ph_4Sn in dioxane, no polymeric product was isolated after 24 h reaction (Table 1, no. 1). Further attempts by using different solvents (THF and toluene) and catalyst (MoCl$_5$–Ph$_4$Sn) all ended up with failure.

Many organorhodium complexes have been synthesized in aqueous solvents. Noticing the remarkable tolerance of the Rh complexes toward polar environments, we tried to polymerize **1a** by [Rh(nbd)Cl]$_2$. Although the concentration of the Rh complex (10 mM) was only half of that of the W catalyst, a yellowish brown polymer was isolated in 40% yield after 24 h of polymerization (Table 1, no. 4). The isolated polymer, however, was only partially soluble. When we changed the solvent from THF/Et$_3$N to DMF/Et$_3$N, polymer with higher molecular weight (M_w 26900) was obtained in 71% yield. Most importantly, the polymer was completely soluble in organic solvents such as DMF, THF, and chloroform, although the dissolution took several days.

Table 1. Polymerization of {4-[({n-[(4'-Cyano-4-biphenylyl)oxy]alkyl}oxy)carboxy]phenyl}acetylenes[a]

No.	Catalyst	Solvent	Yield (%)	M_w	M_w/M_n
		Monomer **1a** ($m = 6$)			
1	WCl$_6$–Ph$_4$Sn	dioxane	0		
2	WCl$_6$–Ph$_4$Sn	THF	0		
3	MoCl$_5$–Ph$_4$Sn	toluene	0		
4	[Rh(nbd)Cl]$_2$	THF/Et$_3$N[b]	40[c]	18600[d]	2.9[d]
5	[Rh(nbd)Cl]$_2$	DMF/Et$_3$N[b]	71	26900	3.8
		Monomer **1b** ($m = 12$)			
6	WCl$_6$–Ph$_4$Sn	dioxane	0		
7	WCl$_6$–Ph$_4$Sn	toluene	29	5400	1.5
8	MoCl$_5$–Ph$_4$Sn	toluene	0		
9	[Rh(nbd)Cl]$_2$	THF/Et$_3$N[b]	59[c]	20300[d]	4.1[d]
10	[Rh(nbd)Cl]$_2$	DMF	70	158000	4.0

[a] Carried out under an atmosphere of dry nitrogen at room temperature for 24 h. [M]$_o$ = 0.2 M. For the W and Mo catalyst systems, [cat.] = [cocat.] = 20 mM; for the Rh catalyst system, [cat.] = 10 mM. [b] Volume ratio of THF or DMF to Et$_3$N: 4:1. [c] Partially soluble in THF. [d] For the THF-soluble fraction. SOURCE: Reproduced from ref 19. Copyright 1997 American Chemical Society.

Polymerization behavior of cyanoacetylene **1b**, which possesses a longer methylene spacer ($m = 12$), was similar to that of **1a**. [Rh(nbd)Cl]$_2$ polymerized **1b** effectively. DMF was found to be a better solvent than THF/Et$_3$N, producing a completely soluble polymer with a high molecular weight (M_w = 158000) in a high yield (70%).

Figure 1 shows the GPC curves of the products of **1b** from the W and Rh catalysts. The product isolated from WCl$_6$–Ph$_4$Sn/toluene system exhibits a sharp peak in the low molecular weight region along with a shoulder in the higher molecular weight region (Figure 1A). Although the polystyrene calibration prevents us from obtaining absolute molecular weights of the products, we suspect that the sharp peak and the shoulder are respectively from trimers and oligomers of **1b**, as WCl_6 can catalyze cyclotrimerization of certain alkynes. In contrast, the product obtained from [Rh(nbd)Cl]$_2$/DMF system shows only one peak in the high molecular weight region (Figure 1B), suggesting that the Rh-catalyzed polymerization has proceeded without complications of side reactions. TGA analysis revealed that a 5% weight loss of the polymers of **1a** and **1b** took place at ~400 °C under nitrogen; that is, the polymers are thermally very stable (*19*).

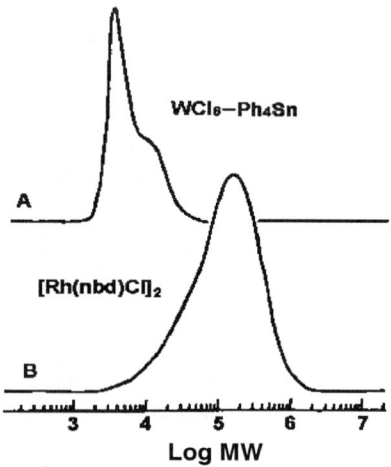

Figure 1. GPC traces of the reaction products of 1b catalyzed by (A) WCl_6–Ph_4Sn (Table 1, no. 7) and (B) [Rh(nbd)Cl]$_2$ (Table 1, no. 10). (Reproduced from ref 19. Copyright 1997 American Chemical Society.)

Since [Rh(nbd)Cl]$_2$ was effective for polymerizing cyanoacetylene **1**, we tried to use it to polymerize cyanoacetylene **2** (*20*). No polymer, however, was produced. All the attempted polymerizations catalyzed by $MoCl_5$(–Ph_4Sn) in toluene and THF and by WCl_6(–Ph_4Sn) in toluene were failed (Table 2, nos. 1–6).

The polymerization catalyzed by WCl_6 in THF, however, was encouraging. After **2a** and WCl_6 were admixed in THF, the color of the mixture changed from red brown to green in a few minutes. The viscosity increased rapidly, and the polymerization solution could not be stirred by magnetic bar in 30 min. After 24 h, we dissolved the reaction product in THF, poured the dilute THF solution into methanol, and isolated the precipitate by filtration. The GPC analysis of the product, however, gave two RI peaks but only one UV peak (Figure 2A). We repeated the purification of the polymer by repeated dissolution–precipitation processes and eventually suppressed the UV-insensitive RI-GPC peak. The suppression of the RI peak thus confirms that the high molecular weight fraction is poly(THF). It is clear that WCl_6 polymerizes **2b** in THF, albeit with the complication of the polymerization of solvent THF. Similar results were obtained when WCl_6–Ph_4Sn was used as the catalysts; that is, the polymerizations of **2b** and THF were simultaneously initiated.

Table 2. Polymerization of 5-{[(4'-cyano-4-biphenylyl)oxy]carbonyl}-1-pentyne (**2b**)[a]

No.	Catalyst	Solvent	Yield (%)	M_w	M_w/M_n
1	$MoCl_5$	toluene	0		
2	$MoCl_5$–Ph_4Sn	toluene	8	4800	2.2
3	$MoCl_5$	THF	0		
4	$MoCl_5$–Ph_4Sn	THF	0		
5	WCl_6	toluene	0		
6	WCl_6–Ph_4Sn	toluene	8	2900	1.4
7[b]	WCl_6	THF	31	25100	1.7
8[b]	WCl_6–Ph_4Sn	THF	49	20200	1.6
9	WCl_6–Ph_4Sn	dioxane	62	22300	1.6

[a] Carried out under an atmosphere of nitrogen at room temperature for 24 h; $[M]_o$ = 0.2 M, [cat.] = [cocat.] = 20 mM. [b] For the **P2b** fraction only, the poly(THF) fraction was removed by repeated precipitation. SOURCE: Reproduced from ref 20. Copyright 1998 American Chemical Society.

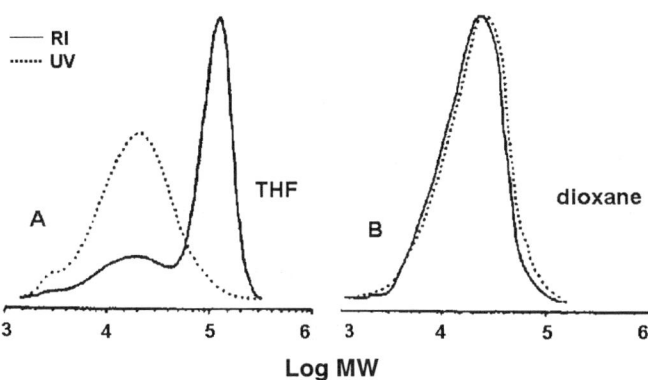

*Figure 2. GPC traces of polymerization products of **2b** catalyzed by (A) WCl_6 in THF (Table 2, no. 7) and (B) WCl_6–Ph_4Sn in dioxane (Table 2, no. 9). (Reproduced from ref 20. Copyright 1998 American Chemical Society.)*

In order to eliminate the solvent polymerization, we tried to use dioxane as the polymerization solvent. Unlike the THF cousin, dioxane is resistant to ring-opening polymerization because of its stable six membered ring structure. When dioxane was used as solvent, the polymerization mixture could be stirred during the whole course of reaction. After 24 h, a red brown polymer was isolated in 62.1% yield. GPC analysis gave only one peak from both UV and RI detectors (Figure 2B), suggesting that the polymerization is "clean" and the solvent dioxane does not undergo the undesired polymerization (20).

The polymerizations of cyanoacetylenes 2a and 2c were similar to that of 2b and can be briefly summarized as follows: $MoCl_5$ is an ineffective catalyst, WCl_6 shows better performance, and dioxane accommodates clean polymerization.

Thus in this study, we have developed two effective catalyst systems for the polymerization of cyanoacetylenes: [Rh(nbd)Cl]$_2$/DMF for the phenylacetylene derivatives 1 and WCl_6–Ph$_4$Sn/dioxane for the alkylacetylene derivatives 2. The polymers of 2 exhibited interesting mesomorphic properties originated from concerted interaction of the mesogenic pendants and the rigid polyacetylene backbones (20,21).

Polymerizations of Substituted Acetylenes in Aqueous Media

Late transition metal catalysts are known to be more tolerant of polar substrate than their early counterparts, and some ruthenium-, osmium-, and iridium-based complexes have been found to be effective ring-opening metathesis polymerization catalysts in water (22). We thus tried to polymerize substituted acetylenes in aqueous media. We first compared catalytic activity of [Rh(nbd)Cl]$_2$ for the polymerization of phenylacetylene (PA) in different solvents. When toluene was used as solvent, the polymerization was very slow, yielding only a trace amount (1 %) of an orange polyphenylacetyelene (PPA) after 1 h reaction. When THF, a polar solvent was used, the polymerization proceeded faster. The polymerization proceeded even faster when it was carried out in water. Unfortunately, however, the PPA obtained in water was insoluble. Similar results were obtained in the polymerization of (p-methylphenyl)-acetylene (p-MePA).

When we used [Rh(cod)Cl]$_2$, another Rh complex, to polymerize PA in water, a high molecular weight PPA was obtained in high yield (Table 3, no. 1). The PPA was completely soluble and the *cis* content was as high as 94%. When triethylamine was used as solvent, soluble PPA was also obtained. We therefore prepared a number of Rh complexes with nitrogen-containing ligands according to published procedures (24) and investigated their catalytic activities in the polymerization of PA. It was found that all the complexes initiate the PA polymerizations in water (Table 3, nos. 3–9).

We have further prepared two water-soluble Rh complexes, Rh(nbd)(tos)(H$_2$O) and Rh(cod)(tos)(H$_2$O), according to Kolle's procedure (25) and examined their catalytic performance in the PA polymerization. Like the [Rh(nbd)Cl]$_2$ complex, Rh(nbd)(tos)(H$_2$O) was a poor catalyst in toluene (Table 4, no. 1). However, in a polar solvent like THF, this complex gave PPA with high molecular weight and high stereoregularity in high yield. A PPA with even higher molecular weight and narrower polydispersity formed when the polymerization was carried out in water (Table 4, no. 3). Similar results were obtained when Rh(cod)(tos)(H$_2$O) was used as catalyst. The difference is that Rh(cod)(tos)(H_2O) yielded PPA with higher stereoregularity. Especially when Rh(cos)(tos)(H_2O) was used in water, highly stereoregular PPA with a 100% *cis* content formed in a short polymerization time. We also conducted a polymerization catalyzed by Rh(cod)(tos)(H_2O) in open air using tap water as solvent. The result was identical to that obtained under nitrogen, demonstrative of the practical value of the catalyst systems.

Table 3. Polymerization of Phenylacetyelene by Organorhodium Catalysts[a]

No.	Catalyst	Solvent	Time (h)	Yield (%)	M_w	M_w/M_n	% cis[b]
1	[Rh(cod)Cl]$_2$	water	0.5	68	46200	2.2	94.0
2	[Rh(cod)Cl]$_2$	Et$_3$N	24.0	92	12500	4.2	79.7
3	Rh(cod)(bbpmt)	water	20.0	63	11400	2.2	83.2
4	Rh(cod)(pip)Cl	water	0.2	72	7600	2.0	86.0
5	Rh(cod)(NH$_3$)Cl	water	0.2	77	23300	1.8	86.4
6	Rh(cod)(t-BuNH$_3$)Cl	water	0,2	57	6500	2.2	87.4
7	Rh(cod)(mid)Cl	water	1.5	75	12500	2.0	87.4
8	[Rh(cod)(mid)$_2$]PF$_6$	water	0.7	98	9900	2.0	87.5
9	[Rh(cod)Cl]$_2$(pda)	water	1.7	39	7300	1.4	89.8

[a] Carried out under nitrogen at room temperature; monomer, 0.5 mL; catalyst, 2 mg; solvent, 5 mL. [b] Determined by ^{1}H NMR analysis. SOURCE: Reproduced from ref 23. Copyright 1997 American Chemical Society.

Table 4. Polymerization of Phenylacetyelene Catalyzed by Water-Soluble Organorhodium Catalysts Rh(diene)(tos)(H_2O)[a]

No.	Catalysts	Solvent	Time (h)	Yield (%)	M_w	M_w/M_n	% cis[b]
1	Rh(nbd)(tos)(H$_2$O)	toluene	1.0	1	30000	3.4	86.0
2	Rh(nbd)(tos)(H$_2$O)	THF	1.0	79	72800	4.2	90.7
3	Rh(nbd)(tos)(H$_2$O)	water	1.0	80	216100	2.8	89.0
4	Rh(cod)(tos)(H$_2$O)	toluene	1.0	2	58600	3.4	94.7
5	Rh(cod)(tos)(H$_2$O)	THF	1.0	70	109000	2.5	99.5
6	Rh(cod)(tos)(H$_2$O)	water	0.5	24	73000	2.5	100.0
7	Rh(cod)(tos)(H$_2$O)	water	1.0	82	31200	5.4	100.0
8	Rh(cod)(tos)(H$_2$O)	(neat)	7.5	20	35400	2.8	82.2

[a] Carried out under an atmosphere of dry nitrogen at room temperature; [M]$_o$ = 0.83 M, [cat.] = 1 mM. [b] Determined by ^{1}H NMR analysis. SOURCE: Reproduced from ref 23. Copyright 1997 American Chemical Society.

Figure 3 shows NMR spectra of the PPAs prepared by Rh(cod)(tos)(H_2O). For comparison, the spectra of PPA prepared by WCl$_6$–Ph$_3$SiH are also given. The absorption peaks in the ^{1}H NMR spectra of the PPA prepared by Rh(cod)(tos)(H_2O) in water are very sharp (Figure 3B), indicative of the high stereoregularity (100% *cis*) of the polymer. In the ^{1}H NMR spectrum of the PPA prepared by the Rh catalyst in neat monomer, a new peak appeared at δ 6.78, which is attributable to the *trans* polyene protons (Figure 3A). The stereostructural inhomogeneity made the spectrum somewhat broad and accounted for the lower *cis* content of the polymer (82.2%). The spectrum of PPA prepared by the classical W catalyst system is very broad. The broadening is a result of an atactic polymer. The high stereoregularity of the PPA prepared by water-soluble Rh catalyst in water is also evidenced by the sharp absorption peaks in the ^{13}C NMR spectrum (Figure 3D), while the peaks of PPA prepared by the W system are much less resolved.

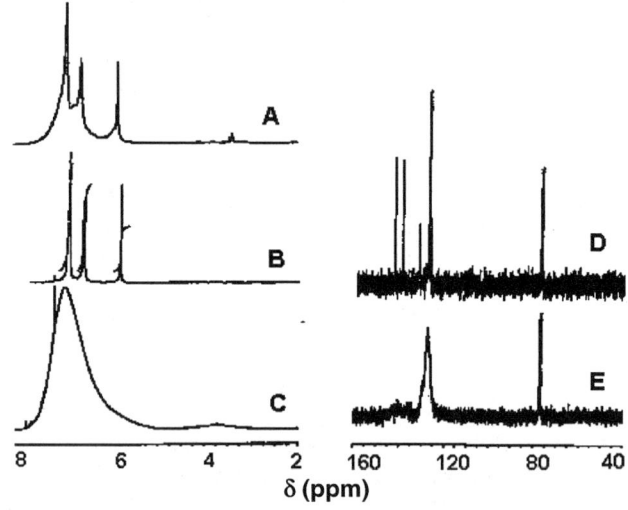

Figure 3. ^{1}H and ^{13}C NMR spectra of PPAs prepared by water-soluble complex Rh(cod)(tos)(H$_2$O) in water (B and D, Table 4, no. 6) and in neat monomer (A, Table 4, no. 8). The spectra for PPA prepared by a "classical" WCl$_6$–Ph$_3$SiH catalyst in toluene (C and E) are also shown for comparison. All the spectra were measured in CDCl$_3$ at room temperature. (Reproduced from ref 23. Copyright 1997 American Chemical Society.)

We tried to use the water-soluble Rh complexes to polymerize other acetylene derivatives. The polymerization behavior of *p*-MePA was very similar to that of PA. However, attempts to polymerize aliphatic acetylene monomers such as HC≡C(CH$_2$)$_5$-CH$_3$, HC≡C(CH$_2$)$_2$OH, HC≡C(CH$_2$)$_3$CO$_2$H, and HO$_2$C≡CCO$_2$H were unsuccessful.

Although much needs to be done to understand the polymerization mechanism, we suspect that the interaction between the Rh complexes and the aromatic rings might be involved in the formation of the active centers. Thus, the lack of the Rh–aromatic interaction might be responsible for the failure in polymerizing the aliphatic acetylene monomers by the Rh catalysts. Toluene is not a good solvent, probably due to the solvent's undesired competitive interaction. The phenyl group in toluene might have partially blocked the way for the PA monomers to coordinate with the Rh complexes.

Catalysts for Polymerizations of Substituted Acetylenes without Cocatalysts and Photoirradiation

Most Mo- and W-based catalysts for polymerizations of substituted acetylenes are air- and moisture-sensitive and many of them require additional Lewis acids as cocatalysts. Mo and W carbonyls are stable complexes but need photoirradiation to generate the active catalytic species. We attempted to explore air-stable catalysts for the polymerization of substituted acetylenes without additional cocatalysts and UV irradiation. We investigated catalytic activity of $MI_2(CO)_3(NCCH_3)_2$ (M = Mo, W) in the polymerization of PA in different solvents without additional cocatalysts and UV irradiation. It was found that the catalysts gave PPAs with M_w of ~20000 in ~20% yields.

Table 5. Polymerization of Phenylacetylene Catalyzed by $M(CO)_xL_y{}^a$

No.	Catalyst	Temp (°C)	Yield (%)	M_w	M_w/M_n
1	$Mo(CO)_4(pip)_2$	60	0		
2	$[Mo(CO)_3cp]_2$	rt	2	41300	1.8
3	$[Mo(CO)_3cp]_2$	60	7	9000	1.7
4	$Mo(CO)_4(nbd)$	rt	52	36400	2.3
5	$Mo(CO)_4(nbd)$	60	58	3200	1.9
6	$W(CO)_4(CH_2PPh_2)_2$	rt	0		
7	$W(CO)_4(CH_2PPh_2)_2$	60	0		
8	$[W(CO)_3cp]_2$	rt	0		
9	$[W(CO)_3cp]_2$	60	0		
10	$W(CO)_3(mes)$	60	100	67400	3.1

[a] The catalysts were stored and handled in open air, while the polymerizations were conducted under nitrogen in CCl_4 for 24 h; $[M]_o = 0.5$ M, $[cat.] = 20$ mM.

In order to prepare PPAs with higher molecular weights in better yields, we checked other Mo and W carbonyl complexes. A halogenated solvent CCl_4 was utilized to provide halogen ligands for the polymerization systems, as it was reported that the halogen atoms from solvent could accelerate the polymerization of acetylenes (26). Table 5 lists the polymerization results. $Mo(CO)_4(nbd)$ and $W(CO)_3(mes)$ show excellent catalytic activity. These catalysts give polymers with higher molecular weights in higher yields even than those catalyst systems with cocatalysts and UV irradiation (8).

Table 6. Polymerization of 1-Chloro-1-Octyne Catalyzed by M(CO)$_x$L$_y$[a]

No.	Solvent	Temp (°C)	Time (h)	Yield (%)	M_w	M_w/M_n
			Mo(CO)$_4$(nbd)			
1	toluene	rt	24	13	102900	2.3
2	toluene	rt	72	28	96600	2.3
3	toluene	80	b	41	309600	2.2
4	toluene	80	1	94	248500	4.1
5	toluene	80	3	100	237900	4.8
6	dioxane	80	72	25	53800	4.4
			Mo(CO)$_6$			
7	toluene	rt	72	0		
8	toluene	80	72	91	316700	2.6
9	dioxane	80	72	trace		
			W(CO)$_3$(mes)			
10	toluene	rt	72	0		
11	toluene	80	72	0		
12	dioxane	80	72	0		

[a] The catalysts were stored and handled in open air, while the polymerizations were conducted under nitrogen; [M]$_o$ = 0.5 M, [cat.] = 20 mM. [b] 3 min (0.05 h).

In the above-mentioned polymerizations, halogen ligands from the solvent are presumed to be involved in the formation of active centers. This motivates us to select 1-chloro-1-octyne, a halogen-containing acetylene for study. As halogen is already contained in the monomer, non-halogenated solvents and catalysts are thus used. The results are given in Table 6. Mo(CO)$_4$(nbd) was found to be the most effective catalyst, with high molecular weight polymers formed within a few minutes. When Mo(CO)$_6$ was used as the catalyst, the polymerization proceeded slowly but gave polymer of high molecular weight and narrow polydispersity. Toluene was found to be an effective solvent. W(CO)$_3$(mes) was however proved to be inactive even though it was very effective for the polymerization of PA. For comparison, an air-sensitive catalyst MoCl$_5$–Ph$_4$Sn was used for the polymerization. The yield and M_w of the obtained polymer were 64% and 5.8 × 10^5, respectively. The yield is lower and molecular weight is somewhat higher than those of the polymers obtained from the carbonyl catalysts. The stereoregularity of the poly(1-chloro-1-octyne) was investigated by ^{13}C NMR analysis. Details of the NMR spectra and assignments of the chemical shifts were reported in ref 27. It was found that the polymers from Mo(CO)$_4$(nbd) and Mo(CO)$_6$ possessed higher *cis* contents than that from MoCl$_5$–Ph$_4$Sn catalyst (27).

New Catalysts for Polymerizations of Diynes

In all the above investigations, we tried to explore new catalysts for the polymerizations of acetylenes with one triple bond, namely, monoynes. It is well known that acetylenes can be cyclized to cyclodimers and cyclotrimers by suitable catalysts (*10*). Diynes, the acetylenes with *two* triple bonds, may undergo intermolecular cyclotrimerization to produce hyperbranched poly(phenylenealkenes) as shown in Scheme 1. Selection of catalysts and control of polymerization would be the most important factors in order to avoid the crosslinking reaction and to prepare soluble hyperbranched poly(phenylenealkenes). For example, some catalysts may initiate the polymerization of triple bonds only to form linear chain polymers as reported by Schrock (*13*) and Choi (*28*), and some may catalyze the cyclotrimerization of diynes to produce branched polymers of polyphenylene architectures (*14–16*). Crosslinking reaction is likely to occur in the polymerization of diynes. At the beginning, the polymers we obtained were crosslinked insoluble products. However, after optimizing the polymerization conditions, we are now able to prepare very soluble hyperbranched polymers of diynes.

Scheme 1

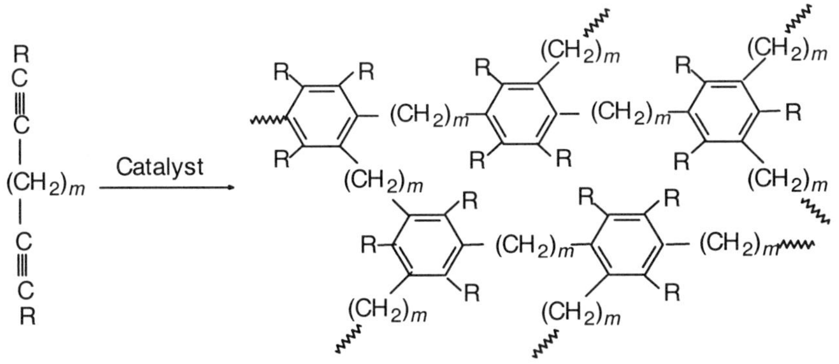

In this investigation, we explored a variety of catalysts for the polymerizations of diynes. Our aim is to prepare polymers of hyperbranched polyphenylenealkene structures via cyclotrimerization of diynes.

Both terminal and internal diynes were investigated. Table 7 summarizes the polymerization results of terminal diynes. Polymerization of 1,8-nonadiyne [HC≡C(CH$_2$)$_5$C≡CH] catalyzed by [Mo(CO)$_3$cp] yielded no polymer in toluene, but gave an insoluble yellow polymer in quantitative yield in CCl$_4$. Similar results were obtained when Mo(CO)$_4$(nbd) was used as catalyst. NbCl$_5$-catalyzed polymerization yielded a partially soluble product. However, TaCl$_5$-catalyzed polymerization gave a soluble polymer at room temperature in a 52% yield, whose M_w and M_w/M_n were 40900 and 4.9, respectively (Table 7, no. 7). The polymer was white powder and was very soluble in common organic solvents (toluene, dichloromethane, THF, etc.). The structure of the polymer was confirmed by different analytical techniques. Its IR spectrum showed a sharp peak at 3008 cm^{-1} (aromatic C-H bonds) and two weak broad peaks at 1905 and 1790 cm^{-1} (characteristics of 1,2,4-trisubstituted aromatic rings). Broad peaks at δ 6.5–6.9 (aromatic protons), 2.4, and 1.5 were observed in its ^{1}H NMR spectrum, and its ^{13}C NMR spectrum consisted of two groups of sharp peaks at δ 125.0–130.0 (unsubstituted aromatic carbons) and 138.8–143.4 (substituted aromatic carbons). The UV spectrum of the polymer shows an absorption peak at 268 nm with a molar absorptivity of 350 L mol^{-1} cm^{-1}, typical of UV absorption of isolated aromatic rings. All these indicate that the cyclotrimerization have taken place and that aromatic rings have formed. Thus we have successfully prepared the polymer with hyperbranched structure as illustrated in Scheme 1 ($m = 5$, R = H), and the formed aromatic rings are of mainly 1,2,4-trisubstituted pattern. TGA data further confirmed the polyphenylenealkene structure. A weight loss of 5% occurred at 480 °C under nitrogen atmosphere, while poly(1-alkynes) of linear structure such as poly(1-hexyne) –[HC=C(C$_4$H$_9$)]$_n$– starts to loss weight at a temperature as low as ~150°C (29).

Table 7. Polymerization of Terminal Diynes[a]

No.	Catalyst	Solvent	Yield (%)	Solubility
		1,8-Nonadiyne		
1	[Mo(CO)$_3$cp]$_2$	toluene	0	
2	[Mo(CO)$_3$cp]$_2$	CCl$_4$	100	insoluble
3	Mo(CO)$_4$(nbd)	toluene	0	
4	Mo(CO)$_4$(nbd)	CCl$_4$	64	insoluble
5	NbCl$_5$	toluene	82	partially soluble
6	TaCl$_5$	toluene	70	partially soluble
7[b]	TaCl$_5$	toluene	52	soluble
		1,9-Decadiyne		
8	NbCl$_5$	toluene	trace	
9[c]	TaCl$_5$	toluene	60	insoluble
10[d]	TaCl$_5$	toluene	72	soluble
11[e]	TaCl$_5$	toluene	88	soluble

[a] Carried out under nitrogen at 65 °C for 24 h; [M]$_o$ = 0.5 M, [cat.] = 20 mM. [b] At room temperature for 24 h; [M]$_o$ = 0.25 M, [cat.] = 15 mM. [c] Polymerized for 8 h. [d] At 0 °C for 2 h; [M]$_o$ = 0.25 M, [cat.] = 80 mM. [e] At room temperature for 30 min; [M]$_o$ = 0.25 M, [cat.] = 3 mM.

The polymerization behavior of 1,9-decadiyne [HC≡C(CH$_2$)$_6$C≡CH] is similar to that of 1,8-nonadiyne. NbCl$_5$-catalyzed polymerization gave only partially soluble products. TaCl$_5$ catalyst gave an insoluble gel in 60% yield at 65 °C. When the polymerization was conducted at 0 °C at a higher catalyst concentration, soluble polymer was isolated (Table 7, no. 10). However, the polymer became partially soluble after dried *in vacuo*. Completely soluble polymer of 1,9-decadiyne was prepared at room temperature at a lower catalyst concentration (Table 7, no. 11). The polymer was white powder and possessed higher molecular weight (M_w = 749700) and polydispersity (M_w/M_n = 25.9). The spectra and properties of the polymer are also similar to those of the soluble polymer of 1,8-nonadiyne discussed above. The only difference is that terminal triple bonds of 1,9-decadiyne polymer (i.e. IR: 3310, 2119 cm^{-1}; ^{1}H NMR: δ 1.9; ^{13}C NMR: δ 68.8, 85.5) can be observed when the polymerization time is less than 2 h. When the polymerization proceeded for longer time, no terminal triple bonds in the polymer could be detected. Thus, hyperbranched polymers from 1,9-decadiyne were also prepared via the cyclotrimerization as illustrated in Scheme 1 ($m = 6$, R = H). The terminal triple bonds may have undergone cyclotrimerization reaction and were then capped by backbiting reactions in the chain termination step. Therefore, no terminal triple bonds could be detected when the termination reaction consumes all the acetylene moieties.

We also tried to polymerize internal diynes by different catalysts. For internal diynes, the steric hindrance of the two end groups connected to the triple bonds may hamper the formation of aromatic rings. Therefore, the cyclotrimerization of internal diynes would be more difficult. Some catalysts for the formation of hexasubstituted aromatic rings such as Pd/C–ClSiMe$_3$ are tested.

Table 8. Polymerization of 1,9-Bis(trimethylsilyl)-1,8-Nonadiyne[a]

No.	Catalyst	Temp (°C)	Yield (%)	Solubility	M_w	M_w/M_n
1[b]	Mo(CO)$_4$(nbd)	65	60	soluble	14200	2.8
2[c]	NbCl$_5$	30	8	soluble	9900	2.0
3[d]	NbCl$_5$	30	10	soluble	16300	3.2
4[e]	Pd/C-ClSiMe$_3$	65	trace			

[a] Carried out under nitrogen for 24 h in toluene; [M]$_o$ = 0.25 M, [cat.] = 20 mM. [b] CCl$_4$ as solvent. [c] Polymerized for 72 h; [M]$_o$ = 0.5 M, [cat.] = 80 mM. [d] Polymerized for 72 h; [M]$_o$ = 0.5 M, [cat.] = 160 mM. [e] Refluxed in THF for 72 h.

We first replaced the terminal hydrogen atoms of 1,8-nonadiyne with trimethylsilyl groups (9). The resulting monomer of 1,9-bis(trimethylsilyl)-1,8-nonadiyne [Me$_3$SiC≡C(CH$_2$)$_5$C≡CSiMe$_3$] showed somewhat lower polymerizability, but completely soluble polymeric products were obtained from this monomer on some catalyst systems (Table 8). However, Pd/C–ClSiMe$_3$, an effective catalyst for cyclotrimerization of internal diynes (30), gave only trace amount of polymer. The polymer of 1,9-bis(trimethylsilyl)-1,8-nonadiyne obtained from NbCl$_5$ catalyst system showed absorption peaks originating from the terminal triple bonds at 2174 cm^{-1} (IR) and δ 108.3 and 85.0 (^{13}C NMR).

The polymerization results of another internal diyne, 3,9-dodecadiyne [EtC≡C-(CH$_2$)$_4$C≡CEt], are given in Table 9. Quite a few catalysts, including Mo(CO)$_4$(nbd), [Mo(CO)$_3$cp]$_2$, NbCl$_5$, PdCl$_2$–ClSiMe$_3$ and Pd/C–ClSiMe$_3$, gave completely soluble polymers under appropriate conditions. Pd/C–ClSiMe$_3$ effectively polymerized the monomer in both nitrogen and air with no obvious difference observed. The polymer of 3,9-dodecadiyne catalyzed by Pd/C–ClSiMe$_3$ and NbCl$_5$ showed strong signals of acetylene triple bonds in the chemical shift region of δ 79.9–82.5 and unsaturated carbons in δ 125–145 in ^{13}C NMR analysis (Figure 4). These data indicate that the polymer possesses a lot of branches or periphery triple bonds.

Table 9. Polymerization of 3,9-Dodecadiyne[a]

No.	Catalyst	Temp (°C)	Yield (%)	Solubility	M_w	M_w/M_n
1	Mo(CO)$_4$(nbd)[b]	65	50	soluble	238300	6.6
2	[Mo(CO)$_3$cp]$_2$[b]	65	24	soluble	40300	8.1
3	NbCl$_5$	65	100	insoluble		
4	TaCl$_5$	65	16	partially soluble		
5	NbCl$_5$	50	79	insoluble		
6	NbCl$_5$	rt	38	soluble	33600	2.3
7	Co(CO)$_2$cp	120	50	partially soluble		
8	PdCl$_2$–ClSiMe$_3$[c]	65	52	soluble	4500	3.0
9	Pd/C–ClSiMe$_3$[c]	65	50	soluble	3600	2.2

[a] Carried out under nitrogen for 24 h in toluene; [M]$_o$ = 0.5 M, [cat.] = 20 mM. [b] CCl$_4$ as solvent. [c] THF as solvent, [Pd] = 55 mM, [ClSiMe$_3$] = 1.0 M.

UV spectra of the polymer showed an absorption peak at 265 nm; that is, there is no high conjugation structure in the polymers. However, the molar absorptivity at the λ_{max} was 3000. This is higher than that of isolated aromatic rings. It is likely that the polymer contains polyene structures. TGA analysis of the polymer showed a gradual initial 20% weight loss in the temperature region of 200–400 °C. This may be induced by the segments with polyene structures. However, the weight loss of the 80% residue took place rapidly in the temperature region of 400–450 °C. This may be contributed from the portion of polymer with polyphenylenealkene structures. Thus, both metathesis polymerization and polycyclotrimerization of the triple bonds may have simultaneously occurred in this system. The obtained polymer is hyperbranched and possesses both polyene and hexasubstituted phenlylene structures.

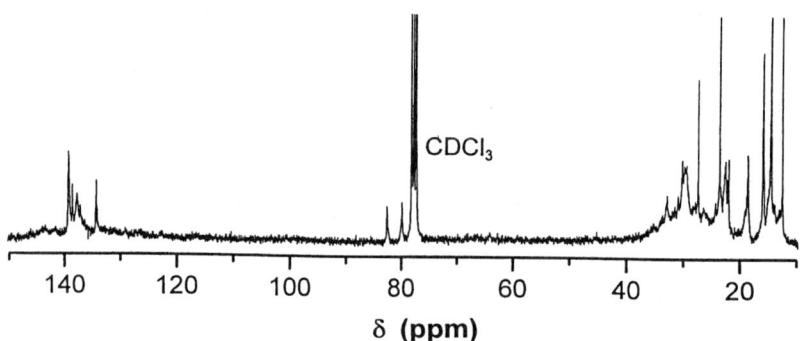

Figure 4. ^{13}C NMR spectrum of chloroform-d solution of poly(3,9-dodecadiyne) obtained from Pd/C–ClSiMe$_3$.

Experimental Section

Materials

Dioxane (Aldrich), THF and toluene (both Lab-Scan) were dried over 4 Å molecular sieves and distilled from sodium benzophenone ketyl prior to use. Dichoromethane, carbon tetrachloride (both Lab-Scan), triethylamine and DMF (both Aldrich) were distilled over calcium hydride. PA, *p*-MePA, 1-octyne, 1,8-nonadiyne, 1,9-decadiyne, and 3,9-dodecadiyne were purchased from Farchan, distilled from calcium hydride, and stored in a dark, cold place. 1-Chloro-1-octyne and 1,9-bis(trimethylsilyl)-1,8-nonadiyne were synthesized in our laboratory, and the synthetic procedures are given below.

Synthesis of 1-Chloro-1-octyne. In a two-necked 500 mL round flask equipped with a nitrogen inlet, a pressure-equalized dropping funnel, and a magnetic stirrer were placed 60 mL anhydrous THF and 14.8 mL (0.1 mol) 1-octyne. To this solution was added 67 mL (0.1 mol) of 1.5 M solution of BuLi in pentane (Aldrich) at −50 °C. After stirring for 1 h, a solution of 20.5 g (0.105 mol) tosyl chloride (Aldrich) in 100 mL anhydrous THF was added into the reactor slowly via a dropping funnel while the reaction mixture was still kept at −50 °C with an acetone/dry ice bath. The reaction mixture was then allowed to warm up to 0 °C and stirred for additional one hour, which was then transferred into a separatory funnel containing ~400 mL ice/water. The organic layer was seperated, and the aqueous phase was extracted with pentane for several times. The combined organic phase was washed with water and dried over anhydrous $MgSO_4$ overnight. The product (1-chloro-1-octyne) was purified by distillation over calcium hydride. Yield: 8.6 g (60%), colorless liquid. IR, v (cm^{-1}): 2952 (CH_3), 2932 (CH_2), 2244 (-C≡C-). ^{1}H NMR (300 MHz, $CDCl_3$), δ (TMS, ppm): 2.2 (t, 2H, CH_2C≡), 1.5 (m, 2H, CH_2-C-C≡), 1.3 (m, 6H, CH_2), 0.9 (t, 3H, CH_3). ^{13}C NMR (75 MHz, $CDCl_3$), δ (TMS, ppm): 70.4 (C≡), 57.6 (≡CCl).

Synthesis of 1,9-Bis(trimethylsilyl)-1,8-nonadiyne. 1,8-Nonadiyne (10 g, 0.0832 mol) and 120 mL anhydrous THF were injected to a two-necked flask equipped with a droping funnel and magnetic stirring bar. Then 98 mL of 1.7 M BuLi in pentane (0.166 mol) was dropped in under an acetone/dry ice bath. After all BuLi was added, the flask was warmed up to 0 °C by changing to an ice bath and kept at that temperature for 1–2 h. During the period, the formed LiC≡C(CH_2)$_5$C≡CLi was precipitated out as white powder. Followed by cooling the flask in an acetone/dry ice bath again, 23.2 mL of 98% Me_3SiCl (Aldrich) (0.179 mol) was added in. The flask was warmed to room temperature and the reaction mixture was stirred overnight. The isolation and purification of 1,9-bis(trimethylsilyl)-1,8-nonadiyne were conducted according to the procedure detailed above. Yield: 18.5 g (85%), colorless liquid. IR, v (cm^{-1}): 2958 (CH_3), 2932 (CH_2), 2174 (C≡C). ^{1}H NMR (300 MHz, $CDCl_3$), δ (TMS, ppm): 2.2 (CH_2-C≡), 1.5 (CH_2), 0.1 (CH_3-Si). ^{13}C NMR (75 MHz, $CDCl_3$), δ (TMS, ppm): 107.8 (≡CSi), 84.9 (C≡).

Rh complexes used in Tables 3 and 4 were prepared according to the procedures published by Zassinovich et al. (24), and Kolle et al. (25). $MI_2(CO)_3(NCCH_3)_2$ (M = Mo, W) were prepared according to Baker's method (31). All other transition metal compounds, i.e. $MoCl_5$, $Mo(CO)_6$, $Mo(CO)_4(pip)_2$, $[Mo(CO)_3cp]_2$, $Mo(CO)_4(nbd)$, $NbCl_5$, $PdCl_2$, Pd/C, $[Rh(nbd)Cl]_2$, $TaCl_5$, WCl_6, $W(CO)_3(mes)$, $[W(CO)_3cp]_2$, $W(CO)_4(CH_2PPh_2)_2$, $Co(CO)_2cp$, and Ph_4Sn, were purchased from Aldrich and used as received without further purification.

Instrumentation

IR spectra were recorded on a Perkin Elmer PC 16 spectrometer. NMR spectra were recorded on a Brucker ARX 300 NMR spectrometer. UV absorption spectra were measured in THF on a Milton Ray 3000 array spectrophotometer. TGA analysis was performed on a Perkin Elmer TGA 7 under nitrogen at a heating rate of 20 °C/min. Molecular weights were estimated by a Waters 510 gel permeation chromatography calibrated by polystyrene standards, using THF as eluent at a flow rate of 1.0 mL/min.

Polymerization

All the polymerization reactions were carried out in septum sealed Schlenk tubes with glass stopcock on side arm. The detailed procedures can be found in refs 9, 19–21, 23, 27, and 32. Typical procedures for each type of polymerization reactions are given below.

Polymerization of Polar Acetylene 1a. Into a baked 20-mL Schlenk tube with a side arm was added 169.4 mg (0.4 mmol) of **1a**. The tube was evacuated under vacuum and then flushed with dry nitrogen three times through the side arm. One milliliter of a DMF/Et_3N mixture (4:1 by volume) was injected into the tube to dissolve the monomer. The catalyst solution was prepared in another tube by dissolving 9.2 mg (0.02 mmol) of $[Rh(nbd)Cl]_2$ in 1 mL of the DMF/Et_3N mixture, which was transferred to the monomer solution using a hypodermic syringe. The reaction mixture was stirred at room temperature under nitrogen for 24 h, and was then quenched with a small amount of methanol. The mixture was diluted with 5 mL of THF and added dropwise to 200 mL of methanol under stirring via a cotton filter. The precipitate was allowed to stay overnight, filtered out, washed with methanol, and dried in vacuum at room temperature to a constant weight. Yield: 120 mg (71%), yellowish brown powder.

Polymerization of Phenylacetylene in Aqueous Media. A catalyst solution (1 mM) was prepared by dissolving 2 mg of $Rh(cod)(tos)(H_2O)$ in 5 mL of distilled water, to which 0.5 mL of distilled PA was added with vigorous stirring at room temperature under nitrogen. An orange solid formed in ~5 min. After 1 h of polymerization, the solid polymer product was separated from the aqueous catalyst solution by filtration and washed with acetone several times. The polymer was dissolved in toluene, and the resulting polymer solution was added dropwise to a large amount of methanol under stirring. The precipitated polymer was filtered out by a Gooch crucible and dried under vacuum at room temperature to a constant weight. Yield: 380 mg (82%), orange powder. The stereostructure of the polymer was elucidated to be 100% *cis*, using the ^{1}H NMR data by the following equation according to ref 33:

$$\% \; cis = [A_{5.9}/(A_{total}/6)] \times 100$$

where A denotes the integrated absorption area and $A_{total} = A_{5.8} + A_{6.6} (+A_{6.8}) + A_{6.9}$.

Polymerization of 1-Chloro-1-octyne Catalyzed by Mo(CO)$_4$(nbd). To a baked Schlenk tube, 15.0 mg Mo(CO)$_4$(nbd) (0.05 mmol) were added in air. The catalyst was mixed with 1.0 mL toluene and aged at 80 °C for 15 min in an oil bath. A solution of 0.2 mL (180 mg, 1.25 mmol) 1-chloro-1-octyne in 1.3 mL toluene was then added dropwise by a syringe. After stirring the polymerization mixture at 80 °C for 1 h, the reaction was quenched by the addition of a small amount of methanol. The polymer solution was then dropped into 200 mL methanol via a cotton filter under stirring. The precipitated polymer was filtered, washed with methanol for several times, and dried in vacuum at room temperature to a constant weight. Yield: 170 mg (94%), yellowish fiber.

Polymerization of 1,8-Nonadiyne. To a baked Schlenk tube, 27.0 mg TaCl$_5$ (0.075 mmol) were added under an inert atmosphere. The catalyst was mixed with 2.0 mL freshly distilled toluene and aged at room temperature for 15 min. A solution of 0.2 mL (160 mg, 1.33 mmol) 1,8-nonadiyne in 2.8 mL toluene was then added dropwise by a syringe over 5–8 min under stirring. An immediate exothermic reaction took place and the color of the reaction mixture changed to dark brown. After stirring at room temperature for 8 h, the reaction was quenched by the addition of a small amount of methanol. The polymer solution was dropped into 200 mL methanol via a cotton filter under stirring. The precipitated polymer was allowed to stand in methanol overnight, which was then filtered out, washed with methanol for several times, and dried in vacuum at room temperature to a constant weight. Yield: 83 mg (52%), white powder.

Acknowledgements. The work described in this chapter was partially supported by the grants from the Research Grants Council of the Hong Kong Special Administrative Region, China (Projects Nos. HKUST595/95P, 6149/97P, and 6062/98P).

Literature Cited

1. Chien, J. C. W. *Polyacetylene*; Academic Press: New York, 1984.
2. Ginsburg, E. J.; Gorman, C. B.; Grubbs, R. H. In *Modern Acetylene Chemistry*; Stang, P. J., Diederich, F., Eds.; VCH: New York, 1995; p 353.
3. *Handbook of Conducting Polymers*, 2nd ed.; Skotheim, T. A., Elsenbaumer, R. L., Reynolds, J. R., Eds.; Marcel Dekker: New York, 1998.
4. Masuda, T.; Higashimura, T. *Adv. Polym. Sci.* **1987**, *81*, 121.

5. Schrock, R. R. *Acc. Chem. Res.* **1990**, *23*, 158.
6. Carlini, C.; Chien, J. C. W. *J. Polym. Sci., Polym. Chem. Ed.* **1984**, 22, 2749.
7. Lee, H. J.; Shim, S. C. *J. Polym. Sci., Polym. Chem. Ed.* **1994**, *32*, 2437.
8. Tamura, K.; Masuda, T.; Higashimura, T. *Polym. Bull.* **1994**, *32*, 289.
9. Tang, B. Z.; Kotera, N. *Macromolecules* **1989**, *22*, 4388.
10. Vollhardt, K. P. C. *Pure Appl. Chem.* **1993**, *64*, 153.
11. Stille, J. K.; Frey, D. A. *J. Am. Chem. Soc.* **1961**, *83*, 1697.
12. Gibson, H. W.; Bailey, F. C.; Epstein, A. J.; Rommelmann, H.; Kaplan, S.; Harbour, J.; Yang, X.-Q.; Tanner, D. B.; Pochan, J. M. *J. Am. Chem. Soc.* **1983**, *105*, 4417.
13. Fox, H.; Wolf, M. O.; O'Dell, R.; Lin, B. L.; Schrock, R. R.; Wright, M. S. *J. Am. Chem. Soc.* **1994**, *116*, 2827.
14. Chalk, A. J.; Gilbert, A. R. *J. Polym. Sci., Polym. Chem. Ed.* **1972**, *10*, 2033.
15. Sergeyev, V. A.; Shitikov, V. K.; Chernomordik, Y. A.; Korshak, V. V. *Appl. Polym. Symp.* **1975**, *26*, 237.
16. Srinrivasan, R.; Farona, M. F. *Polym. Bull.* **1988**, *20*, 359.
17. Ho, T. H.; Katz, T. J. *J. Mol. Catal.* **1985**, *28*, 359.
18. Gorman, C. B.; Vest, R. W.; Palovich, T. U.; Serron, S. *Macromolecules* **1999**, *32*, 4157.
19. Tang, B. Z.; Kong, X.; Wan, X.; Feng, X. D. *Macromolecules* **1997**, *30*, 5620.
20. Tang, B. Z.; Kong, X.; Wan, X.; Lam, W. Y.; Feng, X. D.; Kwok, H. S. *Macromolecules* **1998**, *31*, 2419.
21. Kong, X.; Tang, B. Z. *Chem. Mater.* **1998**, *10*, 3352.
22. Novak, B. M.; Grubbs, R. H. *J. Am. Chem. Soc.* **1988**, *110*, 960.
23. Tang, B. Z.; Poon, W. H.; Leung, S. M.; Leung, W. H.; Peng, H. *Macromolecules* **1997**, *30*, 2209.
24. Zassinovich, G.; Mestroni, G.; Camus, A. *J. Organomet. Chem.* **1975**, *91*, 379.
25. Kolle, U.; Gorissen, R.; Wagner, T. *Chem. Ber.* **1995**, *128*, 911.
26. Tamura, K.; Masuda, T.; Higashimura, T. *Polym. Bull.* **1993**, *30*, 537.
27. Xu, K.; Peng, H.; Tang, B. Z. *Polym. Mater. Sci. Eng.* **1999**, *80*, 485.
28. Choi, D.-C.; Kim, S.-H.; Lee, J.-H.; Cho, H.-N.; Choi, S.-K. *Macromolecules* **1997**, *30*, 176.
29. Masuda, T.; Tang, B. Z.; Higashimura, T.; Yamaoka, H. *Macromolecules* **1985**, *18*, 2369.
30. Jhingan, A. K.; Maier, W. F. *J. Org. Chem.* **1987**, *52*, 1161.
31. Baker, P. K.; Frasher, S. G.; Keys, E. M. *J. Organomet. Chem.* **1986**, *309*, 319.
32. Xu, K.; Tang, B. Z. *Chin. J. Polym. Sci.* **1999**, *17*, 397.
33. Simionescu, C. I.; Percec, V. *J. Polym. Sci., Polym. Chem. Ed.* **1980**, *18*, 147.

Chapter 10

Addition Polymerization of Norbornene: Catalysis of Monocyclopentadienyltitanium Compounds Activated with Methylaluminoxane

Qing Wu, Yingying Lu, and Zejian Lu

Institute of Polymer Science, Zhongshan University, Guangzhou 510275, China

Addition polymerizations of norbornene were carried out with η^5-monocyclopentadienyltitanium compounds, $CpTiCl_3$, $CpTi(OBz)_3$, $Cp*TiCl_3$ and $Cp*Ti(OBz)_3$, in the presence of methylaluminoxane (MAO). Among the catalyst precursors, $CpTi(OBz)_3$ exhibits the highest catalytic activity for the polymerization. It was also found that the content of residual trimethylaluminum (TMA) in MAO has an important effect on the catalytic activity. The catalyst system containing less residual TMA gives optimum results of polymerization. FTIR and ^{13}C-NMR analyses indicated that the polynorbornenes obtained with the catalyst are vinyl-type polymer with relatively low stereoregularity. The polynorbornenes are soluble in hydrocarbons such as cyclohexane, chlorobenzene and tetrahydronaphthalene.

Two modes of reaction have been known for the polymerization of norbornene, ring-opening metathesis polymerization (ROMP, A) and addition polymerization (vinyl-type B) (see Scheme 1). The type of catalyst is a major factor determining the course of reaction. The conventional Ziegler-Natta catalysts, such as the titanium systems $TiCl_4$-AlR_3[1] and $TiCl_4$-$LiAlH_4$[2], can cause ROMP or addition polymerization depending upon the ratio of cocatalyst Al component to Ti. Titanacyclobutanes[3] delivered from the reactions of Tebbe reagent with cycloolefins initiate a living ROMP of norbornene. By thermolysis of dimethyltitanocene and related cyclopentadienyltitanium (IV) derivatives, the titanium complexes afford ring-opening polymers[4].

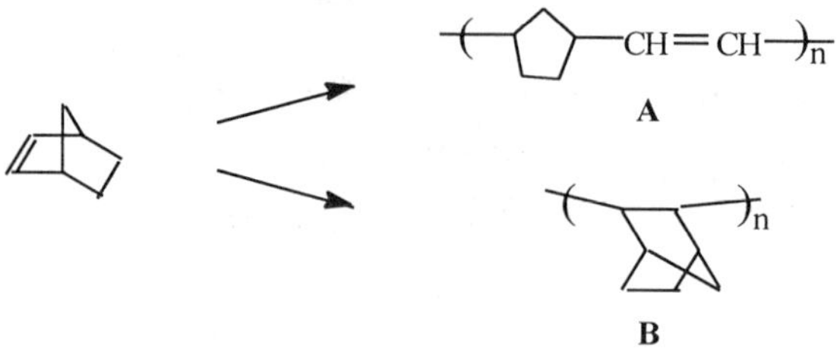

Scheme 1

Since Sinn and Kaminsky[5] reported the direct synthesis of methylaluminoxane (MAO) by controlled hydrolysis of trimethylaluminum (TMA), a variety of metallocene catalyst systems for α-olefin and cyclic olefin polymerizations has been reported[6-10]. The zirconocene-MAO catalyst systems allow the polymerization of strained cyclic olefins such as cyclobutene, cyclopentene, norbornene, dimethanooctahydronaphthalene by double bond opening[11-12]. C_2- and C_s-symmetric catalysts give higher activities for the polymerizations. However, the stereoregular homopolymers of cyclic olefins produced by the catalysts are highly crystalline and decompose upon melting at temperatures over 400 °C, and are completely insoluble in organic solvents. Therefore, it is difficult to study the structure of homopolymers and to process them.

Cationic palladium (II) complex $[Pd(RCN)_4][BF_4]_2$[13] and amine complexes[14] were reported to catalyze norbornene addition polymerizations in a mixture of solvents containing nitrobenzene. The polymer products were found to have high molecular weight with low polydispersities and to be soluble in chlorohydrocarbons. Similar Pd (II) complexes with norbornylnitrile as ligands in CH_3NO_2 or C_6H_5Cl can also affords high-molecular-weight polynorbornene[15]. Recently, a new family of single component cationic nickel and palladium catalysts was used to polymerize norbornene to amorphous and transparent addition polymers having a glass transition temperature of 380-390 °C[16]. The polynorbornenes generated by the palladium catalysts are insoluble in simple hydrocarbons, but those made by using the nickel catalysts are highly soluble in hydrocarbons such as heptane and cyclohexane even at the molecular weights higher than 1,000,000.

Previously, we have reported that the half-titanocene catalyst system, $CpTi(OBz)_3$-MAO, can be used to polymerize α-olefins efficiently and afford polyethylene, atactic polypropylene, syndiotactic polystyrene, and ethylene-styrene and propylene-styrene random copolymers[17-21]. It was found that the dependencies of the catalytic activity on the content of residual TMA in MAO are different for the polymerizations of different monomers. Higher content of residual TMA in MAO favors styrene syndiotactic polymerization, while less residual TMA in MAO favors the polymerizations of ethylene and propylene.

In the present study, we have tried to prepare new polynorbornenes by addition polymerization using the simple catalyst systems, monocyclopentadienyltitanium-MAO.

Experimental Methods

Materials

All work involving air- and moisture-sensitive compounds was carried out in an inert atmosphere. Norbornene and toluene were dried with sodium and distilled under nitrogen.

$CpTiCl_3$, $CpTi(OBz)_3$, $Cp*TiCl_3$, $Cp*Ti(OBz)_3$ (Cp = η-cyclopentadienyl, Cp* = η-pentamethylcyclopentadienyl, Bz = benzyl), was synthesized by the similar procedures as in literature[17, 22].

MAOs were prepared as follows: 200 mL of TMA solution (3.1 M, in toluene) was added dropwise into a flask with appropriate amount of ground $Al_2(SO_4)_3 \cdot 18H_2O$ in toluene at 0 °C. The mixture was gradually heated to 60 °C and stirred for 24 h, and was then filtered. The filtrate was concentrated under reduced pressure to a white solid. Residual TMA content in the MAO was determined by pyridine titration[23]. The MAOs prepared with initial $[H_2O]/[TMA]$ molar ratios of 1.2 and 1.8 contain residual TMA at the levels of 24.9% and 15.4%, respectively, and are referred to MAOI and MAOII.

Polymerization Procedure

Polymerizations were carried out in a 50 mL glass flask equipped with a magnetic stirrer. Ten mL of norbornene solution (4.25M in toluene) and the desired amount of MAO and titanocene compound in toluene were introduced in this order and the solution was stirred at the desired temperature for 1 h. The polymerizations were terminated and the polymers were precipitated by addition of acidified alcohol. The resulting polymers were washed with alcohol and dried in vacuo to constant weight.

Analysis and Characterization

Viscosities of polynorbornenes were measured in monochlorobenzene at 30 °C. The polymer samples were pressed to thin film for IR measurement. The FTIR spectra of the samples were recorded on a Nicolet 205 FTIR spectrometer. ^{13}C-NMR spectra of the polymers in o-dichlorobenzene were recorded on a Bruker-AM400 spectrometer operating at 100 °C under proton decoupling in the Fourier-transform mode. Differential scanning calorimetric

(DSC) curves of the polymers were obtained on a Perkin-Elmer DSC-7c instrument at a scanning rate of 20 °C/min from 30 °C to 400 °C under nitrogen.

Results and Discussion

Polymerizations with Different Titanocene Compounds

We have found that appropriate monocyclopentadienyltitanium compounds combined with MAO can efficiently catalyze norbornene addition polymerization. Four monocyclopentadienyltitanium complexes were used as catalyst precursors in norbornene polymerization with cocatalyst MAOII, the results are shown in Table 1. The Cp(cyclopentadienyl)-bearing titanocene catalysts afford the norbornene polymer at much higher yields than the related Cp*(1, 2, 3, 4, 5, -pentamethylcyclopentadienyl) -bearing one. Bulk of Cp* ligand has certain steric obstruction to the coordinaton of the rigid cyclic monomer onto the vacant orbit of Ti centre. On the other hand, substitution of -OR ligand for chlorine at the Cp-bearing compound causes remarkable increase in catalytic activity. Hence, the system CpTi(OBz)$_3$/MAO was employed in further study.

Table 1. Polymerization of Norbornene Catalyzed by Different Titanocenes Activated with MAOII [a]

Titanocene	Yield, g	A, kg polymer/(molTi.h)
CpTiCl$_3$	0.50	12.0
CpTi(OBz)$_3$	2.74	65.6
Cp*TiCl$_3$	0.25	6.0
Cp*Ti(OBz)$_3$	0.22	5.3

[a] Polymerization: [Ti] = 3.8 mM; [MAOII] = 160 mM; [NBE] = 3.9 M; T_p = 60 °C.

Effect of Cocatalyst MAO

Activity of the titanocene CpTi(OBz)$_3$ for norbornene polymerization is sensitive to the TMA content in MAO, as shown in Table 2. Using MAOI containing plenty of TMA (24.9 mol%), the catalyst shows poor activity (run 1). High catalytic acitivity is achieved with MAOII containing less TMA (15.4 mol%). The addition of external Al(i-Bu)$_3$ into MAOII system causes a dramatic decrease in catalytic activity (run 7).

In previous studies, we have found that the dependences of the catalytic activity on the content of residual TMA in MAO are different for the polymerizations of different monomers. Higher content of residual TMA in MAO favors styrene syndiotactic polymerization[17, 19] while less residual TMA in MAO favors the atactic polymerization of propylene[18]. Combining the monocyclopentadienyltitanium complex with MAO cocatalyst would generate tetravalent Ti species and trivalent Ti species in the ratio depending on the the free TMA concentration. The Ti^{4+} species are active for propylene polymerization, but not for styrene polymerization. Oppositely, the Ti^{3+} species are active for styrene polymerization, but inactive for propylene polymerization. The present results are much like those obtained from propylene polymerization with the same catalyst system. It seems that the norbornene polymerization is catalyzed by the Ti species in tetravalent state. The optimum molar ratio of MAOII to the titanocene for norbornene polymerization is approximately 40, which is much lower as compared to the zirconocene system[11].

Table 2. Polymerization of Norbornene Catalyzed by CpTi(OBz)$_3$ Activated with Various MAOs at Different Al/Ti Ratios [a]

Run	MAO	[Al]/[Ti]	Yield, g	A, kg polymer/(molTi.h)
1	MAOI	50	0.21	5.0
2	MAOII	20	0.50	11.9
3	MAOII	30	2.15	51.4
4	MAOII	40	2.74	65.6
5	MAOII	50	1.25	29.9
6	MAOII	67	0.78	18.7
7	MAOII[b]	40	0.17	4.1

[a] Other conditions are the same as Tab.1. [b] 0.3 mmol Al(i-Bu)$_3$ was added

Polymerization Conditions

Effects of the reaction conditions on norbornene polymerization carried out with the system CpTi(OBz)$_3$/MAOII were studied. As shown in Table 3-5, the catalyst system exhibits high activities in the Ti concentration range of 2-4 mM at 60 °C. Raising the polymerization temperature causes a decrease of the molecular weight of the polymer. The catalytic activity increases linearly with an increase of monomer concentration.

Table 3. Effect of CpTi(OBz)$_3$ Concentration on Norbornene Polymerization with MAOII Cocatalyst [a]

Run	[Ti], mM	Yield, g	A, kg polymer/(molTi.h)
8	0.95	0.22	20.8
9	1.9	0.95	45.5
10	3.8	2.21	52.9
11	7.0	1.50	18.0

[a] [MAOII] = 235 mM, other conditions are the same as Tab.1.

Table 4. Effect of Temperature on Norbornene Polymerization Catalyzed with CpTi(OBz)$_3$/MAOII [a]

Run	T_p, °C	Yield, g	A, kg polymer/(molTi.h)	[η], mL/g
12	0	0.19	4.6	
13	20	0.78	18.8	
14	40	1.87	44.7	131.3
15	60	2.34	56.0	109.2
16	80	1.58	37.9	85.0
17	100	1.00	23.9	

[a] Other conditions are the same as Tab.1.

Table 5. Effect of Monomer Concentration on Norbornene Polymerization with CpTi(OBz)$_3$/MAOII [a]

Run	[NBE], M	Yield, g	A, kg polymer/(molTi.h)
18	1.16	0.23	11.0
19	1.9	0.46	22.0
20	3.9	0.94	45.0

[a] [Ti] = 1.9 mM, other conditions are the same as Tab.1.

Polymer Characterization

The polynorbornenes obtained with the monocyclopentadienyltitanium-MAO catalyst system consist exclusively of saturated macromolecular chains. IR spectrum of the polymer sample (from run 4) is shown in Figure 1. The bands at 730 and 960 cm^{-1}, respectively attributable to the double bonds in cis and trans forms which are generated from ring-opening metathesis polymerization, are not observed. The IR characterization of the polymers produced with the catalyst reveals that norbornene is polymerized only by an insertion mechanism (addition polymerization), not a ring opening mechanism.

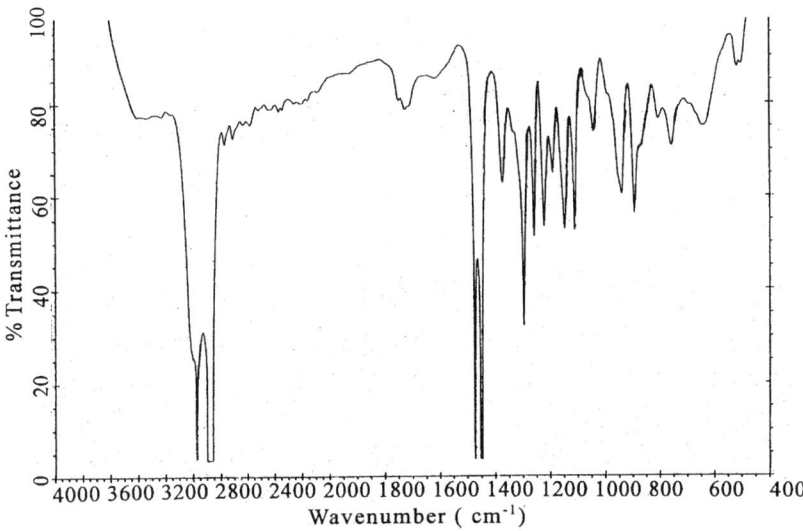

Figure 1. IR spectrum of polynorbornene sample obtained with CpTi(OBz)$_3$-MAOII catalyst (from run4)

Figure 2 shows the ^{13}C-NMR spectrum of polynorbornene sample (from run 4) in o-dichlorobenzene. Comparing the spectrum with those of the hydrodimers and -trimers from hydrooligomerizations of norbornene[24], it is shown that the polymerization of norbornene proceeds by normal 1,2-insertion *via* a *cis-exo* manner. In principle, there are three different stereostructures of the triad sequences in polynorbornene, as shown in Scheme 2. According to the chemical shift assignments for hydrotrimers from hydrooligomerizations of norbornene[24], distribution of the stereo-triad sequences for the present polynorbornene is determined from tertiary-carbon region (C1, C4, mm at 38.1, mr at 38.7~39.5, and rr at 40.1 ppm) of the ^{13}C-NMR spectrum to be [mm] = 14 %, [rr] = 40 %, and [mr] = 46 %.

55
50
45
40
35
30
25
ppm

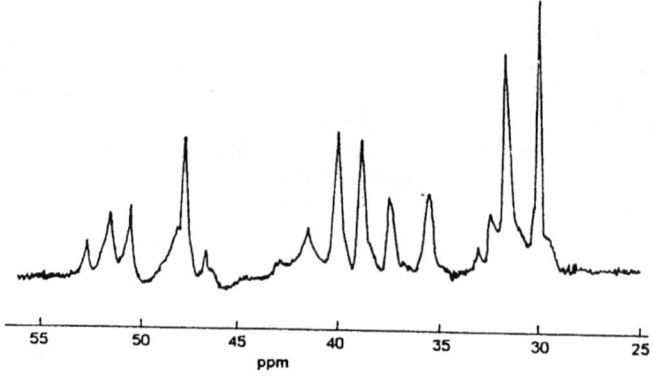

Figure 2. ^{13}C-NMR spectrum of polynorbornene sample obtained with $CpTi(OBz)_3$-MAOII catalyst (from run 4)

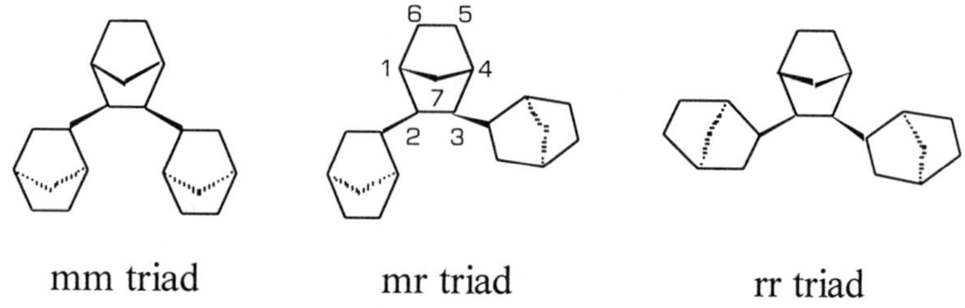

6
5
1
4
7
2
3
mm triad
mr triad
rr triad

Scheme 2

The polynorbornenes are soluble in hydrocarbon solvents such as cyclohexane, monochlorobenzene, *o*-dichlorobenzene and tetrahydronaphthalene. The polymers exhibited a glass transition temperature of about 320 °C in DSC curves.

Conclusions

From these results, it can be concluded that addition polymer of norbornene can be efficiently prepared with the simple half-titanocene catalyst system $CpTi(OBz)_3$/MAO. The content of residual TMA in MAO has a determinative effect on the catalytic activity for the norbornene polymerization. The catalyst system containing less residual TMA gives optimum results for the norbornene polymerization.

Acknowledgement. The authors of this paper would like to thank the National Natural Science Foundation of China for financial support of this research.

References

1. Tsujino, T.; Saegusa, T.; Furukawa, J. *Makromol. Chem.* **1965**, *85*, 71.
2. Boor, J. Jr. *Ziegler-Natta Catalysts and Polymerizations*; Academic Press: New York, 1979; p 523.
3. Gilliom, L. R.; Grubbs, R. H. *J. Am. Chem. Soc.* **1986**, *108*, 733.
4. Petasis, N. A., Fu, D.-K. *J. Am. Chem. Soc.* **1993**, *115*, 7208.
5. Sinn, H.; Kaminsky, W.; Vollmer, H. J.; Woldt, R. *Angew. Chem., Int. Ed. Engl.* **1980**, *19*, 39.
6. Gupta, V. K.; Satish, S.; Bhardwaj, I. S. *J. Macromol. Sci.,-Rew. Macromol. Chem. Phys.* **1994**, *C34(3)*, 439.
7. Brintzinger, H. H.; Fischer, D.; Mülhaupt, R.; Waymouth, R. M. *Angew. Chem., Int. Ed. Engl.* **1995**, *34*, 1143.
8. Huang, J.; Rempel, G. L. *Prog. Polym. Sci.* **1995**, *20*, 459.
9. Olabisi, O.; Atiqullah, M.; Kaminsky, W. *J. Macromol. Sci.,-Rev. Macromol. Chem. Phys.* **1997**, *C37(3)*, 519.
10. Bochmann, M. *J. Chem. Soc., Dalton Trans.* **1996**, 255.
11. Kaminsky, W.; Bark, A.; Steiger, R. *J. Mol. Cata.* **1992**, *74*, 109.
12. Kaminsky, W. *Macromol. Chem. Phys.* **1996**, *197*, 3907.
13. Seehof, N.; Mehler, C.; Breuning, S.; Risse, W. *J. Mol. Cata.* **1992**, *76*, 219.
14. Abu-Surrah, A. S.; Rieger, B. *J. Mol. Cata. A: Chem.* **1998**, *128*, 239.
15. Haselwander, T. F. A.; Heitz, W.; Kriigel, S. A.; Wendorff, J. H. *Macromol. Chem. Phys.* **1996**, *197*, 3435.
16. Goodall, B. L.; Barnes, D. A.; Benedikt, G. M.; McIntosh, L. H.; Rhodes, L. F. *Polym. Mater. Sci. Eng.* **1997**, *76*, 56.
17. Wu, Q.; Ye, Z.; Lin, S.-A. *Macromol. Chem. Phys.* **1997**, *198*, 1823.
18. Wu, Q.; Ye, Z.; Gao, Q.-H.; Lin, S.-A. *J. Polym. Sci., Part A: Polym. Chem.* **1998**, *36*, 2051.
19. Wu, Q.; Gao, Q.-H.; Ye, Z.; Lin, S.-A. *J. Appl. Polym. Sci.* **1998**, *70*, 765.
20. Wu, Q.; Ye, Z.; Gao, Q.-H.; Lin, S.-A. *Macromol. Chem. Phys.* **1998**, *199*, 1715.
21. Wu, Q.; Gao, Q.-H.; Ye, Z.; Lin, S.-A. *Polym. Prepr.* **1998**, *39(1)*, 222.
22. Wade, S. R.; Wallbridge, M. G. H.; Willey, G. R. *J. Chem. Soc., Dalton*, **1982**, 271.
23. Jordan, D. E. *Anal. Chem.* **1968**, *14*, 2150.
24. Arndt, M.; Engehausen, R.; Kaminsky, W.; Zoumis, K. *J. Mol. Cata. A: Chem.* **1995**, *101*, 171.

Chapter 11

Transfer and Isomerization Reactions in Propylene Polymerization with the Isospecific, Highly Regiospecific rac-Me₂C(3-t-Bu-1-Ind)₂ZrCl₂/MAO Catalyst

Isabella Camurati, Anna Fait, Fabrizio Piemontesi,
Luigi Resconi[1], and Stefano Tartarini

Montell Polyolefins, Centro Ricerche G. Natta, 44100 Ferrara, Italy

The influence of polymerization temperature, propylene concentration and hydrogen on the polymerization performance of the isospecific, highly regiospecific rac-Me₂C(3-t-Bu-1-Ind)₂ZrCl₂ / MAO (1/MAO) catalyst has been investigated. Propagation follows the rate law $R_p = (a[M]+b[M]^2)/(c+d[M])$. 1/MAO is more stereospecific compared to unsubstituted bisindenyl complexes, e.g rac-C₂H₄(1-Ind)₂ZrCl₂ / MAO (2/MAO), but growing-chain-end isomerization (epimerization) is faster in 1/MAO (ca 10% of stereoerrors of type mrrm are due to epimerization in liquid monomer at 50 °C) than in 2/MAO. ¹H NMR end group analysis shows that chain transfer occurs by β-methyl (prevalent) and β-hydrogen transfer reactions.

Chiral, C_2-symmetric $ansa$-zirconocenes produce polypropylenes with microstructures ranging from almost atactic to almost perfectly isotactic, and often containing isolated regioirregularities.[1,2] The wide range of molecular properties of these polypropylenes is due not only to the variability of the $ansa$-π-ligand structure, but also to the influence of the polymerization conditions.[3-9] While control over the stereoselectivity of zirconocene catalysts by ligand design has been obtained, a way to prevent secondary insertions has been achieved only recently: we have shown that C_2-symmetric, single-carbon bridged, 3-alkyl-substituted bisindenyl zirconocenes, are highly $regio$specific propylene polymerization catalysts, with stereospecificities and molecular weights that vary with the size of the substituent on C(3).[10]

For example, the catalyst system rac-Me₂C(3-t-Bu-1-Ind)₂ZrCl₂ / MAO (1/MAO, see chart 1) is both fairly isospecific ($mmmm$ ≈ 95 % at T_p = 50 °C)

[1]Corresponding author.

and highly regiospecific (no secondary propylene units being detectable in the 100 MHz ^{13}C NMR spectra), producing i-PP with $\Delta\Delta E^{\ddagger}_{enant}$ = 4.5 kcal/mol and a remarkably high $\Delta\Delta E^{\ddagger}_{transfer}$ of 10.7 kcal/mol (viscosity average molecular weights ranging from ≈ 25,000 at T_p = 70 °C to ≈ 450,000 at T_p = 20 °C).[10]

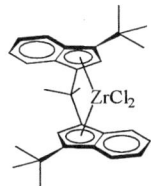

Chart 1. rac-Me$_2$C(3-t-Bu-1-Ind)$_2$ZrCl$_2$

The absence of secondary propylene units either in the chain or as end-groups greatly simplifies kinetic analysis, hence we selected 1/MAO for a study of the influence of polymerization parameters on catalyst performance. Our previous study of the influence of polymerization temperature [10] is here extended to chain transfer reactions. Busico,[3-8] Brintzinger [11,12] and us [9] have shown that the stereospecificity of chiral zirconocenes is lowered, to an extent which is a function of ligand structure, polymerization temperature and propylene concentration, by primary-growing-chain-end isomerization (epimerization), which scrambles the chirality of the last chirotopic methine of the growing chain. Hence, we have investigated the influence of propylene concentration, [M], on the performance of 1/MAO.

Our results are compared to those obtained with Brintzinger's rac-C$_2$H$_4$(Ind)$_2$ZrCl$_2$/MAO (2/MAO),[9,10,13] and are used to discuss the mechanisms of chain transfer and epimerization reactions. Reversible formation of a zirconocene allyl complex is proposed to account for the decrease of catalyst stereospecificity at low monomer concentration, the structure of olefinic end groups arising from primary β-H transfer, and the formation of internal vinylidenes.[14,15] The influence of propylene concentration on catalyst activity and type of chain transfer reactions from 1/MAO is here interpreted in the light of the available kinetic models.[16] The effect of hydrogen addition is also discussed.

^{1}H NMR Analysis of Olefinic Unsaturations

The proton spectra of i-PP samples from 1/MAO show the presence of four different olefinic groups, and the corresponding olefin structures, assigned as previously described,[10,15] are shown in Chart 2. The olefinic region of the proton spectrum of a typical sample prepared in liquid propylene is shown in Figure 1.

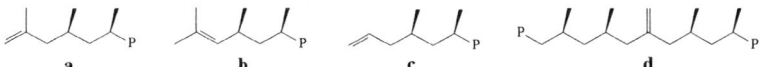

Chart 2. Unsaturated end group and internal vinylidene structures identified in i-PP samples from 1/MAO

Hence the chain transfer reactions operating with 1/MAO are primary: β-hydrogen (vinylidene **a**, peaks at 4.69, 4.77 ppm) and β-methyl transfer (allyl **c**, multiplets around 5.0 and 5.8 ppm), see Scheme 1.[*17-25*] In addition, chain transfer in 1/MAO also occurs via formation of the isobutenyl end group **b** (vinylic hydrogen: doublet at 4.9 ppm).

The allyl end group (**c**) always prevails over the others, confirming that β-CH$_3$ transfer is the most important chain transfer reaction in propylene polymerization with 1/MAO,[*10,15,26*] while β-H transfer gives only a minor contribution to the chain ends. 2-butenyl end groups (at 5.4 – 5.5 ppm) [*27,28*] could not be detected, confirming the high regioregularity of the samples. The inequality of the two vinylidene peaks of the samples prepared in liquid monomer, clearly shows that there is an additional, symmetrical vinylidene overlapping the lower field vinylidene peak at 4.77 ppm. The presence of this internal vinylidene unsaturation has been previously observed in most *i*-PP samples from C_2-symmetric zirconocenes,[*10*] and reassigned to the structure shown in Chart 2, structure **d**.[*15*] The large contribution from the internal vinylidene, and the presence of isobutenyl end groups, **b**, are noteworthy.

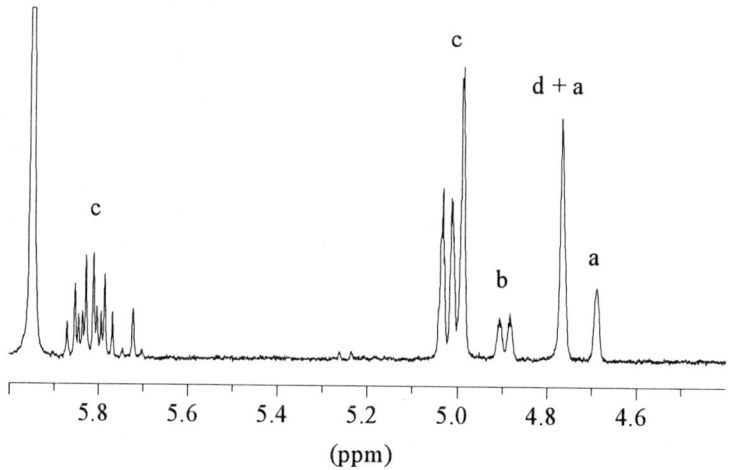

Figure 1. Olefin region of the ^{1}H NMR spectrum ($C_2D_2Cl_4$, 120 °C) of *i*-PP (sample 1 in Table I) prepared with 1/MAO at 70 °C in liquid propylene. Reference C_2HDCl_4 at 5.95 ppm.

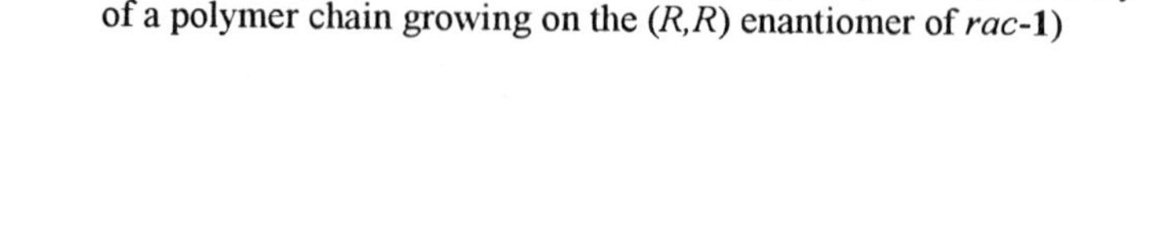

Scheme 1. Primary β-transfer reactions (showing the preferred chirality of a polymer chain growing on the (*R,R*) enantiomer of *rac*-**1**)

Influence of polymerization temperature

The polymerization of liquid propylene with 1/MAO in the temperature range 20 – 70 °C has already been described.[10] Here we complete the description with the ^{1}H NMR analysis of terminal and internal unsaturations of the set of i-PP samples in the T_p range 40 – 70 °C. The polymerization results and polymer analysis are reported in Tables I, II and III. Reaction of red 1 with MAO in toluene gives a violet solution which is remarkably stable under nitrogen, and the catalyst activity increases with aging time.

1. Isotacticity.

The samples of i-PP from 1/MAO show a decrease of isotacticity with increasing polymerization temperature typical of zirconocene catalysts. The samples from 1/MAO are more isotactic than those from 2/MAO at any polymerization temperature.

Table I. Influence of Polymerization Temperature in Propylene Polymerization with 1/MAO.[a]

Sample #	T_p °C	[M][b] mol/L	Activity[b] kg/(mmol$_{Zr}$×h)	$\overline{M}_v$[c]	$(\overline{P}_n)_{[\eta]}$[d]	$mmmm$[e] %
1	70	9.1	108.05	25,300	301	91.2_5
2	70	9.1	109.95	23,400	278	89.0_0
3	60	10.0	160.60	44,400	527	93.0_6
4	50	10.7	124.66[f]	89,400	1062	94.8_0
5	50	10.7	99.06[f]	88,000	1046	94.9_0
6	40	11.3	112.40	130,800	1554	95.9_3[g]
7	40	11.3	116.60	117,500	1396	
8	30	11.7	68.83	221,100	2627	96.7_8
9	20	12.0	32.06	410,400	4876	96.8_9
10	20	12.0	17.78	455,400	5411	

[a] experimental conditions: 0.4 L propene, 1 hour; Al/Zr = 8000, premixed catalyst/cocatalyst in toluene (aging time 10 min); [b] averaged over the polymerization time. [c] Calculated from the intrinsic viscosity [η] (THN, 135 °C), according to the [η] = K×$(\overline{M}_v)^\alpha$ with K = 1.93·10^{-4} and α = 0.74. [d] Estimated assuming $\overline{M}_v = \overline{M}_w = 2\overline{M}_n$. [e] determined by ^{13}C NMR (100 MHz) assuming the enantiomorphic site model. No regiomistakes could be detected. [f] aging time 64h at Al/Zr=1000 + 10min at Al/Zr=8000 (sample 5); 356h at Al/Zr=1000 + 10min at Al/Zr=8000 (sample 4). [g] Average of two measurements.

The Arrhenius plots of $\ln[b_{obs}/(1-b_{obs})]$ versus $1/T_p$, where b_{obs} (obtained through a non-linear least-squares method from the methyl pentad distribution of the sample measured by ^{13}C NMR) is the concentration of correct primary propene units in an i-PP sample prepared at a given [M] and T_p, are compared in Figure 2: 1/MAO shows an apparent difference $\Delta\Delta E^{\ddagger}_{enant}$ of the activation energies for the insertion of the two enantiofaces of propylene, of 4.5 ± 0.5 kcal/mol, compared to 3.3 ± 0.2 kcal/mol for 2/MAO. The $\Delta\Delta E^{\ddagger}_{enant}$ value for **1** approaches that of enantiomorphic site control of highly isospecific Ti catalysts (4.8 kcal/mol),[29] which however show a higher stereospecificity compared to 1/MAO. The melting points of *i*-PP from 1/MAO (142 – 158 °C) are always higher than those of *i*-PP from 2/MAO (125 – 142 °C), confirming the higher isospecificity of **1**.

Table II. Influence of Polymerization Temperature in Propylene Polymerization with 1/MAO: Unsaturated Group Composition from ^{1}H NMR.[a]

Sample #	T_p °C	term. vinylid.	int. vinylid.	isobutenyl	allyl
1	70	16.2 (20.3)	20.1	19.1 (23.9)	44.6 (55.8)
2	70	14.8 (17.6)	16.4	15.8 (18.9)	53.0 (63.5)
3	60	21.0 (25.9)	19.1	20.8 (25.6)	39.2 (48.5)
4,5 [b]	50	15.8 (24.1)	34.1	14.7 (22.3)	35.4 (53.6)
6	40	23.5 (35.6)	34.1	16.1 (24.5)	26.3 (39.9)
7	40	27.8 (39.9)	30.2	17.6 (25.2)	24.4 (34.9)

[a] Composition of unsaturations determined from ^{1}H NMR, 400 MHz, C$_2$D$_2$Cl$_4$, 120 °C. Values are normalized over total unsaturations or (in parenthesis) over end groups only.
[b] Average of four measurements.

Table III. Influence of Polymerization Temperature in Propylene Polymerization with 1/MAO: Molecular Weight Data from ^{1}H NMR.

Sample #	T_p °C	$(\overline{P}_n)$ [η]	$(\overline{P}_n)$ ^{1}H-NMR [a]	$(\overline{P}_n)$ [allyl] [b]	$(\overline{P}_n)$ [vinylid] [c]	internal vinylid. per 1000 units
1	70	301	429	770	2111	0.59
2	70	278	384	605	2175	0.51
3	60	527	802	1655	3099	0.29
4,5 [d]	50	1022	1826	3405	7640	0.29
6	40	1554	2537	6357	7124	0.20
7	40	1396	2065	5918	5177	0.21

[a] Estimated from terminal vinylidene+allyl+2×isobutenyl. [b] Estimated from the allyl end group only. [c] Estimated from the terminal vinylidene end group only. [d] Average of four measurements.

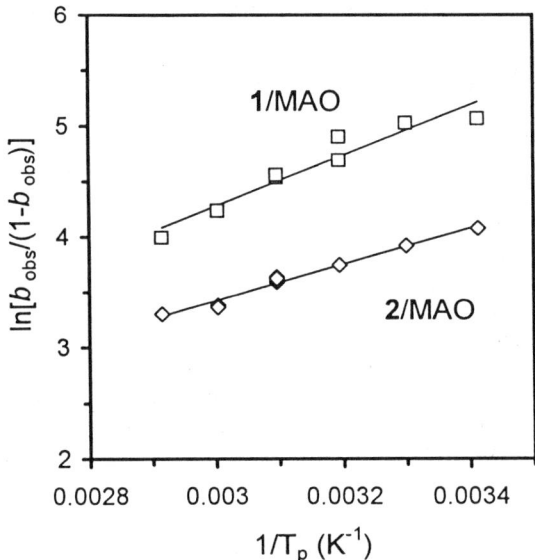

Figure 2. $\ln[b_{obs}/(1-b_{obs})]$ versus $1/T_p$ for i-PP samples from **1**/MAO and **2**/MAO prepared in liquid propylene in the temperature range 20 – 70 °C. □: **1**/MAO, $\Delta\Delta E^{\ddagger}_{enant}$ = 4.5 ± 0.5 kcal/mol, R = 0.961; ◊: **2**/MAO, $\Delta\Delta E^{\ddagger}_{enant}$ = 3.3 ± 0.2 kcal/mol, R = 0.988.

2. Molecular weight

The i-PP molecular weights are much more dependent on the polymerization temperature in the case of 1/MAO than 2/MAO: average viscosity molecular weights decrease from 450,000 at 20 °C to 25,000 at 70 °C for 1/MAO, while they are limited to the range from 56,000 to 20,000 in the case of 2/MAO. End group analysis by ^{1}H NMR (T_p = 40 – 70 °C) allows to recognize the different unsaturated end groups and discern among the contributions of the different chain transfer reactions. While vinylidenes are of too low intensity to allow for a quantitative evaluation, especially for the higher molecular weight samples, both allyl and total end groups show reliable trends. The $\overline{P}_n$ values of i-PP from 1/MAO and 2/MAO, from intrinsic viscosity and ^{1}H NMR, give Arrhenius plots of $\ln(\overline{P}_n)$ versus $1/T_p$ with very good linear correlations, and they are compared in Figure 3, together with $\ln(\overline{P}_{n,allyl})$ calculated for 1/MAO from ^{1}H NMR. The slopes give the overall activation energy barrier for chain transfer, $\Delta\Delta E^{\ddagger}_{tr}$ = 3.9 ± 0.4 kcal/mol for 2/MAO and a much higher 10.7 ± 0.4 kcal/mol for 1/MAO (both set of data from average viscosity molecular weight data). These values can be, in first approximation, attributed to the higher conformational freedom of the growing chain in **2** versus the bulkier and more rigid catalyst **1**. $\overline{P}_n$ data from ^{1}H NMR for 1/MAO can be obtained only in the T_p range 40-70 °C, due to the too high molecular weights obtained at the lower polymerization temperatures, and the obtained $\Delta\Delta E^{\ddagger}_{tr}$ = 12.8 ± 1.5 kcal/mol is hence less reliable than that obtained from viscosity data. From the analysis of end groups we could evaluate the dependence of β-CH$_3$ transfer on T_p, that seems to have a slightly stronger dependence, with a $\Delta\Delta E^{\ddagger}_{tr}$ = 15.6 ± 0.9 kcal/mol. Vinylidenes are of too low concentration to allow the same analysis also for β-H chain transfer.

Influence of Propylene Concentration

The strong influence of the concentration of propylene on i-PP molecular weight, stereoregularity, and end-group structure is a typical feature of zirconocene catalysts. This effect has been described in some detail for Brintzinger's C$_2$-symmetric catalysts rac-C$_2$H$_4$(H$_4$Ind)$_2$ZrCl$_2$/MAO,[3-8] rac-C$_2$H$_4$(Ind)$_2$ZrCl$_2$/MAO (2/MAO),[9] rac-Me$_2$Si(Benz-[e]-Ind)$_2$ZrCl$_2$/MAO and rac-Me$_2$Si(2-Me-Benz-[e]-Ind)$_2$ZrCl$_2$/MAO,[27] for Ewen's C$_s$-symmetric Me$_2$C(Cp)(9-Flu)ZrCl$_2$,[30-32] and for the C$_1$-symmetric zirconocenes developed by Rieger.[33,34] The performance of 1/MAO also is heavily affected by the concentration of propylene, [M]. Propylene was polymerized with 1/MAO at different propylene concentrations at 50 °C ([propylene] from ~0 to ~10 mol/L). The results of the polymerization experiments are reported in Table IV and compared to polymerizations run in liquid propylene. Pentane was chosen as the solvent to better simulate a gas-phase process. The use of pentane also makes polymer isolation easier, compared to toluene or other higher boiling solvents. Run 11 ([M] = 5.1 mol/L) was carried out in a 1:1 propane:propylene mixture.

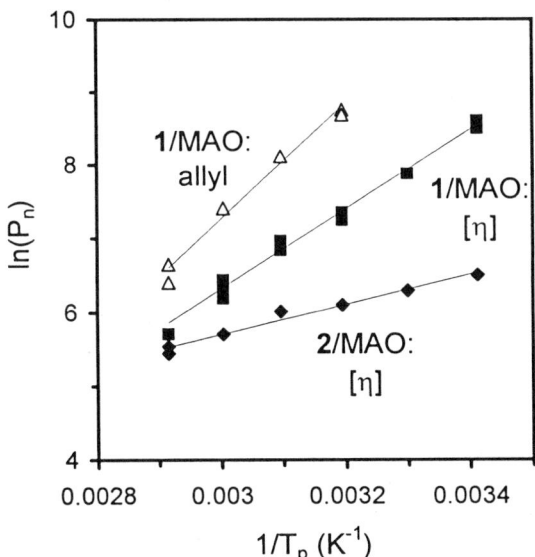

Figure 3. $\ln \overline{P}_n$ versus $1/T_p$ for *i*-PP samples from 1/MAO and 2/MAO prepared in liquid propylene in the temperature range 20 – 70 °C. ◆ = 2/MAO, $\overline{P}_n$ values from intrinsic viscosity; ■ = 1/MAO, $\overline{P}_n$ values from intrinsic viscosity; △ = 1/MAO, $\overline{P}_n$ values from allyl end groups.

1. Activity

The activity of 1/MAO increases in a non-linear fashion with increasing [M], from ca. 1 kg/(mmol$_{Zr}$×h) at [M] ≈ 0.2 mol/L, to ca. 75 kg/(mmol$_{Zr}$×h) in liquid propylene. The activity further increases at longer aging times (Table IV). Compared to 2/MAO in toluene/propylene,[9] 1/MAO in pentane/propylene shows lower activity. Since the two catalysts have similar activities in liquid propylene, we attribute the lower activity of 1/MAO to the different solvents. Different solvents are likely to have a different influence on the rate of catalyst formation (pre-equilibria),[35] or on the rate constant of initiation.[36]

Table IV. Influence of Monomer Concentration in Propylene Polymerization with 1/MAO.[a]

Sample #	solvent	[M] mol/l	Activity [b] kg/(mmol$_{Zr}$×h)	mmmm [c] %	$\bar{P}_n$ [d] ^{1}H NMR
4	propylene	10.7 [b]	124.7 [e]	94.8$_0$	1770
5	propylene	10.7 [b]	99.1 [e]	94.9$_0$	1912
11 [f]	propane	5.1	14.8	93.8$_1$	992
12	pentane	2.73	22.3	-	-
13	pentane	2.21	30.5	92.4$_3$	545
14	pentane	1.69	8.85	90.3$_1$	438
15	pentane	1.18	3.19	89.8$_3$	394
16	pentane	0.68	2.65	87.3$_2$	304
17	pentane	0.2	1.23	80.4$_2$	197

[a] experimental conditions: 0.4 L propene (samples 4,5) or 0.4 L pentane (samples 12-17), 50 °C, 1 hour, premixed catalyst/cocatalyst in toluene (Al/Zr = 8000, aging time 10 min); [b] averaged over the polymerization time. [c] determined assuming the enantiomorphic site model. No regiomistakes could be detected. [d] Estimated from the unsaturated end groups by ^{1}H NMR. [e] see Table I. [f] Al/Zr = 5,000.

The non-linear dependence of activity on [M] (Figure 4) is a common feature of stereospecific zirconocenes [9,30-32] that can be explained by a Single-Center, Two-State catalyst model,[16] where the active metal centers can exist in two energy states (C$_{fast}$ and C$_{slow}$) which interconvert without monomer assistance, and are in equilibrium through K = $k_{fast \rightarrow slow}/k_{slow \rightarrow fast}$. This model leads to Equation 1 for the rate of polymerization,[16] were $k_{p,fast}$ and $k_{p,slow}$ are the propagation rate constants (product of the rate constants of propylene coordination, k_c, and insertion, k_{ins}) of the fast and slow catalyst states respectively.

$$R_p = [C] \frac{(k_{fast \rightarrow slow} k_{p,slow} + k_{p,fast} k_{slow \rightarrow fast})[M] + k_{p,fast} k_{p,slow} [M]^2}{k_{fast \rightarrow slow} + k_{slow \rightarrow fast} + k_{p,slow}[M]} \quad (1)$$

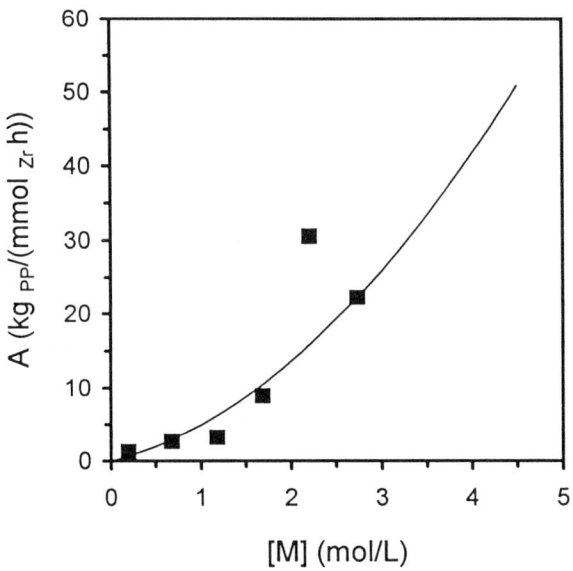

Figure 4. Polymerization activity of 1/MAO. (■): experimental data (50 °C, propylene/pentane, Al/Zr = 8000, 10 min. aging time).
(—): best fit from equation 1. $k_1' = 280$ kg$_{PP}$/(mmol$_{Zr}$×h×mol$_M$/L), $k_2' = 180$ kg$_{PP}$/(mmol$_{Zr}$×h×mol^2_M/L^2), $k_3' = 92$, $k_4' = $ L/mol$_M$.

The experimental activities of 1/MAO reported in Table IV and shown in Figure 4 ([M] = 0-3 mol/L) are averaged over the polymerization time of 1 hour. Since a different [M] also induces a change in the rate of catalyst decay, these values can not be used to estimate the ratios of rate constants in Equation 1. However, the best fit from Eq. 1 in the range 0 – 4 mol/L, is also shown in Figure 4.

As to the nature of the different states of the active centers, some hypothesis have been previously discussed.[16] For example, the two catalyst states can be generated by different conformations of the growing chain, that will have two different agostic interactions: γ-agostic, the kinetic product of insertion (C_{fast}), and β-agostic, expected to be more stable (C_{slow}).[37] A further hypothesis, based on the mechanism we propose for the formation of internal unsaturations and epimerization, is discussed below.

2. Isotacticity.

The isospecificity of both catalysts 1/MAO and 2/MAO is adversely affected by decreasing [M], but 1/MAO is more isospecific than 2/MAO at any [M].

A number of studies have shown that, in the case of C_2-symmetric zirconocenes, isotacticity decreases at lower propylene concentrations,[3-9,11,12] due to unimolecular primary-growing-chain-end epimerization, which scrambles the chirality of the last chirotopic methine of the growing chain. The extent of epimerization at a given [M] depends on the polymerization temperature and on the nature of the *ansa*-π-ligand.[3-8] For example, 1/MAO is more stereospecific than 2/MAO at any [M], although the stereospecificity of 1/MAO shows a more pronounced dependence on [M]. We have shown that the dependence of isotacticity on [M] can be described by Equation 2:[9]

$$\frac{b_{obs}}{1-b_{obs}} = \frac{0.5 + b K_{eq}[M]}{0.5 + (1-b) K_{eq}[M]} \quad (2)$$

where the Bernoullian probability parameter b is the inherent enantioface selectivity which depends on the catalyst structure and T_p, but is independent from [M]. We have observed that, for 2/MAO, in liquid monomer, $b_{obs} \rightarrow b$,[9,10] at least at the polymerization temperature of 50 °C.

The experimental b_{obs} / (1-b_{obs}) values for *i*-PP samples from both 1/MAO and 2/MAO are fitted to Equation 2 in Figure 5. In the case of 1/MAO, Equation 2 gives K_{eq} = 35 L/mol, and b = 0.9906 (corresponding to *mmmm* = 95.4 %), slightly higher than the experimental value obtained in liquid monomer, *mmmm* = 94.9 %. The difference between 1-b (the number of wrong primary insertions due to incomplete catalyst enantioselectivity) and the value of 1-b_{obs} (the total number of insertion errors measured from the ^{13}C NMR spectrum) obtained in liquid monomer, indicates that for 1/MAO about 10% stereoerrors are due to epimerization, already at 50 °C. Hence, since the rate of epimerization increases with temperature,[5,7,8] the value of $\Delta\Delta E^{\ddagger}_{enant}$ = 4.5 kcal/mol for 1/MAO must be overestimated. This explains in part the difference in isospecificity with heterogeneous catalysts.

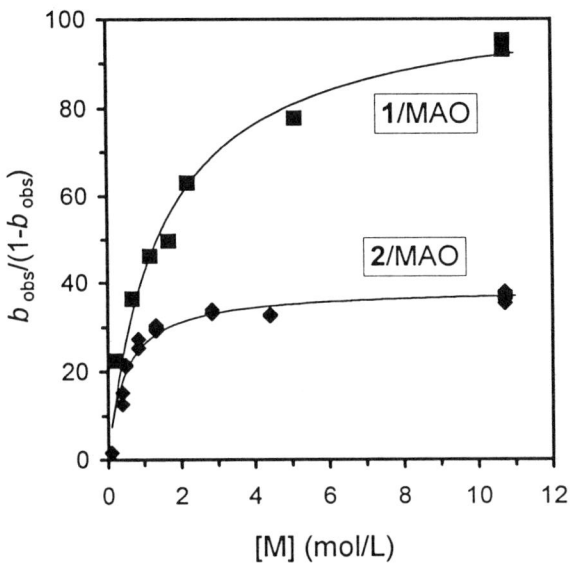

Figure 5. Polypropylene isotacticity ($b_{obs}/(1-b_{obs})$ from ^{13}C NMR) versus [M], for catalysts **1**/MAO (50 °C, propylene/ pentane, ■) and **2**/MAO (50 °C, propylene/ toluene, ◆). Best fit from Equation 2.

Epimerization has been explained by two mechanisms, both requiring formation of a Zr–H(CH$_2$=C(CH$_3$)P) intermediate via unimolecular β-H transfer: Busico's mechanism (I in Scheme 2) involves a sequence of β-H transfers, double bond reorientations, and insertions [3]; we have proposed that epimerization occurs through reversible formation of a zirconocene allyl dihydrogen complex (II in Scheme 2),[14,15] a pathway that stands on literature precedents,[38-45] and also explains the presence of internal unsaturations (see below). Both mechanisms take into account the analysis of Leclerc and Brintzinger,[11,12] who have shown that primary-growing-chain-end epimerization occurs with exchange of the methylene and methyl carbons of the stereoinverted unit (A→B in Scheme 2). Recent work by Bercaw [46] indicates that the allyl can rotate in the wedge plane, as shown in Scheme 2. Mechanism II assumes that H$_2$ does not dissociate during allyl rotation: epimerization here arises from subsequent reinsertion of H$_2$ at either Zr–C(3) in **C** or Zr–C(1) in **C'**.

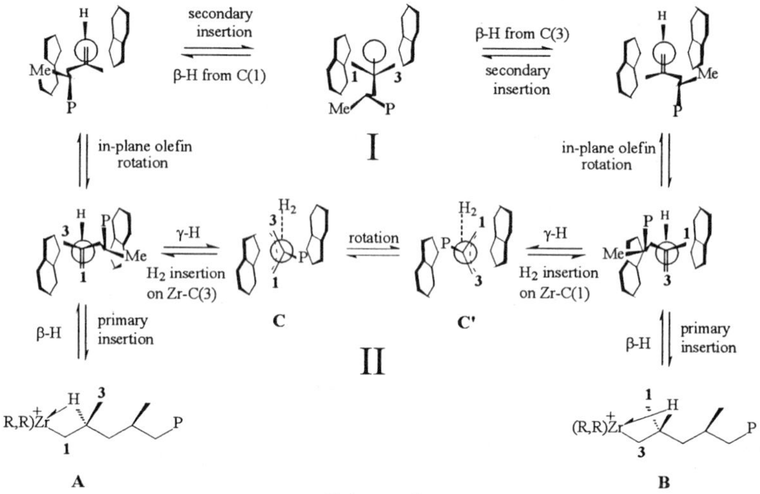

Scheme 2.
Epimerization via double bond reorientation (I) or reversible formation of a (R,R)Zr(allyl)(H$_2$) cation (II) (the ligand bridge is omitted for clarity).

3. Chain Transfer Reactions and Molecular weight.

Along with affecting isotacticity, lowering propylene concentration also induces a non-linear decrease of the molecular weight of *i*-PP. The change of $\overline{P}_n$ (^{1}H NMR) with [M] for catalysts **1**/MAO and **2**/MAO are compared in Figure 6. The higher molecular weights observed in *i*-PP from **1**/MAO compared to **2**/MAO are due to the higher regiospecificity of the former, hence to the absence of chain tranfer to the monomer after a secondary insertion.[27,28] While *i*-PP from **2**/MAO prepared in liquid propylene clearly shows *cis*-2-butenyl end groups due to chain transfer after a secondary insertion, these are undetectable (that is ≤ 1/200,000 units) in the samples

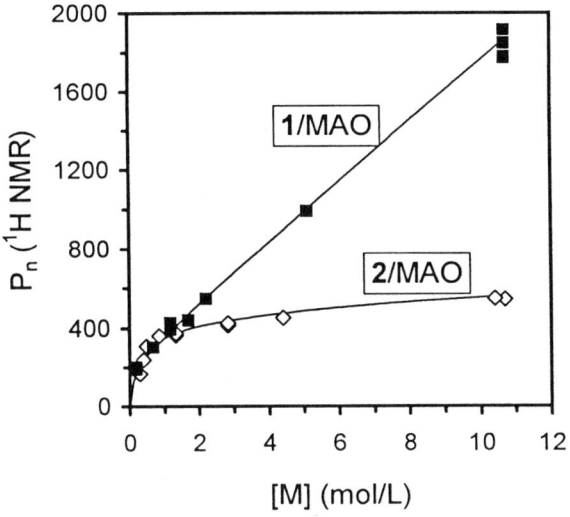

Figure 6. $\bar{P}_n$ from ^{1}H NMR versus [M] for **1**/MAO (■) and **2**/MAO (◊).(—): Best fit from Equation 4. For **1**/MAO: $k_1' = 280$, $k_2' = 180$, $k_3' = 0.12$, $k_4' = 1.17$. For **2**/MAO: $k_1' = 332$, $k_2' = 665$, $k_3' = 0.28$, $k_4' = 0.72$.

prepared with 1/MAO at any value of [M]. Both unimolecular and bimolecular chain transfer mechanisms (that is hydrogen and methyl transfers to the metal, $R_t = k_{t_0}[C]$ and hydrogen transfer to the monomer, $R_t = k_{t_1}[C][M]$) concur in lowering i-PP molecular weights. Equation 3 (which applies for the simple cases in which $R_p = k_p[C][M]$) does not provide a satisfactory description of the molecular weights of i-PP from chiral zirconocenes.

$$\overline{P}_n = \frac{k_p[M]}{k_{t_0} + k_{t_1}[M]} \qquad (3)$$

The dependence of PP molecular weight on propylene concentration is rather more complicated for isospecific catalysts, because of the non-linear dependence of activity on [M]. The change of $\overline{P}_n$ with [M] for both 1/MAO and 2/MAO is described with a very good approximation by Equation 4 (Figure 6):

$$\overline{P}_n = \frac{k_1'[M] + k_2'[M]^2}{k_3' + k_4'[M]} \qquad (4)$$

that has been derived by combining Eq. 1 with the hypothesis that uni- and bimolecular chain transfer occurs only (or prevailingly) at the slow state of the catalyst, or that only unimolecular chain transfer occurs from both the fast and the slow catalyst states. In general, the strong decrease of molecular weight caused by lowering [M] observed with 1/MAO can be ascribed to the increase in the fraction of centers in the slower state, which are then prone to chain transfer reactions.

The formation of an allyl intermediate also accounts for the formation of internal unsaturations: both the isobutenyl end group (structure **b**), and the internal vinylidene (structure **d**), can in fact be generated from Zr(allyl) species (Scheme 3). Based on the above mechanism, the amount of internal vinylidene should be monomer concentration dependent. Indeed, the content of internal vinylidene is largely decreased at lower propylene concentration (higher epimerization rate).[14]

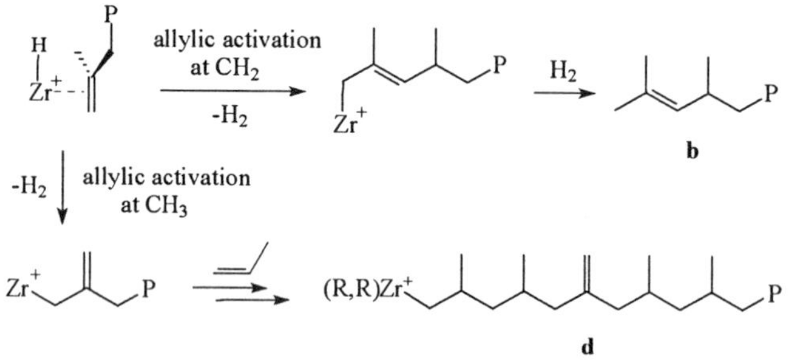

Scheme 3. Proposed reaction scheme for the formation of the internal vinylidene by propylene insertion on a Zr(η^1-allyl)(H$_2$) cation, and the *iso*-butenyl end-group. [14,15]

Influence of hydrogen

Molecular hydrogen is used to regulate the molecular weight of polyolefins in both heterogeneous [47] and metallocene catalysts.[4,48-54] In addition, hydrogen also produces an increase in activity for both catalyst types. Two hypothesis have been described to explain this activating effect: hydrogenation of a secondary growing chain,[4,48-54] or reactivation of an allyl zirconocene species.[55]

The influence of molecular hydrogen on the performance of 1/MAO in the polymerization of liquid propylene was investigated at 50 °C (Table V), with the original purpose to see whether hydrogen activation would be observed also with a regiospecific catalyst. Addition of molecular hydrogen produces a moderate (up to 2-fold) increase in catalyst activity, which decreases at higher levels, together with the expected decrease in molecular weights.

Table V. Propene Polymerization with *rac*-Me$_2$C(3-*t*-Bu-Ind)$_2$ZrCl$_2$/MAO: influence of hydrogen concentration.[a]

Sample #	H$_2$ NmL	Activity[b] kg/(mmol$_{Zr}$×h)	$\overline{M}_v$[c]	$(\overline{P}_n)_{[\eta]}$[d]	$(\overline{P}_n)_{[13C]}$	mmmm[e] %
18	0	44.3	88,000	1046	n.d.	94.9
19	100	47.5	50,100	595	760	95.1
20	200	81.3	42,100	500	451	95.0
21	300	61.7	15,200	181	-	-
22	400	62.8	n.d.	-	55	94.9

[a] experimental conditions: 0.4 L propene, 50 °C, 1 hour; Zr = 0.5 mg (0.92 μmol), Al/Zr = 5000, aging time 10 min; [b] averaged over the polymerization time. [c,d] see Table I. [e] Estimated including the isobutyl end groups, assuming equal chain transfer rates after a correct or wrong insertion.

Since 1/MAO does not show detectable secondary propylene insertions into the growing polypropylene chain, the activation of dormant secondary centers is unlikely. Analysis of the olefin region of the ^{1}H NMR spectra of *i*-PP prepared with 1/MAO in the presence of hydrogen shows that H$_2$ addition in liquid propylene polymerization also suppresses the formation of internal vinylidenes, providing further support to the hypothesis of the allyl intermediate. Interestingly, ^{13}C NMR end group analysis of samples 19-22 shows the presence of the (CH$_3$)$_2$CHCH(CH$_3$)CH$_2$- initiating group, which must arise from secondary propylene insertion into the Zr–H bond, which forms a Zr(*iso*-propyl) species, followed by primary propagation.

This new group is in about 1:5 ratio with the usual *n*-propyl group.[56] Possibly, hydrogen activates the catalyst by reducing the concentration of Zr(allyl) species (**C** in scheme 2), but at the same time the activation is limited by the formation of secondary, slower Zr(*i*-Pr) initiating species, which require either isomerization to *n*-propyl or a new hydrogenolysis to be converted in the faster centers.

Conclusions

The isospecific and highly regiospecific C_2-symmetric zirconocene *rac*-Me$_2$C(3-*t*-Bu-Ind)$_2$ZrCl$_2$/MAO (**1**/MAO) was chosen to study the influence of polymerization temperature, propylene concentration, and hydrogen on the elementary steps of *primary* propylene polymerization: monomer insertion, chain release, and primary growing-chain-end isomerization (epimerization). Secondary (2,1) propylene insertions into the growing PP chain could not be detected by either ^{13}C NMR (2,1 < 0.02 %) or ^{1}H NMR (2-butenyl end groups < 1/200,000 units). The extent of epimerization was measured by ^{13}C NMR, and it is shown that **1**/MAO, although more isospecific than **2**/MAO, undergoes faster epimerization compared to the latter, to some extent already in liquid monomer. Formation of a zirconocene(allyl) cation is proposed to account for growing chain epimerization, and the formation of internal vinylidene unsaturations and isobutenyl end groups, although further investigation is required to fully understand the mechanism of their formation. ^{1}H NMR was used to measure the average polymerization degrees, to determine the dependence of molecular weight on the polymerization conditions. β-methyl transfer is the most important chain transfer reaction. Since a zirconocene allyl cation is effectively deactivated towards propagation,[*40*] we suggest that the nature of the slow catalyst state, which is in equilibrium with the fast propagating catalyst state (equilibrium that generates the apparent exponential dependence of activity on propylene concentration) is indeed a Zr(allyl) cation.

Confirming the nature of the slow catalyst state(s) remains an important target of our research, since it is mainly the slow state(s) which are involved in unimolecular chain transfer and epimerization reactions. Hopefully, a better understanding of these catalyst systems will provide clues for their improvement.

Aknowledgement

We are indebted to Dr. I. E. Nifant'ev for developing the synthesis of **1**, to Mr. Marcello Colonnesi for the NMR spectra, to Dr. L. Cavallo, Prof. G. Guerra and Prof. U. Giannini for helpful discussions, and to Montell for supporting this work.

Experimental Section

The catalyst precursor **1** was prepared by Dr. I. Nifant'ev according to the procedure described in ref. 10. MAO was a commercial 10% solution in toluene (Witco). MAO is a pyrophoric substance that must be handled carefully and under nitrogen. **1** was preracted with MAO in 5-10 mL toluene for 10 minutes and pressurized into the polymerization reactor containing either pentane/propylene, or propylene only. The *i*-PP samples 1-10 were prepared and analyzed as previously described.[9,10] Tests with different propylene concentrations were carried out into a 1-L Büchi glass autoclave: 0.3 L of pentane were charged into the autoclave at 30 °C, the atmosphere changed from nitrogen to propylene by pressurizing with propylene to 3-4 bar-g under stirring and venting the excess pressure (3 times), the temperature of the reactor was brought to 50 °C, the excess pressure was released, the catalyst/cocatalyst mixture was loaded under a nitrogen overpressure, excess pressure was again released, then the polymerization test was carried out for 1 hour at 50 °C continuosly feeding propylene to maintain the desired pressure, and finally stopped with CO. Tests with added hydrogen were carried out as follows: 0.4 L of propylene were charged into the 1-L stainless steel autoclave (Büchi) at 30 °C, the temperature of the reactor was brought to 50 °C, and, in quick sequence: the catalyst/cocatalyst mixture was loaded under a nitrogen overpressure, excess pressure was released, the desired amount of hydrogen (measured into a stainless-steel 100 mL vial) was added; the polymerization test was carried out for 1 hour at 50 °C and then stopped with CO. The polymers were recovered by evaporating the solvents under vacuum, exhaustive extraction with refluxing toluene under nitrogen in a Kumagawa extractor to remove catalyst residues, and finally by evaporating toluene and drying the polymer at 60 °C under high vacuum.

Proton and carbon spectra were obtained using a Bruker DPX-400 spectrometer operating in the Fourier transform mode at 120 °C at 400.13 MHz and 100.61 MHz respectively. The samples were dissolved in $C_2D_2Cl_4$. As reference the residual peak of C_2HDCl_4 in the ^1H spectra (5.95 ppm) and the peak of the *mmmm* pentad in the ^{13}C spectra (21.8 ppm) were used. Proton spectra were acquired with a 45° pulse and 5 seconds of delay between pulsed; 256 transients were stored for each spectrum. The carbon spectra were acquired with a 90° pulse and 12 seconds of delay between pulses. About 3000 transients were stored for each spectrum.

References

1. Ewen, J.; Elder, M.; Jones, R.; Haspeslagh, L.; Atwood, J.; Bott, S.; Robinson, K. *Makromol. Chem., Macromol. Symp.* **1991**, *48/49*, 253.
2. Brintzinger, H. H.; Fischer, D.; Mülhaupt, R.; Rieger, B.; Waymouth, R. M. *Angew. Chem. Int. Ed. Engl.* **1995**, *34*, 1143.
3. Busico, V.; Cipullo, R. *J. Am. Chem. Soc.* **1994**, *116*, 9329.
4. Busico, V.; Cipullo, R.; Chadwick, J. C.; Modder, J. F.; Sudmeijer, O. *Macromolecules* **1994**, *27*, 7538.
5. Busico, V.; Cipullo, R. *J. Organomet. Chem.* **1995**, *497*, 113.
6. Busico, V.; Caporaso, L.; Cipullo, R.; Landriani, L.; Angelini, G.; Margonelli, A.; Segre, A. L. *J. Am. Chem. Soc.* **1996**, *118*, 2105.
7. Busico, V.; Brita, D.; Caporaso, L.; Cipullo, R.; Vacatello, M. *Macromolecules* **1997**, *30*, 3971.
8. Busico, V.; Cipullo, R.; Caporaso, L.; Angelini, G.; Segre, A. L. *J. Mol. Cat.* **1998**, *128*, 53.

9. Resconi, L.; Fait, A.; Piemontesi, F.; Colonnesi, M.; Rychlicki, H.; Zeigler, R. *Macromolecules*, **1995**, *28*, 6667.
10. Resconi, L.; Piemontesi, F.; Camurati, I.; Nifant'ev, I.E.; Ivchenko, P.V.; Kuz'mina, L.G. *J. Am. Chem. Soc.* **1998**, *120*, 2308.
11. Leclerc, M. K.; Brintzinger, H. H. *J. Am. Chem. Soc.* **1995**, *117*, 1651.
12. Leclerc, M. K.; Brintzinger, H. H. *J. Am. Chem. Soc.* **1996**, *118*, 9024.
13. Wild, F.; Wasiucionek, M.; Huttner, G.; Brintzinger, H.H. *J. Organomet. Chem.* **1985**, *288*, 63.
14. Resconi, L. *Polym. Mat. Sci. Eng.* **1999**, 80, 421.
15. Resconi, L.; Camurati, I.; Sudmeijer, O. *Topics in Catalysis*, **1999**, *7*, 145.
16. Fait, A.; Resconi, L.; Guerra, G.; Corradini P. *Macromolecules* **1999**, *32*, 2104.
17. Although we did not unambiguously prove that allyl end groups originate from β-CH_3 transfer in the present system, we assume this to be the case given the large number of precedents established for similar systems.[18-25]
18. Eshuis, J.; Tan, Y.; Teuben, J.H.; Renkema, J. *J. Mol. Cat.* **1990**, *62*, 277.
19. Eshuis, J.; Tan, Y.; Meetsma, A.; Teuben, J.H. *Organometallics* **1992**, *11*, 362.
20. Resconi, L.; Piemontesi, F.; Franciscono, G.; Abis L.; Fiorani, T. *J. Am. Chem. Soc.* **1992**, *114*, 1025.
21. Yang, X.; Stern, C.L.; Marks, T.J., *Angew. Chem. Int. Ed. Engl.* **1992**, *31*, 1375.
22. Mise, T.; Kageyama, A.; Miya, S.; Yamazaki, H. *Chem. Lett.* **1991**, 1525.
23. Yang, X.; Stern, C.L.; Marks, T.J.; *J. Am. Chem. Soc.* **1994**, *116*, 10015.
24. Hajela, S.; Bercaw, J.E. *Organometallics* **1994**, *13*, 1147.
25. Guo, Z.; Swenson D.; and Jordan, R.F. *Organometallics* **1994**, *13*, 1424.
26. Resconi, L.; Piemontesi, F.; Sudmeijer, O. *Polym. Prepr. Am. Chem. Soc., Div. Polym. Chem.* **1997**, *38(1)*, 776.
27. Jüngling, S.; Mülhaupt, R.; Stehling, U.; Brintzinger, H.-H.; Fischer, D.; Langhauser, F. *J. Polym.Sci.: Part A: Polym. Chem.* **1995**, *33*, 1305.
28. Resconi, L.; Piemontesi, F.; Camurati, I.; Balboni, D.; Sironi, A.; Moret, M.; Rychlicki, H.; Zeigler, R. *Organometallics*, **1996**, *15*, 5046-5059.
29. Zambelli, A.; Locatelli, P.; Zannoni, G.; Bovey, F. A. *Macromolecules* **1978**, *11*, 923.
30. Ewen, J. A.; Elder, M. J.; Jones, R. L.; Curtis, S.; Cheng, H. N. in *Catalytic Olefin Polymerization*; Keii, T., Soga, K., Eds.; Kodansha: Tokyo, 1990; p. 439.
31. Herfert, N.; Fink, G.; *Makromol. Chem., Macromol. Symp.* **1993**, *66*, 157.
32. Fink, G.; Herfert, N.; Montag, P. in *Ziegler Catalysts*; Fink, G., Mülhaupt, R., Brintzinger, H.-H., Eds.; Springer-Verlag: Berlin, 1995; p. 159.
33. Rieger, B.; Jany, G.; Fawzi, R.; Steimann, M. *Organometallics* **1994**, *13*, 647.
34. Guerra, G.; Cavallo, L.; Moscardi, G.; Vacatello, M.; Corradini, P. *Macromolecules* **1996**, *29*, 4834.
35. Fischer, D.; Mülhaupt, R. *J. Organomet. Chem.* **1991**, *417*, C7.
36. Richardson, D. E.; Alameddin, N. G.; Ryan, M. F.; Hayes,T.; Eyler, J. R.; Siedle, A. J. *J. Am. Chem. Soc.* **1996**, *118*, 11244.
37. Støvneng, J. A.; Rytter, E. *J. Organomet. Chem.* **1996**, *519*, 277.

38. Jordan, R. F.; LaPointe, R. E.; Bradley, P. K.; Baezinger, N. *Organometallics* **1989**, *8*, 2892.
39. Horton, A. D. *Organometallics* **1992**, *11*, 3271.
40. Eshuis, J.J.W.; Tan, Y.Y.; Meetsma, A.; Teuben, J.H.; Renkema, J.; Evens, G.G. *Organometallics* **1992**, *11*, 362.
41. Tjaden, E. B.; Casty, G. L.; Stryker, J. *J. Am. Chem. Soc.* **1993**, *115*, 9814.
42. Feichtinger, D.; Plattner, D.A.; Chen, P. *J. Am. Chem. Soc.*, **1998**, *120*, 7125.
43. Margl, P.; Woo, T.K.; Ziegler, T. *Organometallics*, **1998**, *17*, 4997.
44. Karol, F.J.; Kao, S.; Wasserman, E.P.; Brady, R.C. *New J. Chem.* **1997**, *21*, 797.
45. Wasserman, E.; Hsi, E.; Young, W.-T. *Polym. Prepr. Am. Chem. Soc., Div. Polym. Chem.* **1998**, *39(2)*, 425.
46. Abrams, M. B.; Yoder, J. C.; Loeber, C.; Day, M. W.; Bercaw, J. E. *Organometallics* **1999**, *18*, 1389.
47. See for example: Chadwick, J. C.; van Kessel, G. M. M.; Sudmeijer, O. *Macromol. Chem. Phys.* **1995**, *196*, 1431 and references therein.
48. Pino, P.; Cioni, P.; Wei, J. *J. Am. Chem. Soc.* **1987**, *109*, 6189.
49. Tsutsui, T.; Kashiwa, N.; Mizuno, A. *Makromol. Chem., Rapid Commun.* **1990**, *11*, 565.
50. Kashiwa, N.; Kioka, M. *Polym. Mat. Sci. Eng.* **1991**, *64*, 43.
51. Kioka, M.; Mizuno, A.; Tsutsui T.; Kashiwa, N. in: Catalysis in Polymer Synthesis, eds. E.J. Vandenberg, J.C. Salamone (ACS Symposium Series 496, American Chemical Society, Washington, DC, 1992) p. 72.
52. Busico, V.; Cipullo, R.; Corradini, P. *Makromol. Chem., Rapid Commun.* **1993**, *14*, 97.
53. Busico, V.; Cipullo, R.; Corradini, P. *Makromol. Chem.* **1993**, *194*, 1079.
54. Carvill, A.; Tritto, I.; Locatelli, P.; Sacchi, M. C. *Macromolecules* **1997**, *30*, 7056.
55. Guyot, A.; Spitz, R.; Journaud, C. in *Studies in Surface Science and Catalysis, vol. 89: Catalyst Design for Tailor-Made Polyolefins*; Soga, K., Terano, M., Eds.; Elsevier: Amsterdam, 1994; p. 43.
56. Moscardi, G.; Piemontesi, F.; Resconi, L. *Organometallics* accepted for publication.

Controlled Radical Polymerization

The discovery of controlled radical polymerization techniques is an extremely important recent development in organometallic polymer synthesis. While free radical polymerization is the most commonly used industrial method for preparing commercial polymers, the high reactivity of the propagating species renders the polymerizations susceptible to many types of unwanted chain transfer and termination reactions.

In controlled radical polymerization, the active radical endgroup is maintained in equilibrium with a dormant organometallic species, supplied to the reaction medium as a transition metal complex capable of undergoing reversible one-electron oxidation. Since the concentration of the actual propagating radical at any given time is low, side reactions are suppressed to the extent that these polymerizations can exhibit "living" behavior under mild conditions. Following the discovery of controlled radical polymerization (often referred to as Atom Transfer Radical Polymerization or ATRP) in 1995, a number of well-controlled polymerizations of styrenic and acrylic monomers based on a variety of transition metals have been reported.

This section first features contributions from two of the original developers of controlled radical polymerization. Sawamoto et al. discuss the use of a new initiator, rhenium(V), for the well-defined polymerization of acrylic monomers. A comprehensive review of ligands used for copper(I)-mediated ATRP is presented by Matyjaszewski et al. Patten and Levy report progress in determining the structure-reactivity relationships of a subset of these initiators, copper(I) bipyridine halides, and Haddleton et al. discuss the application of ATRP to solid-supported transition metal catalysts. Finally, a chapter focusing on mechanistic elucidations of a second type of controlled radical methodology, catalytic chain transfer polymerization, is presented by Davis, Heuts, and Forster.

Chapter 12

Living Radical Polymerization of Acrylates with Rhenium(V)-Based Initiating Systems: $ReO_2I(PPh_3)_2$/Alkyl Iodide

Hiroko Uegaki, Yuzo Kotani, Masami Kamigaito, and Mitsuo Sawamoto[1]

Department of Polymer Chemistry, Graduate School of Engineering, Kyoto University, Kyoto 606–8501, Japan

Rhenium(V) iododioxobis(triphenylphosphine) [$ReO_2I(PPh_3)_2$], a group 7 transition metal complex, induced living radical polymerizations of methyl and n-butyl acrylates in conjunction with an iodide initiator such as $CH_3CH(Ph)I$ and $(CH_3)_2C(CO_2Et)I$ in the presence of $Al(Oi\text{-}Pr)_3$, where the reaction was faster than the $NiBr_2(Pn\text{-}Bu_3)_2$-mediated The number-average molecular weights of the obtained polymers increased in direct proportion to monomer conversion, and the molecular weight distributions were relatively narrow throughout the polymerizations ($M_w/M_n \sim 1.6$). The polymerization most probably proceeds via the rhenium-catalyzed homolytic cleavage of the polymer C–I terminal originated from the iodide initiator, as indicated by the ^1H NMR analysis of the polymer terminal structure and quenching study of the polymerization with a stable nitroxide radical. In contrast, a bromide initiator like CCl_3Br led to an uncontrolled acrylate polymerization with the rhenium(V) complex. The iodide/$ReO_2I(PPh_3)_2$ initiating system also induced a fast polymerization of methyl methacrylate, though the polymerization was not controlled.

Introduction

Transition metal catalysts play important roles in precision control in polymerization. They can permit control of molecular weights, molecular weight distributions (MWDs), steric structure of main chain, and sometimes sequence of monomer units in various modes of polymerizations for many monomers. Radical polymerizations is now among these metal-catalyzed precision processes, although it has been believed difficult to control because of highly reactive radical intermediates and their side reactions such as bimolecular termination (*1*), as with radical addition reactions for small molecule synthesis (*2*). Thus, the emergence of transition metal catalysts has changed the recognition of radical reactions from uncontrollable to

[1]Corresponding author.

controllable. Owing to their variable oxidation states, these complexes can generally serve as effective oxidants or reductants to generate radical species from organic precursors. For example, a ruthenium complex, $RuCl_2(PPh_3)_3$, induces controlled radical addition reactions between CCl_4 and an alkene, where one of the C–Cl bonds in the former is cleaved with the aid of the oxidation of Ru(II) into Ru(III) to generate $CCl_3\cdot$, which subsequently adds to the double bond of the olefin to give a 1:1 adduct effectively along with the reduction of the Ru(III) center (3). During the addition reaction, therefore, the ruthenium center undergoes a reversible redox reaction between the divalent and trivalent states. Such addition reactions are widely employed for inter- and intramolecular reactions for organic synthesis (2).

Employing similar catalysts, we have recently developed a series of transition metal-mediated living radical polymerizations that precisely control molecular weights and molecular weight distributions (MWDs) of polymers for methacrylates, acrylates, styrenes, etc (4). The processes were initially reported for the initiating systems that consist of an organic halide like CCl_4 as an initiator and a metal catalyst such as $RuCl_2(PPh_3)_3$ (5, 6). Following the incipient works, such metal-catalyzed processes vastly developed in terms of the scope of usable catalysts (metals and ligands) and monomers. Effective metals thus far reported include not only Ru (5–9) but also Cu (10–14), Fe (15–17), Ni (18–21), Rh (22–24), and Pd (25).

Therein the metals complexes mediate a controlled sequence of the formation of radicals from the C–halogen terminals derived from the initiator, their addition to the monomers, and the regeneration of the analogous C–halogen terminals (eq 1). In these living processes, the equilibrium between the dormant and the active or radical species is shifted to the dormant side, and thereby the concentration of the growing radical species is most probably kept so low as to diminish bimolecular radical termination and other side reactions.

$$R-X \xrightleftharpoons{Ru(II)} R^\bullet \; XRu(III) \xrightarrow[-Ru(II)]{CH_2=C(R^1)(R^2)} R-CH_2-C(R^1)(R^2)-X \xrightleftharpoons{CH_2=C(R^1)(R^2)}_{Ru(II)}$$

$$R\text{\textasciitilde\textasciitilde\textasciitilde}CH_2-C(R^1)(R^2)-X \xrightleftharpoons{Ru(II)} R\text{\textasciitilde\textasciitilde\textasciitilde}CH_2-C^\bullet(R^1)(R^2) \; XRu(III) \qquad (1)$$

Dormant Species   **Active Species**

As evident from eq 1, the dormant-active species equilibrium depends on the nature of halogens, monomers, and metal complexes along with other factors. This in turn means that the halogens and the metal complexes should be carefully selected for a particular monomer so as to affect the reversible homolytic cleavage of the dormant carbon–halogen bonds. The halogens thus far generally employed were chlorine and bromine, and on the other hand, most of the complexes were based on late transition metals (4). Quite recently, we have reported a $ReO_2I(PPh_3)_2$-catalyzed living radical polymerization of styrene in conjunction with an iodo-initiator $CH_3CH(Ph)I$ (26). This system is unique in that it involves rhenium, a group 7 metal that clearly differs from the group 8–10 counterparts. Specifically for the styrene polymerizations (*26*), for example, the Re(V) complex is more active and gives narrower MWDs of polystyrene than does $RuCl_2(PPh_3)_3$ under similar conditions.

In this study, we employed $ReO_2I(PPh_3)_2$ for possible living radical polymerization of methyl and *n*-butyl acrylates (MA and BA, respectively) as well as methyl methacrylate (MMA), in conjunction with an alkyl iodide or bromide as an initiator (**1–5**; Scheme 1). Specifically, the iodide (**1** or **2**)/$ReO_2I(PPh_3)_2$ initiating system proved effective in living radical polymerization of acrylates in the presence of $Al(Oi-Pr)_3$.

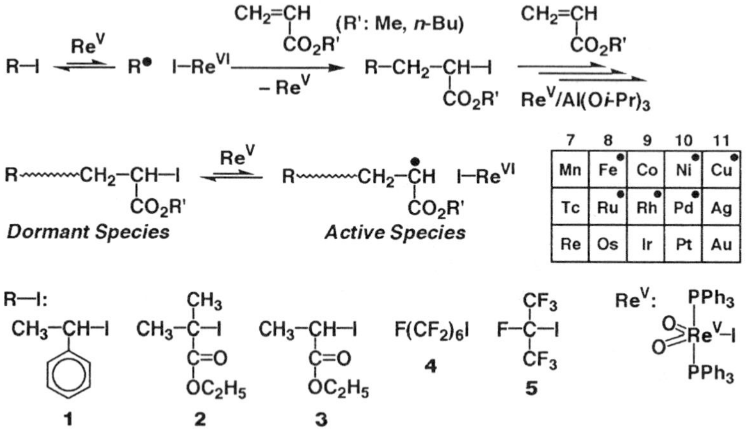

Scheme 1. Living radical polymerization of acrylates with R–I/Re(V) complex. The dots in the attached periodic table show the metals whose complexes have thus far been shown effective for living radical polymerization..

Results and Discussion

Living Radical Polymerization of Methyl Acrylate

Methyl acrylate was polymerized with $ReO_2I(PPh_3)_2$ in conjunction with a bromide (CCl_3Br) or an iodide [$CH_3CH(Ph)I$] in the presence of $Al(Oi-Pr)_3$ in toluene at 80 °C (Figure 1). The bromide has proved effective for MA living polymerization with $NiBr_2(Pn-Bu_3)_2$ (*20*), while the iodide for styrene with $ReO_2I(PPh_3)_2$ (*26*). Polymerizations occurred smoothly without an induction phase, and the bromide initiator induced a faster polymerization than the iodide. With the former initiator, $ReO_2I(PPh_3)_2$ is apparently more active than a nickel complex [$NiBr_2(Pn-Bu_3)_2$], which induced living radical polymerization of MA to give relatively narrow MWDs ($M_w/M_n \sim 1.4$) (*20*); note that therein the concentrations of the rhenium complex was lower (10 vs 80 mM). Thus, the rhenium complex is highly active for the polymerization of MA as well as styrene (*26*).

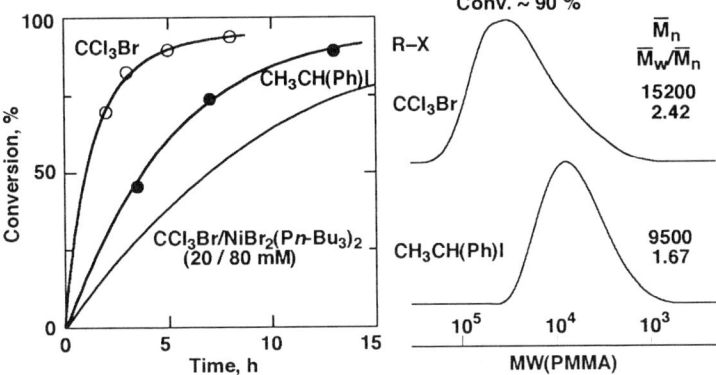

Figure 1. Polymerization of MA with R–X/ReO$_2$I(PPh$_3$)$_2$/Al(Oi-Pr)$_3$ in toluene at 80 °C: [M]$_0$ = 2.0 M; [R–X]$_0$ = 20 mM; [ReO$_2$I(PPh$_3$)$_2$]$_0$ = 10 mM; [Al(Oi-Pr)$_3$]$_0$ = 40 mM. R–X: (●) CH$_3$CH(Ph)I; (○) CCl$_3$Br.

Figure 1 also shows that, for the Re(V) catalyst, the iodide is better suited than the bromide, giving narrower MWDs and controlled molecular weights (see below also). With the bromide, the M_n slightly decreased with conversion [M_n = 21600 (70%); 22500 (82%); 15200 (94%)].

Figure 2 shows the M_n, M_w/M_n, and MWDs of poly(MA) obtained with the iodide at varying monomer conversion. The M_n increased in direct proportion to conversion and stayed close to the calculated values, one polymer chain per CH$_3$CH(Ph)I molecule. The slight deviation in M_n is probably due to the molecular weight determination by size-exclusion chromatography (SEC) with poly(MMA) standard samples. The polydispersity ratios, M_w/M_n, were around 1.7 throughout the polymerizations.

Iodides (2–5) other than CH$_3$CH(Ph)I (1) were also employed as possible initiators with the Re(V) complex (Figure 3). These iodides were unimer-models of poly(methacrylates) (2) and poly(acrylates) (3) with a C–I terminal or perfluoroalkyl iodides (4 and 5). They led to smooth MA polymerizations at nearly the same rates.

The polymers obtained with 2 were similar to those with 1, whose molecular weights increased with conversion and were close to the calculated values. In contrast, 3 gave higher M_n values (open triangles) than the calculated, although it is a unimer-model of poly(acrylates). Probably, the C–I bond of 3 is more difficult to be dissociated to radicals than that of the dormant polymer terminal, which results in a slow initiation from 3, as observed in the RuCl$_2$(PPh$_3$)$_3$-catalyzed MMA living radical polymerization with a unimer-model chloride (27). The perfluoroalkyl iodides 4 and 5 were not effective as initiators, giving higher M_n and broader MWDs, although their carbon–iodine linkages seem to generate radicals efficiently as degenerative chain-transfer agents (28, 29). These results indicate that ReO$_2$I(PPh$_3$)$_2$ is a good catalyst for living radical polymerization of MA with an appropriate iodide as an initiator.

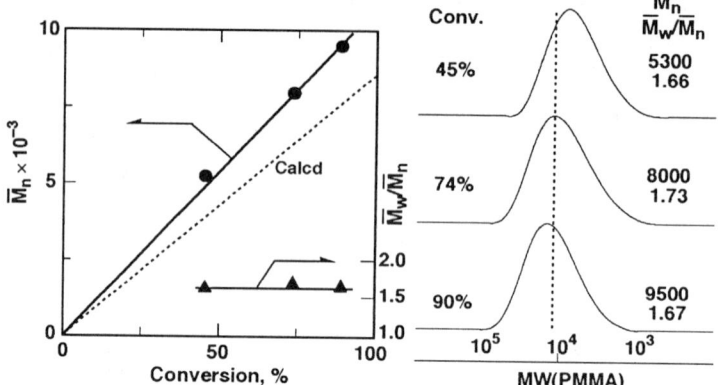

Figure 2. M_n, M_w/M_n, and SEC curves of poly(MA) obtained with $CH_3CH(Ph)I$/ $ReO_2I(PPh_3)_2$/$Al(Oi-Pr)_3$ in toluene at 80 °C: $[M]_0$ = 2.0 M; $[CH_3CH(Ph)I]_0$ = 20 mM; $[ReO_2I(PPh_3)_2]_0$ = 10 mM; $[Al(Oi-Pr)_3]_0$ = 40 mM. The diagonal dashed line indicates the calculated M_n assuming the formation of one living polymer per $CH_3CH(Ph)I$ molecule.

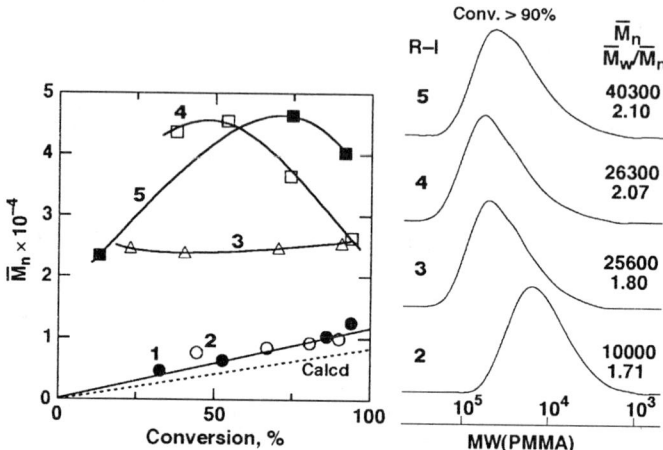

Figure 3. M_n, M_w/M_n, and SEC curves of poly(MA) obtained with R–I/$ReO_2I(PPh_3)_2$/ $Al(Oi-Pr)_3$ in toluene at 80 °C: $[M]_0$ = 2.0 M; $[R-I]_0$ = 20 mM; $[ReO_2I(PPh_3)_2]_0$ = 10 mM; $[Al(Oi-Pr)_3]_0$ = 40 mM. (●) 1; (○) 2; (△) 3; (□) 4; (■) 5 (see Scheme 1). The diagonal dashed line indicates the calculated M_n assuming the formation of one living polymer per R–I molecule.

Living Radical Polymerization of *n*-Butyl Acrylate

The rhenium(V) complex was also employed for BA polymerizations in conjunction with $CH_3CH(Ph)I$ and $Al(Oi\text{-}Pr)_3$ in toluene at 80 °C. The polymerization occurred smoothly to reach 91% conversion in 20 h. As shown in Figure 4, the M_n increased in direct proportion to monomer conversion, similarly to those of poly(MA) obtained under similar conditions (cf. Figure 2), although the molecular weights were higher than the calculated values. The MWDs were relatively narrow throughout the polymerization and became narrower with conversion. Thus, the $CH_3CH(Ph)I/ReO_2I(PPh_3)_2$ initiating system also induced living radical polymerization of BA.

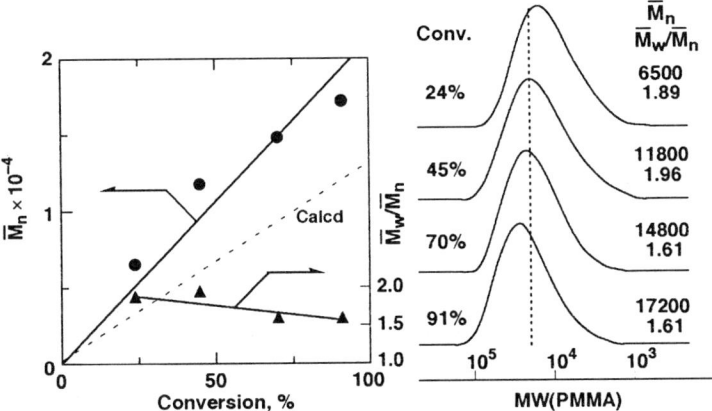

Figure 4. M_n, M_w/M_n, and SEC curves of poly(BA) obtained with $CH_3CH(Ph)I/ReO_2I(PPh_3)_2/Al(Oi\text{-}Pr)_3$ in toluene at 80 °C: $[M]_0 = 2.0$ M; $[CH_3CH(Ph)I]_0 = 20$ mM; $[ReO_2I(PPh_3)_2]_0 = 10$ mM; $[Al(Oi\text{-}Pr)_3]_0 = 40$ mM. The diagonal dashed line indicates the calculated M_n assuming the formation of one living polymer per $CH_3CH(Ph)I$ molecule.

Polymerization of Methyl Methacrylate

We then used the Re(V) complex for MMA polymerization in conjunction with CCl_3Br or $CH_3CH(Ph)I$ in the presence of $Al(Oi\text{-}Pr)_3$ in toluene at 80 °C. The catalyst was also active for MMA to induce faster polymerizations than the nickel-based system under the same conditions (Figure 5). The molecular weights of the obtained polymers were higher than the calculated values and decreased with conversion. The MWDs were broad throughout the polymerizations with both initiators. Other rhenium complexes such as $ReOCl_3(PPh_3)_2$ and $ReCl_3(PMe_2Ph)_3$ also resulted in uncontrolled polymerizations of MMA.

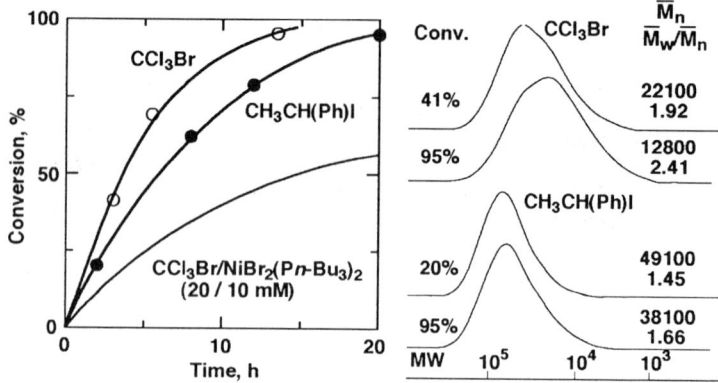

Figure 5. Polymerization of MMA with R–X/ReO$_2$I(PPh$_3$)$_2$/Al(Oi-Pr)$_3$ in toluene at 80 °C: [M]$_0$ = 2.0 M; [R–X]$_0$ = 20 mM; [ReO$_2$I(PPh$_3$)$_2$]$_0$ = 10 mM; [Al(Oi-Pr)$_3$]$_0$ = 40 mM R–X: (●) CH$_3$CH(Ph)I; (○) CCl$_3$Br.

Quenching Experiments

The effects of potential terminators [TEMPO (2,2,6,6-tetramethyl-1-piperidinoxyl), methanol, and water] were examined to clarify the nature of the Re(V)-mediated living polymerization of acrylates. Thus, the polymerization was first run without additives, where conversion reached 49% in 5.5 h and 71% in 17.5 h (open circles in Figure 6). At 49% conversion, the inhibitors (10 equiv each to the initiator) were added to the reaction mixtures. The polymerizations were not affected by methanol or water, proceeding without significant changes in rate, molecular weights, and MWDs relative to the additive-free system. In contrast, TEMPO quenched MA consumption almost completely, where the conversion remained unchanged (51%) in an additional 12 h. There were almost no effects on the molecular weights and MWDs, either. These may suggest that the polymerization proceeds via a radical mechanism.

Polymer Terminal Structure

Figure 7 shows the ^{1}H NMR spectrum of a typical sample of the poly(MA) obtained with CH$_3$CH(Ph)I/ReO$_2$I(PPh$_3$)$_2$/Al(Oi-Pr)$_3$. It shows the characteristic signals of the initiator moiety; e.g., peaks c–e were attributed to the aromatic protons of the phenylethyl group at the α-end. The number-average degree of polymerization based on the absorptions of the α-end terminal and the main-chain methyl ester was 90 [DP$_n$(NMR) = (h + h')/3e], which is higher than the calculated value [DP$_n$(calcd) = 71] based on the initial ratio of the monomer to the initiator and conversion by gas chromatography. The difference is apparently due to some loss of low molecular weight fractions in purification of the polymers by preparative SEC, as indicated by

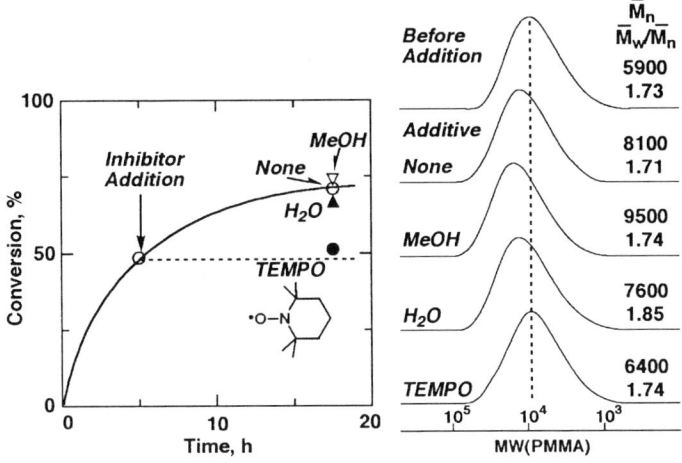

Figure 6. Effects of additives (200 mM) on polymerization of MA with $CH_3CH(Ph)I/ReO_2I(PPh_3)_2/Al(Oi-Pr)_3$ in toluene at 80 °C: $[M]_0 = 2.0$ M; $[CH_3CH(Ph)I]_0 = 20$ mM; $[ReO_2I(PPh_3)_2]_0 = 10$ mM; $[Al(Oi-Pr)_3]_0 = 40$ mM. Additives: (●) TEMPO; (∇) MeOH; (▲) H_2O; (○) none. Each additive was added to the polymerization mixture when conversion reached 49% in 5.5 h.

the increase in M_n(SEC) from 8100 to 8800. Another cause may be a low initiation efficiency of the iodide due to some side reactions. The iodo ω-end terminal was seen at 4.3 ppm (peak *i*), which supports the proposed reversible activation of the C–I terminal derived from $CH_3CH(Ph)I$. However, the ratio of the ω- to the α-ends was 0.63, suggesting a partial loss of the living end during the polymerization or the work-up process. Along with this, there appeared some olefinic protons around 6 ppm. These observations indicate that the C–I terminal is less stable than the corresponding C–Br or C–Cl terminal.

Conclusion

This study shows that the complex based on a group 7 metal is also effective in acrylates living radical polymerization with the use of an appropriate initiator. Specifically, the rhenium(V)-complex, $ReO_2I(PPh_3)_2$, induced living radical polymerization of acrylates specifically in conjunction with an iodide initiator such as $CH_3CH(Ph)I$ in the presence of $Al(Oi-Pr)_3$ to give polymers with controlled molecular weights and MWDs, similarly to the polymerization of styrene (26). In contrast, the system is not effective in controlling methacrylate polymerization. This is probably due to that such an active C–halogen terminal as C–I is not suited for the polymerizations of α-methyl-substituted monomers. $ReO_2I(PPh_3)_2$ shows a higher catalytic activity than $NiBr_2(Pn-Bu_3)_2$ but gives broader MWDs.

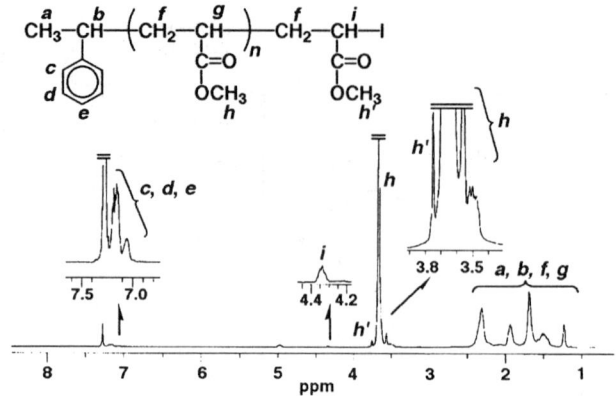

Figure 7. ^1H NMR spectrum of poly(MA) obtained with $CH_3CH(Ph)I/ReO_2I(PPh_3)_2/Al(Oi-Pr)_3$ in toluene at 80 °C: $[M]_0 = 2.0$ M; $[CH_3CH(Ph)I]_0 = 20$ mM; $[ReO_2I(PPh_3)_2]_0 = 10$ mM; $[Al(Oi-Pr)_3]_0 = 40$ mM. $M_n(NMR) = 8000$; $M_n(SEC) = 8800$.

Experimental

Materials

MA, BA, and MMA (all Tokyo Kasei; purity >99%) was dried overnight over calcium chloride and distilled twice over calcium hydride under reduced pressure before use. $ReO_2I(PPh_3)_2$ (Aldrich), $ReOCl_3(PPh_3)_2$ (Aldrich), $ReCl_3(PMe_2Ph)_3$ (Aldrich, purity >97%), $NiBr_2(Pn-Bu_3)_2$ (Aldrich; purity >97%) and $Al(Oi-Pr)_3$ (Aldrich; purity >99.99%) were used as received and handled in a glove box under a moisture- and oxygen-free nitrogen atmosphere (H_2O <1 ppm; O_2 <1 ppm). $CH_3CH(Ph)I$, $(CH_3)_2C(CO_2Et)I$, and $(CH_3)CH(CO_2Et)I$ were prepared as reported (26). $F(CF_2)_6I$ and $(CF_3)_2CFI$ (both Daikin Industries) were used as received. CCl_3Br (Wako Chemicals; purity > 99%) was doubly distilled over calcium chloride before use. Toluene (solvent) and n-octane (internal standard for gas chromatography) were dried overnight over calcium chloride, distilled twice over calcium hydride, and bubbled with dry nitrogen for more than 15 min immediately before use.

Polymerization Procedures

Polymerization was carried out by the syringe technique under dry nitrogen in baked glass tubes equipped with a three-way stopcock or in baked and sealed glass vials. In a 50-mL round-bottomed flask was placed $ReO_2I(PPh_3)_2$ (87 mg), and toluene (3.43 mL), n-octane (1.17 mL), MA (1.80 mL), and solutions of $CH_3CH(Ph)I$ (500 mM in toluene; 0.40 mL) and $Al(Oi-Pr)_3$ (125 mM in toluene; 3.20 mL) were added sequentially in this order at room temperature under dry nitrogen. The total volume of the reaction mixture was thus 10.0 mL. Immediately after mixing, aliquots (1.20 mL each) of the solution were injected into baked glass tubes, which were then sealed (except when a stopcock was used) and placed in an oil bath kept at 80 °C. In predetermined intervals, the polymerization was terminated by cooling the reaction mixtures to –78 °C. Monomer conversion was determined from the concentration of residual monomer measured by gas chromatography with n-octane as an internal standard. The quenched reaction solutions were diluted with toluene (ca. 20 mL) and rigorously shaken with an absorbent [KYOWAAD-2000G-7 ($Mg_{0.7}Al_{0.3}O_{1.15}$); Kyowa Chemical Industry] (ca. 5 g) to remove the metal-containing residues. After the absorbent was separated by filtration (Whatman 113V), the filtrate was washed with water and evaporated to dryness to give the products, which were subsequently dried overnight under vacuum at room temperature.

Measurements

The MWD, M_n, and M_w/M_n ratios of the polymers were measured by size-exclusion chromatography (SEC) in chloroform at 40 °C on three polystyrene gel columns (Shodex K-805L × 3) that were connected to a Jasco PU-980 precision pump and a Jasco RI-930 refractive index detector. The columns were calibrated against twelve standard poly(MMA) samples (Polymer Laboratories; M_n = 630–1200000; M_w/M_n = 1.06–1.22) as well as the monomer. ^{1}H NMR spectra were recorded in $CDCl_3$ at 25 °C on a JEOL JNM-LA500 spectrometer, operating at 500.16 MHz. Polymers for NMR analysis were fractionated by preparative SEC (column: Shodex K-2002) to be freed from low molecular impurities originated from the catalysts.

Acknowledgment

With appreciation M.S. and M.K. acknowledge the support from the New Energy and Industrial Technology Development Organization (NEDO) under the Ministry of International Trade and Industry (MITI), Japan, through the grant for "Precision Catalytic Polymerization" in the Project "Technology for Novel High-Functional Material" (1996–2000).

References and Notes

1. For a recent review on radical polymerizations, see: Moad, G.; Solomon, D. H. *The Chemistry of Free Radical Polymerization*; Elsevier Science: Oxford, U.K., 1995.
2. For recent reviews on metal-mediated controlled radical reactions, see: (a) Curran, D. P.; Porter, N. A.; Giese, B. *Stereochemistry of Radical Reactions*; VCH: Weinheim, Germany, 1996. (b) Iqbal, J.; Bhatia, B.; Nayyar, N. K. *Chem. Rev.* **1994**, *94*, 519.
3. Matsumoto, H.; Nakano, T.; Nagai, Y. *Tetrahedron Lett.* **1973**, *51*, 5147.

4. For recent reviews on the transition-metal mediated living radical polymerizations, see: (a) Sawamoto, M.; Kamigaito, M. *Trends Polym. Sci.* **1996**, *4*, 371. (b) *Controlled Radical Polymerization;* Matyjaszewski, K., Ed.; ACS Symposium Series 685; American Chemical Society: Washington, DC, 1998. (c) Sawamoto, M.; Kamigaito, M. *Synthesis of Polymers (Materials Science and Technology Series)*, Schlüter, A.-D., Ed.; Wiley-VCH, Weinheim, Germany, 1999; Chapter 6. (d) Sawamoto, M.; Kamigaito, M. *CHEMTECH* **1999**, *29* (6), 30.
5. Kato, M.; Kamigaito, M.; Sawamoto, M.; Higashimura, T. *Macromolecules* **1995**, *28*, 1721.
6. Ando, T.; Kato, M.; Kamigaito, M.; Sawamoto, M. *Macromolecules* **1996**, *29*, 1070.
7. Takahashi, H.; Ando, T.; Kamigaito, M.; Sawamoto, M.; *Macromolecules* **1999**, *32*, 3820.
8. Takahashi, H.; Ando, T.; Kamigaito, M.; Sawamoto, M.; *Macromolecules*, in press.
9. Simal, F.; Demonceau, A.; Noels, A. F. *Angew. Chem. Int. Ed.* **1999**, *38*, 538.
10. Wang, J.-S.; Matyjaszewski, K. *J. Am. Chem. Soc.* **1995**, *117*, 5614.
11. Patten, T. E.; Xia, J.; Abernathy, T.; Matyjaszewski, K. *Science* **1996**, *272*, 866.
12. Percec, V.; Barboiu, B. *Macromolecules* **1995**, *28*, 7970.
13. Percec, V.; Barboiu, B.; Kim, H.-J. *J. Am. Chem. Soc.* **1998**, *120*, 305.
14. Haddleton, D. M.; Jasieczek, C. B.; Hannon, M. J.; Shooter, A. J. *Macromolecules* **1997**, *30*, 2190.
15. Ando, T.; Kamigaito, M.; Sawamoto, M. *Macromolecules* **1997**, *30*, 4507.
16. Matyjaszewski, K.; Wei, M.; Xia, J.; McDermott, N. E. *Macromolecules* **1997**, *30*, 8161.
17. Kotani, Y.; Kamigaito, M.; Sawamoto, M.; *Macromolecules*, in press.
18. Granel, C.; Dubois, Ph.; Jérôme, R.; Teyssié, Ph. *Macromolecules* **1996**, *29*, 8576.
19. Uegaki, H.; Kotani, Y.; Kamigaito, M.; Sawamoto, M. *Macromolecules* **1997**, *30*, 2249.
20. Uegaki, H.; Kotani, Y.; Kamigaito, M.; Sawamoto, M. *Macromolecules* **1998**, *31*, 6756.
21. Uegaki, H.; Kamigaito, M.; Sawamoto, M. *J. Polym. Sci., Part A, Polym. Chem.* **1999**, *37*, 3003.
22. Percec, V.; Barboiu, B.; Neumann, A.; Ronda, J. C.; Zhao, M. *Macromolecules* **1996**, *29*, 3665.
23. Moineau, G.; Granel, C.; Dubois, Ph.; Jérôme, R.; Teyssié, Ph. *Macromolecules* **1998**, *31*, 542.
24. Petrucci, M. G. L.; Lebuis, A.-M.; Kakkar, A. K. *Organometallics* **1998**, *17*, 4966.
25. Lecomte, Ph.; Draiper, I.; Dubois, Ph.; Teyssié, Ph.; Jérôme, R. *Macromolecules* **1997**, *30*, 7631.
26. Kotani, Y.; Kamigaito, M.; Sawamoto, M. *Macromolecules* **1999**, *32*, 2420.
27. Ando, T.; Kamigaito, M.; Sawamoto, M. *Polym. Prepr. Jpn.* **1997**, *46* (8), 1502
28. Ueda, N.; Kamigaito, M.; Sawamoto, M. *Polym. Prepr. Jpn.* **1996**, *45* (7), 1267.
29. Gaynor, S. G.; Wang, J.-S.; Matyjaszewski, K. *Macromolecules* **1995**, *28*, 8051.

Chapter 13

The Effect of Ligands on Copper-Mediated Atom Transfer Radical Polymerization

Jianhui Xia, Xuan Zhang, and Krzysztof Matyjaszewski[1]

Department of Chemistry, Carnegie Mellon University,
4400 Fifth Avenue, Pittsburgh, PA 15213

The use of different ligands in copper-mediated atom transfer radical polymerization (ATRP) is reviewed. Particular emphasis is placed on nitrogen-based ligands. Ligands are classified according to the coordination number. Tridentate and tetradentate ligands generally provide faster polymerizations than bidentate ligands, while monodentate nitrogen ligands yield redox-initiated free radical polymerization. In addition, ligands with an ethylene linkage between the nitrogens are more efficient than those with a propylene or butylene linkage. Electronic and steric effects are important and reduced catalytic activity or efficiency is observed when there is excessive steric hindrance around the metal center or the ligand has strongly electron-withdrawing substituents.

Introduction

Controlled/'living' radical polymerization has been a field of intensive research in recent years. Several new methodologies have been developed to gain better control over molecular weight, molecular weight distribution, and architectures of polymers *(1)*. Transition metal catalyzed atom transfer radical polymerization (ATRP) is one of the most successful systems *(2)*. The basic concept of ATRP is based on atom transfer radical addition reaction (ATRA) in organic chemistry, which is sometimes referred to as Kharasch addition. In a typical ATRP system, alkyl halides are used as the initiators, transition metals with multiple accessible oxidation states are used as the catalysts, and ligands, typically amines or phosphines, are used to increase the solubility of the transition metal salts in solution and to tune the reactivity of the metal towards halogen abstraction. The polymerization involves the activation of the dormant species, P_n-X, through halogen abstraction by the transition metal in its lower oxidation state, M_t^n, to form propagating species, carbon-centered radical P•, and the transition metal in its higher oxidation state, $X-M_t^{n+1}$ (Scheme 1).

---

[1]Corresponding author.

The radical can add to alkene to form a new carbon-centered radical, PM•, or it can be deactivated through halogen transfer from the transition metal in its higher oxidation state to form dormant alkyl halide, PMX. If PMX can be reactivated by halogen abstraction, the alkene addition and halogen transfer reactions are repeated, and a polymer is obtained. Of course, more than one monomer unit can be incorporated in one activation step.

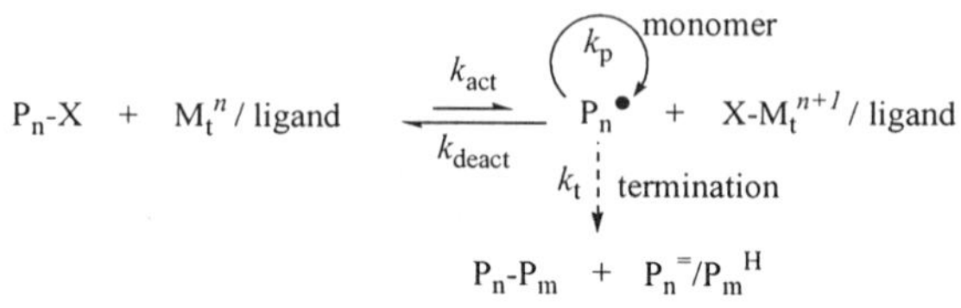

$$P_n\text{-}X + M_t^n/\text{ligand} \underset{k_{deact}}{\overset{k_{act}}{\rightleftarrows}} P_n^\bullet \overset{k_p\ (\text{monomer})}{} + X\text{-}M_t^{n+1}/\text{ligand}$$

$$\downarrow k_t\ \text{termination}$$

$$P_n\text{-}P_m + P_n^=/P_m^H$$

Scheme 1. Proposed mechanism for transition metal mediated ATRP.

The key to the control in ATRP is the dynamic equilibrium between the active species (propagating radical) and dormant species (alkyl halide) via halogen transfer. Several factors are important in order to achieve good polymerization control. First, fast and quantitative initiation ensures that all polymer chains start to grow at about the same time. When the contribution from chain termination and transfer reactions is significantly small, the degree of the polymerization (DP_n) can be predetermined by the ratio of consumed monomer and the initiator ($DP_n = \Delta[M]/[I]_o$). Second, the equilibrium between the alkyl halide and transition metal is strongly shifted to the dormant species. This equilibrium position will render most of growing polymer chains stored in the dormant species reservoir and a low radical concentration. As a result, the contribution of radical termination reactions to the overall polymerization is minimized. Third, fast deactivation of propagating radicals by halogen transfer assures that all polymer chains are growing at approximately the same rate leading to narrow molecular weight distributions. Fourth, a relatively fast activation of dormant polymer chains gives rise to a reasonable polymerization rate. Fifth, there should be no side reactions such as β-H abstraction or reduction/oxidation of radicals.

Originally, two ATRP systems were reported in 1995. Sawamoto et al. reported the controlled polymerization of methyl methacrylate (MMA) with $Ru(PPh_3)_3Br_2$ as the catalyst in conjunction with aluminum alkoxide as the activator *(3)*. Matyjaszewski et al. reported the controlled polymerizations of styrene and (meth)acrylates using CuX (X = Cl, Br) complexed by 2,2'-bipyridine (bpy) as the catalyst.*(4,5)* Since then, many new catalytic systems employing other transition metals, such as Ni *(6,7)* and Fe *(8,9)*, have been reported. Copper system is one of the most extensively studied and also one of the most promising in terms of cost and versatility. It has been applied toward the preparation of a large variety of well-defined polymers such as styrenics, (meth)acrylic esters, and acrylonitrile *(10)*. This brief overview describes the recent development in new catalytic systems in the copper-mediated ATRP, focusing on the design, synthesis and application of new nitrogen ligands.

One of the challenges in ATRP is to extend the scope of monomers that can be polymerized in a controlled manner. This can be accomplished through the development of more active catalysts. For a specific metal, catalytic activity is varied by the employment of different ligands. Thus, detailed studies of the effect of ligands can provide more insightful information in catalyst design.

Effect of the Ligand

Transition metal catalysts are the key to ATRP since they determine the position of the atom transfer equilibrium and the dynamics of exchange between the dormant and active species. There are several basic prerequisites for the transition metal. First, the metal center must have at least two readily accessible oxidation states separated by one electron. Second, the metal center should have reasonable affinity towards halogen. Third, the metal coordination sphere should be expandable upon oxidation to selectively accommodate a (pseudo)halogen.

The role of the ligand in ATRP is to solubilize the transition metal salts in organic media and to adjust the metal center for an appropriate reactivity and dynamics in halogen exchange. The electron donating ability of the ligand can greatly affect the redox potential of the transition metal complex and influence the reactivity of the metal center in halogen abstraction and transfer. In general, transition metal catalysts with low redox potentials favor the formation of copper(II) species and are active catalysts in ATRP with fast polymerization rate. In contrast, transition metal catalysts with high redox potentials disfavor the formation of copper(II) species and are not reactive in halogen abstraction. These transition metal catalysts tend to lead to slow or not well controlled polymerization. However, even in the most thoroughly studied copper/bpy catalytic system, the exact structure of the active species is not completely clear. The literature data on the coordination chemistry of copper species in nonpolar solvents are limited. In addition, both copper(I) and copper(II) species are quite labile in solution which renders structural studies difficult. Preliminary EPR and UV-Vis studies suggested the presence of quite complex species in polymerization solution. ^{1}H NMR studies indicate copper(I) coordinated by bpy is in fast exchange with the free ligand in solution. Combined with the literature data on the structure of copper/bpy complex in polar solvents, such as methanol and acetonitrile, it can be assumed that during ATRP the majority of the copper species can be best represented by a tetrahedral $Cu(I)/(bpy)_2$ *(11)* and a trigonal bipyramidal $XCu(II)/(bpy)_2$ *(12)* (Scheme 2). Other transition metal complexes with only halides as the ligands or bridged structures are also possible.

The effectiveness of the catalyst is greatly influenced by the choice of ligand. Nitrogen ligands generally work well for copper-mediated ATRP. In contrast, sulfur, oxygen or phosphorous ligands are less effective due to the inappropriate electronic effects or unfavorable binding constants. Many monodentate nitrogen ligands can provide coordination for copper; however, the resulting copper complexes do not promote successful ATRP. Our studies have been focused on the design and investigation of nitrogen ligands which will be the central topic in this report.

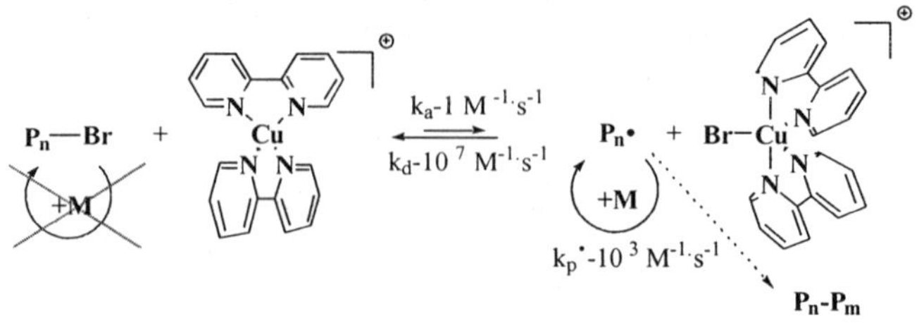

Scheme 2. Proposed copper(I) and copper (II) species using bpy as the ligand

The nitrogen donors in multidentate ligands can be classified in several different ways. 1. According to the electronic property of the coordinating nitrogen atom, ligands can be separated into those with aromatic nitrogen, aliphatic nitrogen, imine nitrogen, and mixed nitrogen ligands. 2. According to geometry, ligands can be separated into linear, cyclic and branched (e.g., tripodal) structures. 3. According to the coordination number, ligands can be separated into monodentate, bidentate, tridentate, tetradentate ligands, etc. Our studies have indicated that the coordination number of the nitrogen ligand plays the most important role in determining the control and rate of the polymerization (*vide infra*). Thus, ligands in this paper will be discussed in the following three major categories: bidentate, tridentate and tetradentate nitrogen ligands. In most cases, the polymerization conditions were kept the same or similar to facilitate comparison; however, readers are cautioned not to place too much emphasis on comparing the rates of the polymerizations as these are all apparent and are affected by many intrinsic factors, such as copper(I) and copper(II) concentrations in the solution and the rate of deactivation of the growing radicals.

Bidentate Nitrogen Ligands

Bidentate nitrogen ligands are readily available and have been most systematically studied for copper-mediated ATRP. X-ray structures have been solved for some of the copper-ligand complexes. In general, complexes of copper(I) and bidentate ligands exhibit a distorted tetrahedral arrangement of two coordinating ligands surrounding the metal center. Kinetic studies of the homogeneous polymerizations of both styrene and MA suggest a 2/1 ratio of ligand to metal is needed to achieve maximum rate of polymerization *(13,14)*; however, a 1/1 ratio is enough for MMA under homogeneous conditions *(15)*. According to their electronic properties, bidentate ligands are discussed in the following subgroups: bipyridine ligands, diamine ligands, diimine ligands, and mixed ligands.

Bipyridine Derivatives

Bpy was the first ligand used to successfully promote copper-mediated ATRP of styrene and (meth)acrylates *(4,5)*. Linear semilogarithmic kinetic plots and linear increase of molecular weights with monomer conversion have been observed. The experimental molecular weights were close to the theoretical values as predicted by the following equation: $M_{n,Cal} = ([M]_o/[In]_o) \times (MW)_o \times$ conversion, where $[M]_o$ and $[In]_o$ represent the initial concentrations of monomer and initiator, and $(MW)_o$ is the molecular weight of the monomer. Polydispersities were quite narrow ($M_w/M_n \sim 1.2$ - 1.5). These results suggest that initiator efficiency was high, and the number of chains was approximately constant, which was consistent with the criteria for a controlled polymerization. However, the $CuX(bpy)_2$ catalytic system is heterogeneous in non-polar media which rendered further kinetic studies difficult. Reverse ATRP, which employs a conventional radical initiator such as azobisisobutyronitrile (AIBN) in the presence of copper(II) species to promote a controlled polymerization, was difficult with bpy as the ligand due to the low solubility of copper(II) species *(16)*. A large excess of bpy was used in an attempt to increase the solubility of copper(II) in the polymerization and the control of the polymerization was unsuccessful for (meth)acrylates under the reaction conditions.

4,4'-Di-*n*-heptyl-2,2'-bipyridine (dHbpy), 4,4'-di(5-nonyl)-2,2'-bipyridine (dNbpy) and 4,4'-di-*n*-nonyl-2,2'-bipyridine (bpy9) were synthesized to provide homogeneous copper complexes in ATRP (Figure 1) *(17,18)*. Polymers with very low polydispersities ($M_w/M_n \sim 1.05$) were prepared using the homogeneous catalysts *(17)*. This was attributed to the increased rate of deactivation of the growing radicals in solution. Kinetic studies were possible for these systems.*(13-15)* Reverse ATRP with AIBN as the initiator have been successfully carried out under homogeneous conditions for the polymerizations of styrene, MA and MMA *(19)*. When perfluoroalkyl side chains were attached to bpy (dR6*f*bpy, Figure 1), controlled polymerizations were performed in supercritical carbon dioxide *(20)*.

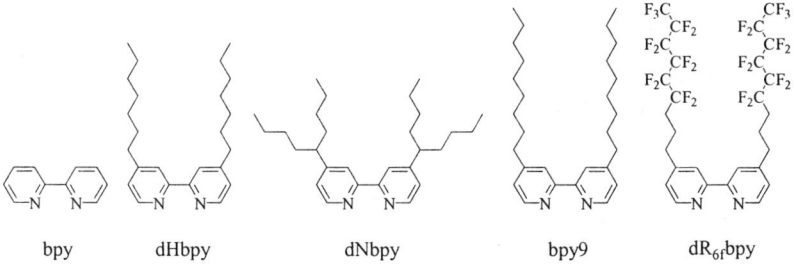

Figure 1. Structures of some bpy derivatives

The substitution on the 4,4' position of bpy has been found to greatly effect the rate and control of ATRP (Table 1).

Table 1. Cu-Mediated ATRP of Styrene in Bulk Using Different Bpy Ligands

entry	ligand	time (h)	conv (%)	$M_{n,Cal}$	$M_{n,SEC}$	M_w/M_n
1[a]	(bpy)	3.0	65	6500	6600	1.21
2[b]	($H_{15}C_7$, C_7H_{15} substituted bpy)	3.0	62	6200	6300	1.05
3[a]	(dialkyl bpy)	4.0	13	1300	52900	1.90
4[a]	(MeO_2C, CO_2Me substituted bpy)	18.5		No Polymer		

[a] 110 °C; [styrene]$_0$/[PEBr]$_0$ = 96; [PEBr]$_0$/[CuBr]$_0$/[CuBr]$_0$ = 1/1/3.
[b] 110 °C; [styrene]$_0$/[PEBr]$_0$ = 96; [PEBr]$_0$/[CuBr]$_0$/[CuBr]$_0$ = 1/1/2.

Electron-donating groups on bpy can form copper complexes with lower redox potentials which shift the equilibrium more towards radical formation, and hence lead to a faster polymerization rate. In contrast, electron-withdrawing groups on bpy provide copper complexes with higher redox potentials which shift the equilibrium to the dormant species. For example, copper(I) complexed by 4,4'-dicarboxylate-2,2'-bipyridine was so stable that it was difficult to abstract halide from alkyl halide or polymeric halide to form radicals. Steric effect is another important consideration in ligand design. Substitutions on the 6,6' positions of bpy leads to uncontrolled polymerization, largely due to the unfavorable steric effect for the formation of copper(II) species since it becomes more difficult for copper(II) to accept incoming halogens and to adopt the sterically more demanding triagonal bipyramidal configuration (12). Similar result has been previously reported (18).

1,10-Phenanthroline and its derivatives have also been used successfully as the ligands for copper-mediated ATRP of styrene (21). Solubilities of the copper complexes were found to be an important factor in controlling the polymerization. The lowest polydispersities were obtained under homogeneous conditions using 4,7-diphenyl-1,10-phenanthroline as the ligand and 1,2-dimethoxybenzene as the solvent.

Diamine Ligands

The most common diamine ligand is ethylenediamine (EDA) and its derivatives. They are inexpensive and readily available. Generally, copper complexes coordinated by unalkylated ligands have low solubilities in common organic solvents and lead to ill-controlled polymerizations. Alkylation of EDA greatly improves its ability to solubilize transition metals in organic solvents. Sterics plays a significant role in the catalytic activities of alkylated ethylenediamines. Compared to bpy, N,N,N',N'-tetramethylethylenediamine (TMEDA) leads to relatively controlled polymerizations of styrene, MA and MMA with relatively high polydispersities ($M_w/M_n \sim 1.4$) and low rates of polymerization *(22)*. Computer modeling illustrates the significant steric hindrance between the methyl groups on nitrogen upon halogen abstraction by the copper(I)/(TMEDA)$_2$ complex. When the sterics on the nitrogen atoms was increased further, as in the case of 2-(dimethylamino)ethyl morpholine or tetraethylethylenediamine, ill-controlled polymerizations were observed yielding polymers with much higher molecular weights than the predicted values and high polydispersities (Table 2).

Table 2. Steric Effect with Bidentate Amines in Cu-Mediated ATRP

ligand[a]	monomer	time (h)	conv (%)	$M_{n,Cal}$	$M_{n,SEC}$	M_w/M_n
TMEDA	styrene	3.65	48	4800	6300	1.37
TMEDA	MA	3.00	43	8600	10100	1.31
TMEDA	MMA	2.75	66	13200	12100	1.39
DMAEM	styrene	7.50	61	6100	14500	1.98
DMAEM	MA	1.40	46	9200	44100	2.53
DMAEM	MMA	6.15	52	10400	21600	1.51
TEEDA	styrene	1.00	42	4200	4200	1.98
TEEDA	MA	0.50	47	9400	30400	1.90
TEEDA	MMA	1.55	54	10800	17000	1.54

[a] Conditions: styrene: bulk; 110 °C; [styrene]$_0$/[PEBr]$_0$ = 96; [PEBr]$_0$/[CuBr]$_0$/[ligand]$_0$ = 1/1/2. MA: bulk; 110 °C; [MA]$_0$/[EBP]$_0$ = 232; [EBP]$_0$/[CuBr]$_0$/[ligand]$_0$ = 1/1/2. MMA: 50 vol % in anisole; 90 °C; [MMA]$_0$/[EBiB]$_0$ = 200; [EBiB]$_0$/[CuBr]$_0$/[ligand]$_0$ = 1/1/2. For experimental details regarding polymerizations, see Reference *23*.

It has to be recognized that ATRP equilibrium constants and propagation rate constants of styrene, MA and MMA are quite different. For example, the values of equilibrium constant using CuBr/(dNbpy)$_2$ as the catalyst are approximately $K_{eq} \sim 4 \times 10^{-8}$ (styrene at 110 °C), $K_{eq} \sim 1.2 \times 10^{-9}$ (MA at 90 °C), and $K_{eq} \sim 7 \times 10^{-7}$ (MMA at 90 °C). Under the same conditions, the values of rate constant are $k_p \sim 1.6 \times 10^3$ M^{-1}S^{-1} (styrene), $k_p \sim 5.0 \times 10^4$ M^{-1}S^{-1} (MA), and $k_p \sim 1.0 \times 10^3$ M^{-1}S^{-1} (MMA). Thus, in order to have a reasonable polymerization rate (e.g., 50% monomer conversion in ca. 3 h), it is necessary to adjust concentrations, temperature and dilution. The following conditions were typically used for the three standard monomers: styrene: bulk, 110 °C, [styrene]$_0$/[initiator]$_0$/[catalyst]$_0$ = 96/1/1; MA: bulk; 90 °C; [MA]$_0$/[initiator]$_0$/[catalyst]$_0$ = 232/1/1.; MMA: 50 vol % anisole; 90 °C;

$[MMA]_o/[initiator]_o/[catalyst]_o = 200/1/1$. In addition, initiators resembling dormant species were used. PEBr (PEBr = 1-phenylethyl bromide or (1-bromoethyl)benzene), EBP (EBP = ethyl 2-bromopropionate) and EBiB (EBiB = ethyl 2-bromoisobutyrate) were used for styrene, MA and MMA, respectively. Effect of the spacer length between the nitrogen ligands was also studied using TMEDA, N,N,N',N'-tetramethylpropylenediamine (TMPDA) and N,N,N',N'-tetramethylbutylenediamine (TMBDA). As shown in Table 3, with an increase of the number of carbons between the two nitrogens, the control of the polymerization progressively degraded with higher molecular weight polymers formed at very low monomer conversions. In the case of TMBDA, the polymerizations displayed behavior typical for redox-initiated free radical polymerization.

Table 3. Effect of Linker with Bidentate Amines in Cu-Mediated ATRP

ligand[a]	monomer	time (h)	conv (%)	$M_{n,Cal}$	$M_{n,SEC}$	M_w/M_n
TMEDA	styrene	3.65	48	4800	6300	1.37
TMEDA	MA	3.00	43	8600	10100	1.31
TMEDA	MMA	2.75	66	13200	12100	1.39
TMPDA	styrene	4.90	43	4300	16300	1.60
TMPDA	MA	1.30	59	11800	18100	1.77
TMPDA	MMA	1.40	30	6000	16500	1.44
TMBDA	styrene	0.50	14	1400	53000	2.23
TMBDA	MA	0.20	19	3800	106800	2.20
TMBDA	MMA	0.50	< 5	< 1000	65400	2.00
N(nBu)$_3$	styrene	1.30	< 5	< 500	85200	2.45

[a] See Table 2 for conditions.

Diimine Ligands

Diimines usually do not function well as ligands in copper-mediated ATRP due to the high steric hindrance (23,24). For example, polymerizations employing N,N'-di-(2,4,6-trimethylphenyl) ethylenediimine as the ligand yielded polymers with much higher molecular weights than theoretical values (23).

Bidentate Ligands with Mixed Nitrogen Types

Pyridine-imine mixed nitrogen ligands have been prepared and successfully used for the ATRP of MMA (24-26). The ligands can be easily prepared by the condensation reactions between the appropriate aldehyde, or ketone, with a primary amine. In one paper, several N-alkyl-(2-pyridyl)methanimine, where alkyl = n-butyl (nBPMI), isobutyl (iBPMI), sec-butyl (sBPMI), n-propyl (nPPMI), and N-(n-propyl)-1-(2-pyridyl)ethanimine (PPEI) were synthesized and used in ATRP (Figure 2) *(26)*. In all cases, the experimental molecular weights increased linearly with conversion and were slightly higher than the theoretical values. Polydispersities remained low for polymers produced using copper complexed by nBPMI, sBPMI and nPPMI ($M_w/M_n \sim 1.3$); however, polydispersities were relatively high for polymers produced using copper complexed by sBPMI and nPPEI ($M_w/M_n > 1.5$). Polymerizations using nBPMI, nPPMI and nPPEI are significantly faster than those using iBPMI and sBPMI. In general, it was concluded that catalysts which contain n-alkyl substituents are more effective than those containing branched alkyl substituents in the imine group. The cause of these effects is not well understood and is presumed to be related to the steric and electronic effect upon oxidation of copper(I). Similar results were obtained in a more recent paper *(24)*.

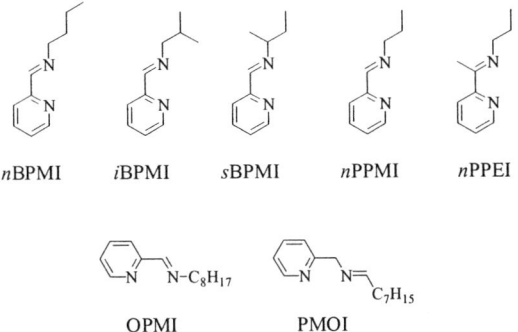

Figure 2. *Structures of pyridine-imines*

Other unconjugated pyridine-imine type ligands, such as N-(2-pyridylmethyl)-1-octanimine (PMOI), have also been studied *(27)*. Compared with its conjugated analog, N-(n-octyl)-(2-pyridyl)methanimine (OPMI), PMOI led to significantly faster polymerizations (Table 4). Both ligands provided reasonably controlled polymerizations where the experimental molecular weights were close to the theoretical values with relatively low polydispersities. The difference in the rate of polymerization can partially be ascribed to the different electronic properties of the ligands. The LUMO π* orbital in the conjugated π system in the case of OPMI provides lower lying empty orbitals for delocalizing the electron density away from the transition metal center than PMOI and thus better stabilizes the metal center in its low oxidation state. As a result, the copper(I)/copper(II) redox potential is increased which can lead to a slower polymerization.

Tridentate Nitrogen Ligands

Tridentate nitrogen ligands are fairly common and have found many applications in various areas of organometallic chemistry. With the increase in coordination number, larger varieties of ligands are possible by different combinations of nitrogen types and geometries. In contrast to bidentate ligands which form cationic tetrahedral complexes, tridentate ligands form 1:1 uncharged complexes with copper. As a result, one equivalent of the tridentate ligand (to metal) is sufficient to achieve maximum polymerization rate. According to the type of nitrogens, the tridentate nitrogen ligands are discussed in the following subgroups: tripyridine ligands, triamine ligands, and mixed ligands.

Table 4. Cu-Mediated ATRP Using Pyridine-Imine Ligands

ligand[a]	monomer	temp (°C)	time (h)	conv (%)	$M_{n,Cal}$	$M_{n,SEC}$	M_w/M_n
OPMI	styrene	110	15.0	92	9200	10400	1.34
OPMI	MA	90	15.0	84	16800	14700	1.23
OPMI[b]	MMA	90	3.5	53	10600	15800	1.33
PMOI	styrene	90	1.05	75	7500	11500	1.33
PMOI	MA	110	0.5	58	11600	12100	2.03
PMOI	MMA	90	0.5	62	12400	15100	1.50

[a] See Table 2 for general conditions. [b] 50 vol % in anisole; $[MMA]_0/[EBiB]_0 = 200$; $[EBiB]_0/[CuCl]_0/[OPMI]_0 = 1/1/2$.

Tripyridine Ligands

Bulk ATRP promoted by CuX complexed by unsubstituted 2,2':6'',2''-terpyridine (tpy) was heterogeneous for most monomers. Uncontrolled polymerizations were observed with molecular weight evolution resembling redox-initiated free radical polymerization. High molecular weight polymers were obtained within 0.5 h and polydispersities were close to 2 for the ATRP of styrene and MA. ATRP of styrene and MA promoted by CuX complexed by 4,4',4''-tris(5-nonyl)-2,2':6'',2''-terpyridine (tNtpy) was initially homogeneous. As the polymerization proceeded, green precipitate formed. The polymerizations of styrene and MA were well-controlled with molecular weights evolving linearly with conversion of monomers and polydispersities less than 1.2 *(28)*. The difference in the results between the polymerization using tpy and tNpy is attributed to the fact that tNtpy can better solublize copper complexes in nonpolar organic solvents and thus provide sufficient amount of the deactivating copper(II) species in the solution.

Triamine Ligands

Tridentate linear amine, *N,N,N',N'',N''*-pentamethyldiethylenetriamine (PMDETA), has been used successfully as a ligand to promote ATRP under similar conditions as dNbpy *(22)*. In general, employment of multidentate aliphatic amines as ligands have the advantage of being commercially available and less expensive, causing less coloring to the polymerization solution and promoting a faster rate when compared with the bpy system. The polymerization media is heterogeneous for both styrene and MMA, but initially homogeneous for MA. With PMDETA as the ligand, the polymerization for styrene was faster than that using dNbpy, while the rate of polymerization of MA was comparable to that of copper/dNbpy. Linear semilogarithmic kinetic plots were observed for both styrene and MA polymerizations; however, a severe curvature in kinetic plot was observed for MMA, suggesting the occurrence of a significant amount of termination reactions presumably due to a higher equilibrium constant. For all three monomers, molecular weight increased linearly vs conversion. The polydispersity for styrene, MA and MMA were ca. 1.3, 1.05 and 1.15, respectively.

Cyclic aliphatic amines, such as 1,4,7-trimethyl-1,4,7-triazacyclononane (TMTACN) and 1,5,9-trimethyl-1,5,9-triazacyclodeodecane (TMTACDD), were also used in ATRP (A and B in Table 5). A similar spacer effect was observed as discussed previously in the case of bidentate ligands. Copper coordinated by a ligand with an ethylene linker (i.e., TMTACN) promoted well-controlled ATRP of styrene, MA and MMA. In contrast, copper coordinated by a ligand with a propylene linker (i.e., TMTACDD) yielded polymers with higher polydispersities under similar reaction conditions.

Table 5. Use of Cyclic Amines in Copper-Mediated ATRP

ligand[a]	monomer	time (h)	conv (%)	$M_{n,Cal}$	$M_{n,SEC}$	M_w/M_n
A	styrene	1.50	63	6300	8100	1.11
A	MA	5.00	63	12600	13700	1.10
A	MMA	1.25	65	13000	18000	1.23
B	styrene	1.00	51	5100	7100	1.23
B	MA	1.00	43	8600	15700	2.65
B	MMA	1.00	65	13000	14000	1.43

[a] See Table 2 for conditions.

Tridentate Ligands with Mixed Nitrogen Types

There are many possible structures for tridentate ligands with mixed nitrogen types, and they can be formed by the combinations of any two of pyridine, imine, and amine or all three. The most studied is the combination of two pyridine and one amine, such as *N,N*-bis(2-pyridylmethyl)octylamine (BPMOA) which was synthesized by the coupling of 2-picolyl chloride with *n*-octylamine (Figure 3).

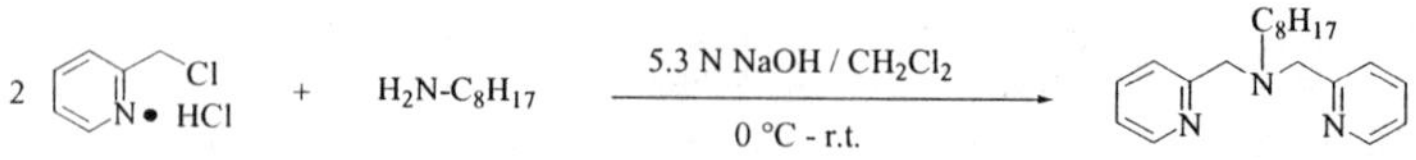

Figure 3. Synthesis of BPMOA

With BPMOA as the ligand, well-controlled polymerizations were carried out for styrene, MA and MMA. Polydispersities for all three monomers remained quite low ($M_w/M_n < 1.2$) *(29)*. Different from PMDETA, BPMOA yielded a linear kinetic plot for MMA, and this was attributed to the better solubility of the catalyst.

Other mixed nitrogen tridentate ligands containing a pyridine-imine moiety, such as *N*-(2-pyridylmethyl)-(2-pyridyl)methanimine (PMPMI) and *N*-(2-*N*-(dimethyl)-ethyl)-(2-pyridyl)methanimine (DMEPMI) were synthesized by the condensation reactions using bidentate primary amines with 2-pyridinecarbox-aldehyde *(27)*. Table 6 displays the results of the ATRP using these two ligands. With PMPMI as the ligand, well-controlled polymerizations were obtained for both styrene and MA. However, less efficient control was observed for the polymerization of MMA yielding polymers with polydispersities of 2.8. When DMEPMI was used as the ligand, well-controlled polymerizations were also obtained for both styrene and MA, and the control of MMA polymerization was slightly improved. °

Table 6. Cu-Mediated ATRP Using PMPMI or DMEPMI as the Ligand

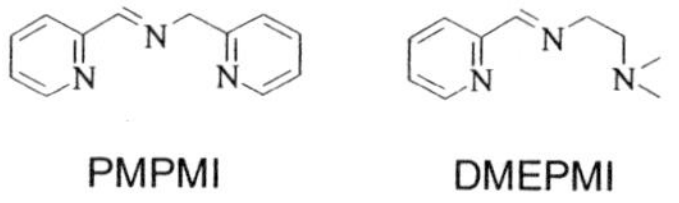

ligand[a]	monomer	temp (°C)	time (h)	conv (%)	$M_{n,Cal}$	$M_{n,SEC}$	M_w/M_n
PMPMI	styrene	110	1.5	60	6000	8100	1.14
PMPMI	MA	90	2.0	57	11400	12300	1.06
PMPMI	MMA	90	1.5	40	8000	17700	2.77
DMEPMI	styrene	90	3.2	61	6100	7299	1.17
DMEPMI	MA	110	5.2	47	9400	15000	1.37
DMEPMI	MMA	90	1.0	29	5800	10500	1.67

[a] For styrene: bulk; [styrene]$_0$/[PEBr]$_0$ = 96; [PEBr]$_0$/[CuBr]$_0$/[ligand]$_0$ = 1/1/1. For MA: bulk; [MA]$_0$/[EBP]$_0$ = 232; [EBP]$_0$/[CuBr]$_0$/[ligand]$_0$ = 1/1/1. For MMA: 50 vol % in anisole; [MMA]$_0$/[EB*i*B]$_0$ = 200; [EB*i*B]$_0$/[CuCl]$_0$/[ligand]$_0$ = 1/1/1.

Tetradentate Nitrogen Ligands

For tetradentate ligands, ligand geometries other than linear or cyclic geometry are possible. Ligands with a tripodal structure have been found to form some of the most reactive catalytic systems in copper-mediated ATRP.

Tetramine Ligands

Similarly to PMDETA, linear tetramine, 1,1,4,7,10,10-hexamethyltriethylene-tetramine (HMTETA) (Figure 4), has also been used successfully as the ligand to promote ATRP of Sty, MA and MMA *(22)*. In contrast to the copper/PMDETA system, a linear semilogarithmic plot of conversion vs time was observed for MMA. Molecular weights increased linearly with conversion, and were close to the theoretical values for all three monomers. The polydispersities of the obtained polymers were very low ($M_w/M_n \sim 1.1$).

Copper complexed by a cyclic tetramine, 1,4,8,11-tetramethyl-1,4,8,11-tetraazacyclotetradecane (TMTACTD, Figure 4), was found to afford polymers with much higher molecular weights than the predicted values and high polydispersities. High molecular weight polymers were obtained at relatively low monomer conversions, suggesting a slow deactivation process.

Controlled ATRP of acrylates can be carried out at ambient temperature using a tripodal tetramine, i.e., tris[2-(dimethylamino)ethyl]amine (Me$_6$TREN) (Figure 4) *(30)*. The polymers obtained have molecular weights close to the theoretical values and narrow molecular weight distributions. The rate of polymerization was significantly faster than that with dNbpy or its linear counterpart. Conversion of 80% of MA was reached within 3 h at 22 °C with a catalyst to initiator to monomer ratio of 0.1/1/232. The polydispersities were as low as 1.09. It should be recognized that increased polymerization rate as a result of an increased radical concentration may adversely affect the polymerization control; however, this is less pronounced for monomers with high rate constants of propagation, e.g., MA. The strong copper binding and fast rate of polymerization using Me$_6$TREN as the ligand was further applied to the controlled polymerization of 4-vinylpyridine *(31)*.

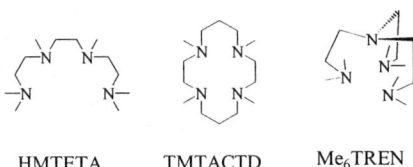

HMTETA   TMTACTD   Me$_6$TREN

Figure 4. Linear, cyclic and branched (tripodal) tetradentate amine ligands

Tetradentate Ligands with Mixed Nitrogen Types

As in the case of tridentate mixed nitrogen ligands, there are many possible combinations and geometry possible. However, they have not be extensively studied so far. Two tetradentate ligands with mixed nitrogen types are shown in Figure 5.

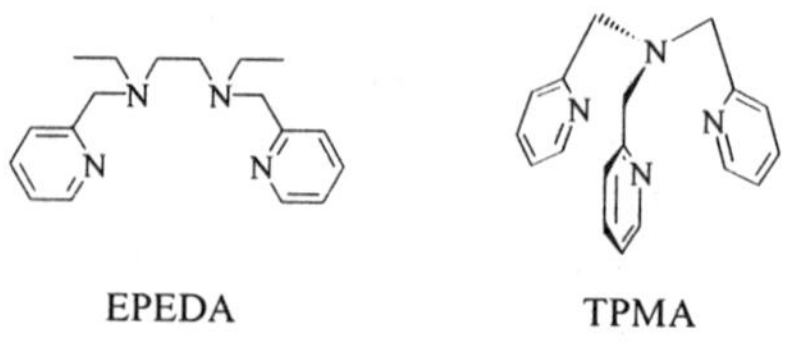

EPEDA   TPMA

Figure 5. Two examples of tetradentate ligand with mixed nitrogen types

N,N'-diethyl-N,N'-bis[(2-pyridyl)methyl]ethylenediamine (EPEDA) and tris[(2-pyridyl)methyl]amine (TPMA) were synthesized by the coupling of two equivalents of 2-pycolyl chloride with N,N'-diethyl-ethylenediamine and (2-aminometyl)pyridine, respectively. Similar to its amine analog, Me$_6$TREN, TPMA as the ligand promoted well-controlled polymerizations of Sty and MA at 110 °C and 50 °C, respectively *(29)*. For MA, a catalyst to initiator ratio of 0.2 was sufficient to produce well-defined polymers (M_n = 15200 and M_w/M_n = 1.05) within 1 h.

Summary of Ligand Effect

A variety of multidentate nitrogen ligands have been successfully employed in copper-mediated ATRP. Although the comprehensive understanding of the effect of various structural parameters on ATRP should require more complete kinetic studies, it is possible to reach the following preliminary conclusions. (1). In general, tetradentate ligands of analogous structures provide similar polymerization rates as tridentate ligands, but yield faster polymerizations than bidentate ligands. Monodentate nitrogen ligands are not effective for copper-mediated ATRP, leading to redox-initiated free radical polymerization. (2). The redox potential of a metal complex serves as useful guideline for catalyst design. Since aliphatic amines are more electron rich and generally better σ donors than pyridines or imines, their copper complexes usually have lower redox potentials and favor the formation of copper(II) species. As a result, copper complexes with aliphatic amines tend to have faster polymerization rates than their pyridine and imine analogs. In addition, electron-donating substituents on the ligand should lead to increased catalytic activity. (3). The spacer between the coordinating nitrogens is another important factor. Ethylene linkage seems to provide the best coordination angle and affords the most reactive catalysts. In contrast, propylene linkage leads to catalysts with reduced molecular weight control. Spacers longer than three carbons generally yield polymerizations similar to monodentate nitrogen ligands. (4). Conformational change of the ligand complexing transition metal during atom transfer is crucial. The catalytic cycle of ATRP requires the transition metal to be able to expand the coordination number by one upon halogen abstraction. In the case of complexes with two bidentate ligands, the coordination geometry of copper undergoes a transition from tetrahedral to trigonal bipyramidal when copper(I) is oxidized to copper(II). The change in the coordination sphere requires the flexibility of the coordinating ligand and precludes the presence of sterically demanding substituents near the coordination sphere. Copper complexes with tripodal tetradentate ligands are among the most active catalytic systems reported and provide much faster polymerizations than their linear analogs. This may be related to their specific type of geometry which does not require a significant conformational change during atom transfer. (5). Stability of the metal complex should be sufficient. This is especially important for ATRP of monomers that compete for the binding of transition metals, e.g., 4-vinylpyridine.*(31)* Additionally, if the metal complex is unstable, it may release a nucleophilic ligand which may undergo side reaction, e.g., S_N2 reaction with terminal alkyl halides.*(32)*

Outlook

The field of ATRP has grown significantly since its discovery in the early 1995. Many advances have been made in the improvement of the catalyst and the application of ATRP towards the synthesis of novel materials. It seems that copper-based catalysts are among the most efficient, selective, robust and affordable for the widest range of monomers. The discovery of the use of multidentate amines in ATRP has greatly improved catalyst efficiency and reduced the cost. Considering the endless combinations of ligand composition, coordination number, coordinating atom type and geometry, the potential for discovering new, highly reactive and highly effective polymerization systems is tremendous. An increasing effort in the elucidation of reaction mechanisms and active catalytic species is necessary for a more effective and logical approach to ligand design and evaluation.

Acknowledgments

Financial support from NSF, EPA and the industrial members of the ATRP Consortium at Carnegie Mellon University is gratefully acknowledged.

References

(1) *Controlled Radical Polymerization*; Matyjaszewski, K., Ed.; American Chemical Society: Washington, DC 1998; Vol. 685.

(2) Matyjaszewski, K. *ACS Symp. Ser.*, **1998**, *685*, 258-283.
(3) Kato, M.; Kamigaito, M.; Sawamoto, M.; Higashimura, T. *Macromolecules* **1995**, *28*, 1721-1723.
(4) Wang, J. S.; Matyjaszewski, K. *J. Am. Chem. Soc.* **1995**, *117*, 5614-5615.
(5) Wang, J. S.; Matyjaszewski, K. *Macromolecules* **1995**, *28*, 7901-7910.
(6) Granel, C.; Dubois, P.; Jérôme, R.; Teyssié, P. *Macromolecules* **1996**, *29*, 8576-8582.
(7) Uegaki, H.; Kotani, Y.; Kamigaito, M.; Sawamoto, M. *Macromolecules* **1997**, *30*, 2249-2253.
(8) Ando, T.; Kamigaito, M.; Sawamoto, M. *Macromolecules* **1997**, *30*, 4507-4510.
(9) Matyjaszewski, K.; Wei, M.; Xia, J.; McDermott, N. E. *Macromolecules* **1997**, *30*, 8161-8164.
(10) Patten, T. E.; Matyjaszewski, K. *Adv. Mater.* **1998**, *10*, 901-915.
(11) Munakata, M.; Kitagawa, S.; Asahara, A.; Masuda, H. *Bull. Chem. Soc. Jpn.* **1987**, *60*, 1927-1929.
(12) Tyagi, S.; Hathaway, B.; Kremer, S.; Stratemeier, H.; Reinen, D. *J. Chem. Soc., Dalton Trans.* **1984**, 2087-2091.
(13) Matyjaszewski, K.; Patten, T. E.; Xia, J. *J. Am. Chem. Soc.* **1997**, *119*, 674-680.
(14) Davis, K. A.; Paik, H.-J.; Matyjaszewski, K. *Macromolecules* **1998**, *32*, 1767-1776.
(15) Wang, J.-L.; Grimaud, T.; Matyjaszewski, K. *Macromolecules* **1997**, *30*, 6507-6512.
(16) Wang, J. S.; Matyjaszewski, K. *Macromolecules* **1995**, *28*, 7572-7573.
(17) Patten, T. E.; Xia, J.; Abernathy, T.; Matyjaszewski, K. *Science* **1996**, *272*, 866-868.
(18) Percec, V.; Barboiu, B.; Neumann, A.; Ronda, J. C.; Zhao, M. *Marcromolecules* **1996**, *29*, 3665-3668.
(19) Xia, J.; Matyjaszewski, K. *Macromolecules* **1997**, *30*, 7692-7696.
(20) Xia, J.; Johnson, T.; Gaynor, S.; Matyjaszewski, K.; DeSimone, J. *Macromolecules* **1999**, *32*, 4802-4805.
(21) Destarac, M.; Bessière, J.-M.; Boutevin, B. *Macromol. Rapid Commun.* **1997**, *18*, 967-974.
(22) Xia, J.; Matyjaszewski, K. *Macromolecules* **1997**, *30*, 7697-7700.
(23) DiRenzo, G. M.; Messerschmidt, M.; Mülhaupt, R. *Macromol. Rapid Commun.* **1998**, *19*, 381-384.
(24) Haddleton, D. M.; Martin, C. C.; Dana, B. H.; Duncalf, D. J.; Heming, A. M.; Kukulj, D.; Shooter, A. J. *Macromolecules* **1999**, *32*, 2110-2119.
(25) Haddleton, D. M.; Jasieczek, C. B.; Hannon, M. J.; Shooter, A. J. *Macromolecules* **1997**, *30*, 2190-2193.
(26) Haddleton, D. M.; Duncalf, D. J.; Kukulj, D.; Martin, C. C.; Jackson, S. G.; van, S. A. B.; Clark, A. J.; Shooter, A. J. *Eur. J. Inorg. Chem.* **1998**, 1799-1806.
(27) Xia, J.; Matyjaszewski, K. *Polym. Prepr. (Am. Chem. Soc., Div. Polym. Chem.)* **1999**, *40(2)*, 442-443.
(28) Kickelbick, G.; Matyjaszewski, K. *Macromol. Rapid Commun.* **1999**, *20*, 341-346.

(29) Xia, J.; Matyjaszewski, K. *Macromolecules* **1999**, *32*, 2434-2437.
(30) Xia, J.; Gaynor, S. G.; Matyjaszewski, K. *Macromolecules* **1998**, *31*, 5958-5959.
(31) Xia, J.; Zhang, X.; Matyjaszewski, K. *Macromolecules* **1999**, *32*, 3531-3533.
(32) Coessens, V.; Matyjaszewski, K. *J. Macromol. Sci. Pure Appl. Chem.* **1999**, *A36*, 653-666.

Chapter 14

Structural Chemistry and Polymerization Activity of Copper(I) Atom Transfer Radical Polymerization Catalysts

Amy T. Levy and Timothy E. Patten[1]

Department of Chemistry, University of California at Davis,
One Shields Avenue, Davis, CA 95616-5295

The use of structurally characterized copper(I) halide bipyridine complexes in styrene ATRP was examined. It was found that a preformed copper(I) bromide complex with dNEObipy of a 1-to-1 ligand to metal stoichiometry, complex-**1**, was an effective ATRP catalyst for styrene. Molecular weights evolved with conversion in correlation with the relation, $DP_n = ([M]_o - [M]_t) / [initiator]_o$, and the molecular weight distributions after early polymerization conversions remained low, $M_w / M_n = 1.04$ to 1.06. Added equivalents of free ligand had very little effect upon either the rate of polymerization or the evolution of molecular weights and molecular weight distributions with conversion. Thus, the kinetically optimum ratio of ligand to metal in these polymerizations was 1-to-1.

In addition to their use in organic synthesis *(1-3)*, copper(I) halide complexes with bidentate and polydentate amines play an important role as catalysts for controlled / living radical polymerizations. Atom transfer radical polymerization (ATRP) employs atom transfer from an organic halide polymer end group to a transition-metal complex to generate and deactivate propagating radicals. Control in ATRP is established via the persistent radical effect *(4,5)*, in which a small amount of radical coupling serves to build up a concentration of a higher oxidation state complex until the rate at which the radicals are deactivated in an atom transfer step becomes faster than the rate at which the radicals can terminate through coupling and disproportionation *(6)*. ATRP is applicable to a range of vinyl monomers and reaction conditions, and complexes of copper(I) *(7-9)*, ruthenium(II) *(10)*, iron(II) *(11,12)*, and nickel(II) *(13,14)* with various ligands all demonstrate activity as ATRP catalysts. Additionally, polymers can be prepared using ATRP with good synthetic control over the final molecular weights, molecular weight distributions, and end group structures *(15)*.

Much spectroscopic and kinetic information has been acquired on ATRP catalyzed by *in situ* catalyst combinations of one equivalent of copper(I) halide to two

[1]Corresponding author.

equivalents 4,4'-disubstituted-2,2'-bipyridines *(16-18)*. The structure(s) and fundamental reactivity(ies) of the copper center(s) that effect atom transfer are not known and usually are inferred from previous literature crystallographic studies of 2,2'-bipyridine (bipy) complexes of copper(I) halides. Thus, direct structure-activity relationships have not been elucidated for these catalyst systems. Such chemical information, if known, would allow for deeper insights into the chemistry of copper(I) halide / 2,2'-bipyridine ATRP systems, and a number of outstanding issues in ATRP (the polymerization of vinyl esters and dienes, extension of polymerization control to even higher molar mass polymers, and the development of more active catalysts that can be used in smaller concentrations) could be solved rationally through knowledge about the structure, solution dynamics, and chemistry of the copper centers involved in the atom transfer steps.

Coordination Chemistry of Copper(I) Halides with 2,2'-Bipyridine and 1,10-Phenanthroline Ligands

There is a good deal of literature information on the synthesis and solid-state structures of copper(I) halide complexes with 2,2'-bipyridines and an analogous group of ligands, 1,10-phenanthrolines *(19,20)*. All of the complexes may be categorized into a few general structural types with the understanding that complexes within a category may have unique bond lengths, bond angles, and geometric distortions due to differences in ligand substituents or counterions (Figure 1). The copper(I) ion has a d^{10} electron count and, therefore, only needs 8 electrons via ligand coordination to satisfy the 18 electron rule. Copper(I) complexes usually are tetracoordinate, and because the d-shell is filled the ideal copper(I) complex adopts a tetrahedral geometry such that the ligands are as far apart as possible.

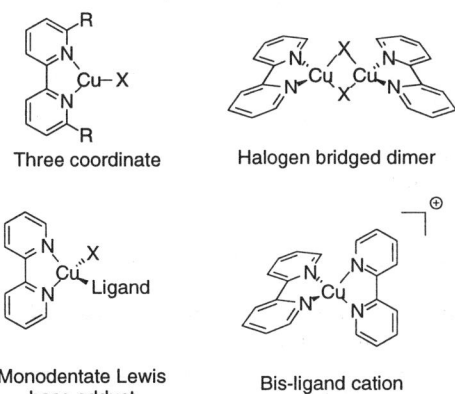

Figure 1. Structural types for copper(I) halide complexes with 2,2'-bipyridines and 1,10-phenanthrolines. Example structures are shown using the bipy ligand.

The first and rarest of the structural types is the three coordinate complex containing one equivalent of bidentate ligand and the halide counterion ligand. These complexes are coordinatively unsaturated, and they typically form when sterically bulky substituents crowd the metal center and prevent dimerization or addition of a second equivalent of ligand. The next general structural type is the halogen bridged dimer. Conceptually, these are three-coordinate complexes without bulky substituents near the metal center and are able to fill the vacant fourth coordination site via donation from an electron pair of the halide ligand on another complex. When the counterion is incapable of donating a pair of electrons (such is the case with BF_4^- and PF_6^-) these dimeric species will not form. The bridging interaction in the dimeric complexes is not very strong and can be broken up via addition of certain monodentate ligands (i.e., a pyridine or a phosphine). This reaction forms the third structural type, monodentate Lewis base adducts. The final structural type is the bis-ligand cation, which can form from reactions that have either a 1-to-1 or 2-to-1 stoichiometry of ligand to copper(I) halide. Under 1-to-1 stoichiometry conditions, a bis-ligand copper(I) cation forms along with a dihalocuprate counterion. Under 2-to-1 stoichiometries, displacement of the halogen ligand can occur to form a bis-ligand copper(I) cation and a halide counterion. Usually, polar or hydrogen-bonding solvents are used for the latter reactions in order to facilitate the halide ligand acting as a leaving group, although it is not known if such species will form in nonpolar media such as bulk monomer.

Dynamic Behavior of Copper(I) Halide Complexes with 2,2'-Bipyridine and 1,10-Phenanthroline Ligands

Few studies have been performed on the solution behavior of copper(I) halide complexes with 2,2'-bipyridines or 1,10-phenanthroline ligands. Pallenberg, et al. *(21)* examined both the solution and solid state structures of a range of 1-to-1 complexes of 2,9-dialkyl-1,10-phenanthrolines with copper(I) halides. For those ligands that possessed small alkyl substituents, the authors found that while different types of solid-state structures could be observed, each of the complexes displayed solution structures comprised of a bis-ligand copper(I) cation and a dihalocuprate anion. This study indicates that the solid state structures of these complexes do not necessarily translate into actual solution structures and that these complexes can adopt a range of structures at the same stoichiometry.

Structural Chemistry of Copper(I) ATRP Catalysts

The synthesis, isolation, and structural characterization of ATRP catalysts and the study of their chemistry and mechanisms through which they effect atom transfer is a present direction of research in ATRP. Haddleton, et al. *(22-24)* have structurally characterized several copper(I) and copper(II) complexes of bidentate 2-iminopyridines and have examined their activities in the ATRP of methyl methacrylate, which has provided much needed insight into that system. Studies such as these are needed to help build a framework for understanding the polymerization data provided in the literature and for developing predictive models for designing new catalyst systems.

Experimental

Materials and Characterizations

The syntheses and characterizations of dNEObipy, dTHEXbipy, dTBDMSbipy, and complex-1 have been reported elsewhere *(25)*. Styrene was distilled under vacuum before use. CuBr (Acros) was purified until colorless by grinding into a fine powder using a mortar and pestle, stirring in glacial acetic acid, consecutive washes of absolute ethanol and diethyl ether, and removal of volatile materials under vacuum. All syntheses and polymerizations were performed under nitrogen atmospheres using standard Schlenk techniques. Unless stated otherwise, all other materials were purchased from commercial sources and used without further purification. Number-averaged molecular weights (M_n), weight-averaged molecular weights (M_w), and molecular weight distributions (M_w/M_n) were determined using gel-permeation chromatography in THF at 30 °C. Three Polymer Standards Sevices columns (100 Å, 1000 Å, and linear) were connected in series to a Thermoseparation Products P-1000 isocratic pump, autosampler, column oven, and Knauer refractive index detector. Calibration was performed using polystyrene samples (Polymer Standard Services; M_p = 400 to 1,000,000; M_w/M_n < 1.10).

Procedure for the Solution Polymerization of Styrene using Complex-1

A 25 mL Schlenk flask was charged with 1.5 mL of *p*-xylene, 3.5 mL (30.5 mmol) of styrene, and 42 µL (0.308 mmol) of 1-phenylethyl bromide. The flask was sealed with a ground glass stopper and three cycles of freezing, evacuation, and thawing were performed to remove any oxygen from the polymerization system. Next, 0.248 g (0.350 mmol) of complex-1 was added to the flask and two more cycles between vacuum and nitrogen gas were performed. The flask was immersed in an oil bath at 110 °C ± 1 °C, and when all the solids had dissolved a time = 0 data point was taken. At time intervals, 0.50 mL sample of solution was withdrawn using a purged syringe and added to 5.00 mL of THF. The THF solutions were injected into the GC, and percent conversions were calculated relative to the time = 0 data point using the xylene signal integration as an internal standard. The samples were then filtered through a small column of alumina and a 0.45 micron filter and then injected into the GPC for analysis. Similar procedures were used for other polymerization experiments; concentrations of reagents are listed in the figure legends.

Results and Discussion

Three 4,4'-disubstituted-2,2'-bipyridines were prepared and used for these investigations: dTBDMSbipy, dTHEXbipy, and dNEObipy *(25)*. Of the three ligands dNEObipy, 4,4'-di-(neophyldimethylsilylmethyl)-2,2'-bipyridyl, yielded complexes that were both soluble in organic solvents and easily crystallized. A 1-to-1

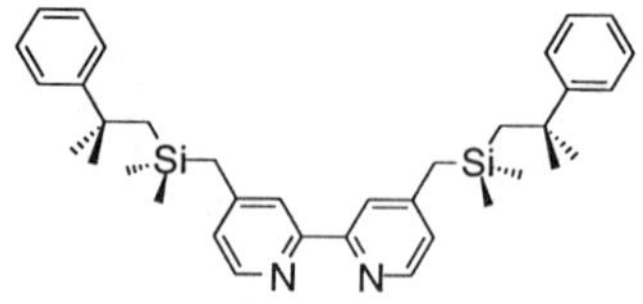

dNEObipy

complex of dNEObipy and CuBr, complex-1, was prepared, and the crystal structure of complex-1 was solved *(25)*. The structure consisted of a bis-ligand copper(I) cation and dibromocuprate anion (Figure 2). The ^{1}H NMR spectrum of complex-1 in

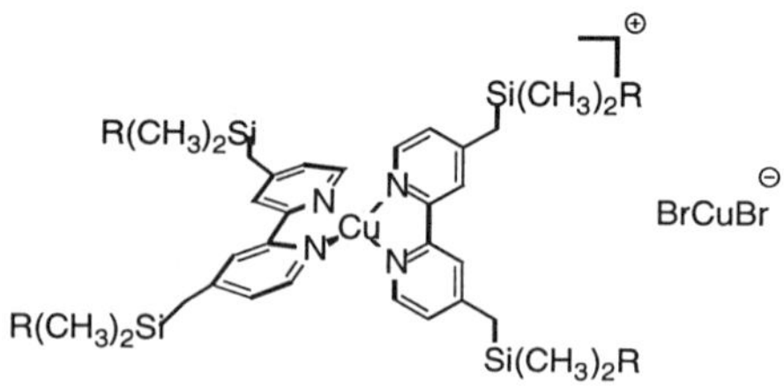

Complex-1 (R = neophyl)

Figure 2. The structure of complex-1 as determined in reference 25.

CD_2Cl_2 showed sharp resonances at 30 °C, and at -80 °C only the pyridyl ring proton signal at 8.15 ppm showed broadening (Figure 3). This behavior suggests that either no dynamic processes occurred in CD_2Cl_2 solutions of complex-1 or that any dynamic processes occurring were fast on the NMR timescale. Elevated temperatures were not investigated, although Haddleton and coworkers *(26)* studied the variable temperature ^{1}H NMR spectra of CuBr/2,2'-bipyridine d_6-DMSO solutions and observed broadening of the pyridyl ring proton signals at elevated temperatures consistent with ligand dissociation or ligand exchange. Interestingly, ^{1}H NMR spectra of the corresponding solutions of CuBr/2-pyridinal-propylimine in d_6-DMSO showed no signal broadening at elevated temperatures *(26)*. The ^{1}H NMR chemical shifts of the pyridyl ring protons of complex-1 matched those found in the spectrum of a CD_2Cl_2 solution of 1 equivalent of dNEObipy and 1 equivalent of CuBr. These results suggest that in relatively polar solvents such as acetonitrile and dichloromethane (and speculatively acrylates and methacrylates) complex-1 is the predominant species formed in solution at this stoichiometry. In contrast, the ^{1}H NMR spectrum of

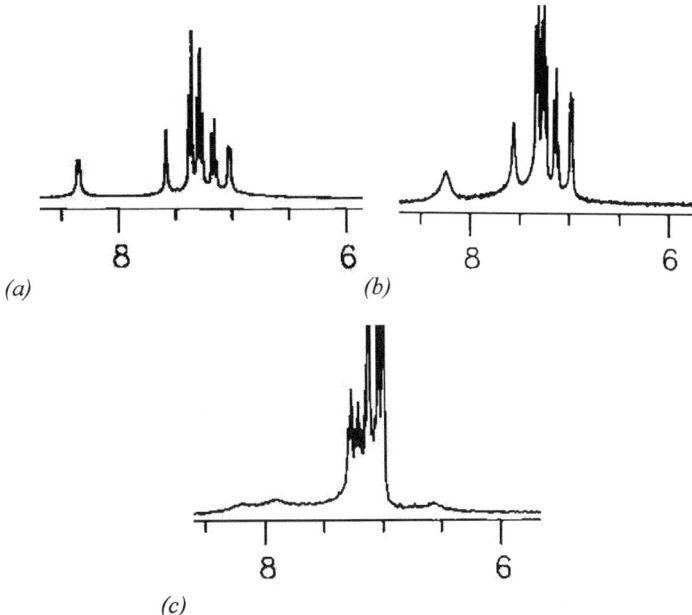

Figure 3. Aromatic proton regions of the ^1H NMR spectra of complex-1 in CD_2Cl_2 at room temperature (a), in CD_2Cl_2 at -80 °C (b), and in d_8-toluene at room temperature (c).

complex-1 in d_8-toluene (Figure 3) showed broadened resonances for the pyridyl ring protons at room temperature that were not due to the presence of paramagnetic copper(I) species incidentally formed in solution. Thus, in this less polar solvent dynamic processes occurred at a rate observable on the NMR timescale. Such a solvent effect on the appearance of the NMR spectrum might be due to a dynamic equilibrium that results in the elimination of charges: for example, between complex-1 and a halogen bridged dimer. We have not yet identified what species comprise this equilibrium in toluene, because variable temperature ^1H NMR spectra recorded at temperatures up to 90 °C or down to -70 °C do not reveal simple coalescence-decoalescence behavior of the resonances, possibly indicating the presence of more than two species in equilibria.

The polymerization activity of complex-1 was assessed using a standard styrene polymerization (Equation 1). The initial ratio of monomer and 1- phenylethyl

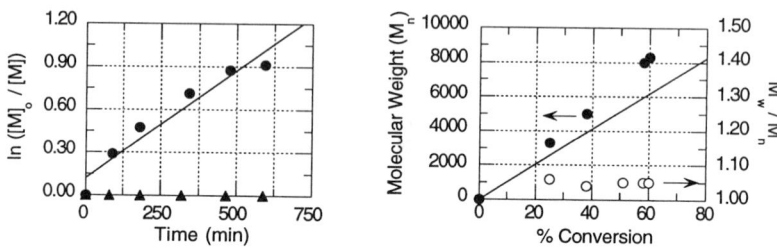

(1)

bromide initiator, was 100-to-1, and the initial ratio of initiator and complex-1 was 1-to-1. The results of this polymerization are shown in Figure 4. From the

Figure 4. Semilogarithmic kinetic plot (left) and molecular weight and molecular weight distribution plot for the ATRP of styrene using complex-1 (filled and open circles) and n-Bu$_4$N$^+$ CuBr$_2^-$ (triangles). The line in the molecular weight versus conversion plot is the theoretical molecular weight line. Conditions: solvent = p-xylene, T = 105 ± 1 °C; [1-PEBr]$_o$ = 61.5 mM; [Styrene]$_o$ = 6.11 M. Complex-1: [Cu(I)]$_o$ = [(dNEObipy)$_2$Cu$^+$]$_o$ + [CuBr$_2^-$]$_o$ = 61.0 mM. nBu$_4$N$^+$ CuBr$_2^-$: [CuBr$_2^-$]$_o$ = 30.5 mM.

semilogarithmic kinetic plots one can conclude that the active species in this polymerization must be either the bis-ligand copper(I) cation, (dNEObipy)$_2$Cu$^+$, or a species created from an equilibrium between the bis-ligand cation and the diahalocuprate counterion, because no conversion of monomer was observed when tetra-*n*-butylammonium dibromocuprate was used as the copper(I) catalyst for the polymerization. Additionally, there is curvature in the semilogarithmic kinetic plot which we do not understand at the present time. We can rule out chain termination as a cause for this effect, because the addition a small amount of free ligand to the polymerization eliminates this curvature (*vide infra*). A linear correspondence between molecular weight (M_n) and conversion was observed, and the data points lay somewhat above the theoretical line (DP_n = ([M]$_o$ - [M]$_t$) / [initiator]$_o$). The latter observation can be rationalized by the fact that complex-1 is a pure copper(I) species, and consequently, some radical coupling must occur in the initial stages of the polymerization in order to establish the persistent radical effect. Through conversion of some of the initiator into inactive dimer, the effective molecular weights should be somewhat higher than the theoretical molecular weights, as is observed. The molecular weight distributions followed a concave curve typically observed for controlled / living radical polymerizations and, after approximately 30 % conversion, were quite low, M_w / M_n = 1.04 to 1.06.

Two standard styrene polymerizations were performed using CuBr / 1 dNEObipy and CuBr / 2 dNEObipy as the copper(I) catalysts. In Figure 5, the resulting data are overlayed onto the polymerization data obtained using complex-1. While the data for the three polymerizations were largely similar, a few clear differences were observed. First, a pronounced curvature in the semilogarithmic plot was observed for the polymerization conducted using CuBr / 1 dNEObipy, whereas the data curvature for the polymerization conducted using complex-1 was less

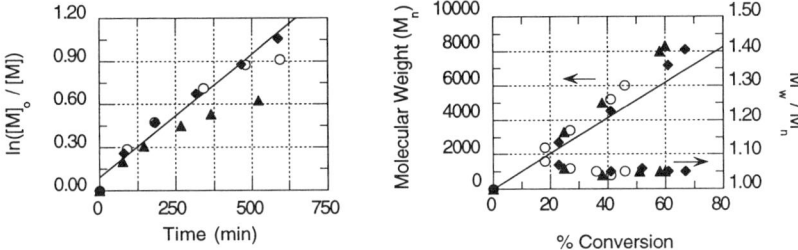

Figure 5. Comparisons of the semilogarithmic kinetic plots (left) and molecular weight and molecular weight distribution plots for the ATRP of styrene using complex-1 (hollow circles), CuBr + 1 eq. dNEObipy (triangles), and CuBr + 2 eq. dNEObipy (diamonds). The line in the semilogarithmic kinetic plots is a best fit for the polymerization data for CuBr + 2 eq. dNEObipy, and the line in the molecular weight versus conversion plot is the theoretical molecular weight line for all three polymerizations Conditions: solvent = p-xylene, T = 105 ± 1 °C; $[1\text{-}PEBr]_o$ = 61.5 mM; $[Styrene]_o$ = 6.11 M. Complex-1: $[Cu(I)]_o = [(dNEObipy)_2Cu^+]_o + [CuBr_2^-]_o$ = 61.0 mM. CuBr/2dNEObipy: $[CuBr]_o = 1/2[dNEObipy]_o$ = 61.0 mM. CuBr/1dNEObipy: $[CuBr]_o = [dNEObipy]_o$ = 61.0 mM.

pronounced and the data for the polymerization conducted using CuBr / 2 dNEObipy was even less so. Very little difference was apparent between the three molecular weight and molecular weight distribution plots. Interestingly, the polymerization catalyzed by complex-1 behaved more like the polymerization catalyzed by CuBr / 2 dNEObipy than that by CuBr / 1 dNEObipy. Thus, both the stoichiometry of ligand to metal and how the catalyst is formed apparently have an effect upon the kinetics of polymerization but no effect upon the evolution of molecular weights or polydispersities as a function of conversion. The deceleration observed in semilogarithmic kinetic plots could be a result of three processes: termination via radical coupling, termination via end group loss (27), or a catalyst deactivation unrelated to irreversible chain termination. It has not been determined yet which of these three processes occurred in these polymerizations.

To investigate the effect of added ligand on the activity of the polymerization independent of the variable of complex formation, we performed a series of styrene polymerizations using complex-1 as the catalyst with added equivalents of dNEObipy. Overlay plots of the results are shown in Figure 6. Within experimental error there was no difference in the plots of molecular weight and molecular weight distribution as a function of conversion. In the semilogarithmic kinetic plots, the only difference between the data sets was observed for polymerizations performed using complex-1 alone, which exhibited a deceleration curvature towards the end of the polymerization. With 0.5 equivalents of added ligand or more, this curvature was no longer observed in the semilogarithmic kinetic plots and the apparent rates were, within experimental error, identical. This trend is best observed in Figure 7, which shows the apparent rate of polymerization plotted as a function of the total ratio of moles of ligand to moles of copper(I) in the polymerization. Essentially there is no

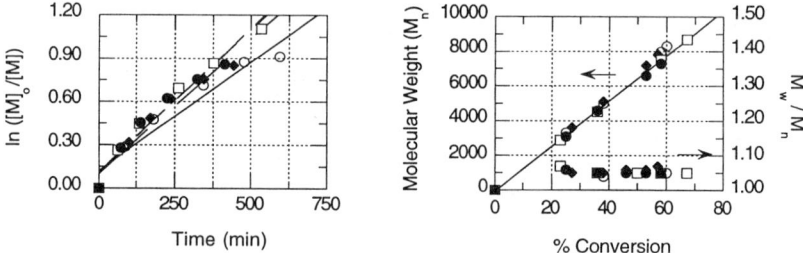

*Figure 6. Comparisons of the semilogarithmic kinetic plots (left) and molecular weight and molecular weight distribution plots for the ATRP of styrene using complex-1 plus n eq. of dNEObipy relative to the total initial concentration of copper(I) (i.e., $[Cu(I)]_o = [(dNEObipy)_2Cu^+]_o + [CuBr_2^-]_o)$: n = 0 eq. (hollow circles), 0.5 eq. (diamonds), 1 eq. (hollow squares), and 2 eq. (filled circles). Note: the line in the molecular weight plot is a fit to the data and not a theoretical molecular weight line. Conditions: solvent = p-xylene, T = 105 °C; $[1\text{-}PEBr]_o$ = 61.5 mM; $[Styrene]_o$ = 6.10 M. Complex-1: $[Cu(I)]_o = [(dNEObipy)_2Cu^+] + [CuBr_2^-]$ = 61.0 mM. Added dNEObipy: n eq. * 61.0 mM.*

change in the apparent rate of polymerization with increasing ligand-to-metal ratio. The apparent rate at 1.0 mol dNEObipy to mol Cu(I) is slightly less than the other apparent rates, but if one considers the corresponding data in Figure 6, one can see that this data point was obtained from the slope of a line fitted to data exhibiting a deceleration curvature. If one considers only the initial stages of the polymerization, one sees that the initial rate is within experimental error of the other polymerizations.

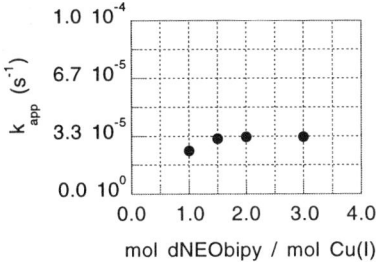

Figure 7. Plot of apparent rate constants from Figure 6 as a function of the ratio of moles of dNEObipy to the moles of copper(I) centers.

The dependence of the apparent rate of styrene polymerization on the ligand-to-metal ratio as seen in Figure 7 is different than that observed for styrene polymerizations using *in situ* catalyst formulations of 2 dNbipy / CuBr *(6)*. With the *in situ* catalyst formulations a clear increase in apparent rate of styrene ATRP was observed until a ligand-to-copper(I) halide stoichiometry of 2-to-1 was reached, the kinetically optimum ratio, but in Figure 7 the rate appeared to be saturated at ratio of 1-to-1 when a preformed complex was used. These observations of the kinetically optimum ratios for styrene ATRP can be rationalized on the basis of copper(I) halide solubility. First, the initial concentrations of ligand and copper(I) halide in these polymerizations are quite small. Second, styrene is a much less polar solvent than, say, methyl methacrylate and is a noncoordinating solvent unlike methyl methacrylate. Thus, under such conditions either (a) the solubility product for CuBr and 4,4'-dialkyl bipyridines is such that two equivalents are necessary to bring all of the copper halide into solution, even though only about one equivalent may be necessary to form the active species or (b) the rate at which the copper halide is brought into solution is slow on the polymerization timescale. Either way, apparently two equivalents of ligand is necessary for a saturation in the apparent rate of polymerization. If the active species or catalyst precursor is preformed, as in complex-**1** or if a more polar, coordinating solvent is used, such as methyl methacrylate, then only about one equivalent of ligand is necessary for a saturation in the apparent rate of polymerization. Thus, the kinetically optimum ratio of ligand-to-metal, independent of the variable of complex formation for styrene ATRP appears to be 1-to-1. Additionally, for styrene ATRP using complex-**1**, a fractional equivalent of free ligand serves to eliminate the small deceleration in kinetics observed at later stages of the polymerization. The reasons for this effect are not yet understood. The variable of complex formation and its effect upon the apparent rate of polymerization can be seen in Figure 5, in which two copper(I) catalyst systems with 1-to-1 ligand to metal stoichiometries, one preformed and one formed *in situ*, exhibited different polymerization kinetics.

The situation when comparing kinetically optimum ratios of styrene to that of different monomers appears to be more complex, so similar types of studies need to be performed for methyl methacrylate and methyl acrylate ATRP. In the meantime, the reader is referred to kinetic studies on *in situ* catalyst formulations. In the case of methyl methacrylate ATRP catalyzed by *in situ* combinations of dNbipy and CuBr, a kinetically optimum ratio of 1-to-1 was also observed *(28)*, and for methyl acrylate ATRP a 2-to-1 ratio was observed *(29)*. In similar studies using alkylpyridylmethanimine ligands a 2-to-1 ratio was observed for methyl methacrylate ATRP *(30)*. A study of methyl methacrylate ATRP that examined the kinetically optimum ratio as a function of the initiator-to-copper(I) concentrations showed different ratios for high and low initiator concentrations *(18)*.

Conclusions

In summary, the use of structurally characterized copper(I) halide bipyridine complexes in styrene ATRP was examined. It was found that a preformed 1-to-1 ligand-to-metal stoichiometry complex of dNEObipy and copper(I) bromide, complex-1, was an effective ATRP catalyst for styrene. Molecular weights evolved with conversion in correlation with the relation, $DP_n = ([M]_o - [M]_t) / [initiator]_o$, and the molecular weight distributions after early polymerization conversions remained low, $M_w / M_n = 1.04$ to 1.06. Added equivalents of free ligand to polymerizations using complex-1 had very little effect upon the rate of polymerization and the evolution of molecular weights and molecular weight distributions with conversion. Thus the variable of complex formation has important effects on the kinetics of polymerization and should be considered when examining the kinetics of ATRP. Further understanding of the dynamic solution behavior of complex-1 and similar complexes will allow for a direct correlation of copper(I) coordination environment to polymerization activity and for a more detailed understanding of the atom transfer step in these polymerizations.

Acknowledgments

We would like to acknowledge the UC Davis Committee on Research for financial support for this work.

References

1. Minisci, F., *Acc. Chem. Res.* **1975**, *8*, 165-171.
2. Bellus, D., *Pure Appl. Chem.* **1985**, *57*, 1827-1838.
3. Iqbal, J.; Bhatia, B.; Nayyar, N. K., *Chem. Rev.* **1994**, *94*, 519-564.
4. Fischer, H., *J. Am. Chem. Soc.* **1986**, *108*, 3925-3927.
5. Daikh, B. E.; Finke, R. G., *J. Am. Chem. Soc.* **1992**, *114*, 2938-2943.
6. Matyjaszewski, K.; Patten, T. E.; Xia, J. H., *J. Am. Chem. Soc.* **1997**, *119*, 674-680.
7. Wang, J. S.; Matyjaszewski, K., *J. Am. Chem. Soc.* **1995**, *117*, 5614-5615.
8. Percec, V.; Barboiu, B., *Macromolecules* **1995**, *28*, 7970-7972.
9. Haddleton, D. M.; Jasieczek, C. B.; Hannon, M. J.; Shooter, A. J., *Macromolecules* **1997**, *30*, 2190-2193.

10. Kato, M.; Kamigaito, M.; Sawamoto, M.; Higashimura, T., *Macromolecules* **1995**, *28*, 1721-1723.
11. Matyjaszewski, K.; Wei, M.; Xia, J.; McDermott, N. E., *Macromolecules* **1997**, *30*, 8161-8164.
12. Ando, T.; Kamigaito, M.; Sawamoto, M., *Macromolecules* **1997**, *30*, 4507-4510.
13. Granel, C.; Dubois, P.; Jerome, R.; Teyssie, P., *Macromolecules* **1996**, *29*, 8576-8582.
14. Uegaki, H.; Kotani, Y.; Kamigaito, M.; Sawamoto, M., *Macromolecules* **1997**, *30*, 2249-2253.
15. Patten, T. E.; Matyjaszewski, K., *Adv. Mater.* **1998**, *10*, 1-15.
16. Matyjaszewski, K., *Macromolecules* **1998**, *31*, 4710-4717.
17. Matyjaszewski, K.; Woodworth, B. E., *Macromolecules* **1998**, *31*, 4718-4723.
18. Percec, V.; Barboiu, B.; Kim, H. J., *J. Am. Chem. Soc.* **1998**, *120*, 305-316.
19. Massey, A. G. "Copper" in *Comprehensive Inorganic Chemistry*; Bailar, J. C., Emeleus, H. J., Nyholm, R. and Trotman-Dickenson, A. F., Ed.; Pergammon: Oxford, 1973; Vol. 3, pp 1-78.
20. McKenzie, E. D., *Coord. Chem. Rev.* **1971**, *6*, 187-216.
21. Pallenberg, A. J.; Koenig, K. S.; Barnhart, D. M., *Inorg. Chem.* **1995**, *34*, 2833-2840.
22. Haddleton, D. M.; Duncalf, D. J.; Kukulj, D.; Crossman, M. C.; Jackson, S. G.; Bon, S. A. F.; Clark, A. J.; Shooter, A. J., *Eur. J. Inorg. Chem.* **1998**, 1799-1806.
23. Haddleton, D. M.; Duncalf, D. J.; Kukulj, D.; Heming, A. M.; Shooter, A. J.; Clark, A. J., *J. Mater. Chem.* **1998**, *8*, 1525-1532.
24. Haddleton, D. M.; Clark, A. J.; Duncalf, D. J.; Heming, A. M.; Kukulj, D.; Shooter, A. J., *J. Chem. Soc., Dalton Trans.* **1998**, 381-385.
25. Levy, A. T.; Olmstead, M. M.; Patten, T. E., *Inorg. Chem.* , submitted.
26. Haddleton, D. M.; Shooter, A. J.; Heming, A. M.; Crossman, M. C.; Duncalf, D. J.; Morsley, S. R. "Atom Transfer Radical Polymerization of Methyl Methacrylate: Effect of Phenols" in *Controlled Radical Polymerization*; Matyjaszewski, K., Ed.; American Chemical Society: Washington, D. C., 1998; Vol. 685, pp 284-295.
27. Matyjaszewski, K.; Davis, K.; Patten, T. E.; Wei, M. L., *Tetrahedron* **1997**, *53*, 15321-15329.
28. Wang, J. L.; Grimaud, T.; Matyjaszewski, K., *Macromolecules* **1997**, *30*, 6507-6512.
29. Davis, K. A.; Paik, H.-J.; Matyjaszewski, K., *Macromolecules* **1999**, *32*, 1767-1776.
30. Haddleton, D. M.; Crossman, M. C.; Dana, B. H.; Duncalf, D. J.; M., H. A.; Kukulj, D.; Shooter, A. J., *Macromolecules* **1999**, *32*, 2110-2119.

Chapter 15

Atom Transfer Polymerization Mediated by Solid-Supported Catalysts

David M. Haddleton, Arnaud Radigue[1], Dax Kukulj[2], and David J. Duncalf[3]

Department of Chemistry, University of Warwick,
Coventry CV4 7AL, United Kingdom

Atom transfer polymerization of methyl methacrylate mediated by heterogeneous supported catalyst is described. Primary amino functional silica and crosslinked polystyrene are transformed to supported Schiff base ligands by reaction with 2-pyridine carbaldehyde. These supported ligands react with copper(I) bromide to produce active catalysts for living polymerization of methyl methacrylate with an activated alkyl bromide initiator. All catalysts lead to efficient polymerization with a linear increase in Mn with conversion, acceptable values of Mn and polymerization following first order kinetics, as is typical for this chemistry. Molecular masses are higher than targeted and PDi is approximately 1.6–2.0 and 1.4–1.6 for both silica and polystyrene supported catalysts respectively. Block co-polymers have been prepared using a macro-initiator approach. In all cases there are undetectable levels of copper present in the products and the catalyst may be re-used in subsequent polymerization reactions.

Living free radical polymerization of vinyl monomers is of increasing interest for the controlled polymerization of a wide-range of vinyl monomers to polymers with

---

[1]Current address: Elf Atochem, Centre d'application de Levallois, 95 rue Danton, 92300 Levallois-Perret, France.
[2]Current address: Unilever Research, Quarry Bank Road, Port Sunlight, United Kingdom.
[3]Current address: Avecia, Blackley, Manchester, United Kingdom.

designed architecture (1-4). The initial breakthrough in this area came in the form of using stable organic free radicals, such as TEMPO, to regulate chain growth as pioneered by Rizzardo, Solomon and Moad (5) and greatly developed by Georges (6, 7) and Hawker (8). Although Hawker has made major advances in optimizing this stable free radical polymerization for styrenes and acrylates, methacrylates still remain an elusive goal. The use of transition metal complexes in an atom transfer process has been the focus of attention since Sawamoto (2) and Matyjaszewski (1) realized the potential of this approach developed primarily for atom transfer cyclization reactions. Sawamoto has been mainly developing a $Ru(PPh_3)_3Cl_2$ catalyst in conjunction with aluminum triisopropoxide whilst Matyjaszewski has carried out a tremendous amount of work using copper(I) bipyridine and aliphatic amine complexes. An ever increasing number of other metal catalysts have been developed based on nickel (9, 10), copper(0) (11), palladium (12), rhodium (13), iron (14), etc. We have been concentrating on a copper(I) bromide catalyst stabilized by complexation with Schiff base ligands of type 1 where R = alkyl chain with no branching at the α-position, primarily for the controlled polymerization of methacrylates (15-18), Figure 1.

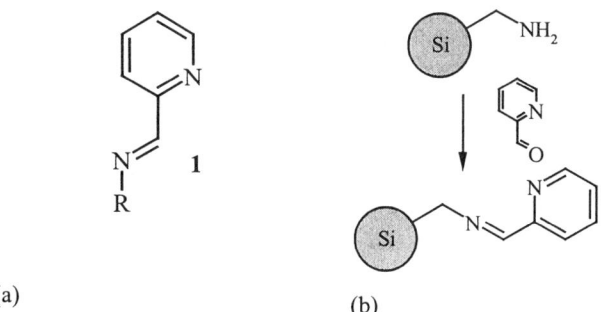

(a)   (b)

Figure 1 (a) Schiff base ligand for homogeneous atom transfer polymerization (b) condensation of pyridine 2-carbaldehyde with aminated support to give supported ligand

Although catalytic activity has been improving with new generation ligands and additives, resulting in lower levels of catalyst being required for acceptable rates of polymerization, the residual catalyst in the products remains problematic. For example, atom transfer polymerization mediated by copper(I) Schiff base complexes in toluene solution typically results in an extremely dark brown/black solution. The color can be removed from the product by passing the solution through a column of basic alumina and celite followed by precipitation, typically into aqueous media. However, this raises two major issues, namely cost of catalyst removal and inability to re-use the catalyst in subsequent polymerization reactions. A potential way to overcome these problems is to support the catalyst on an insoluble support that can be subsequently separated from the product by a simple filtration/decanting step. We have previously reported our initial experiments in this area where a copper (I) catalyst is covalently linked to a cross-linked polystyrene particle or aminated silica gel and also the adsorption/complexation of $Ru(PPh_3)_3Cl_2$ to primary amine functional insoluble supports (19-20), Figure 1. This approach has the potential to leave no, or very small amounts of, catalyst in the product and also for re-use of the catalyst in either repeated batch or a continuous process. This paper reports further our results for supported copper(I) complexes for atom transfer polymerization of methacrylates. Since submission of this paper Matyjaszewski has published on the use of immobilized catalysts for atom transfer polymerization (21).

EXPERIMENTAL

GENERAL METHODS AND ANALYSIS

All reagents were used as received without further purification. 2-Pyridine carbaldehyde was obtained from Avocado. Potassium phthalimide, hydrazine monohydrate, and 3-aminopropyl silica gel (~9% functionalised, 1.05 mmol of NH_2/g of support) and ethyl 2-bromoisobutyrate (98+%, E2BI) were obtained from Aldrich and used as received without further purification. CuBr (Avocado, 98 %) was purified according to the method of Keller and Wycoff. (22) Cross-linked chloromethylated polystyrene beads (~ 4 mmol of Cl g^{-1}, 1 % DVB), para pure and mixture of para/meta chloromethylated beads, were obtained from Avecia.

Molecular weight distributions were measured using size exclusion chromatography (SEC), on a system equipped with a guard column and one mixed E column (Polymer Laboratories) with differential refractive index detection using tetrahydrofuran at 1 mL min^{-1} as an eluent. Poly(MMA) standards in the range (6 × 10^4 to 200 g mol^{-1}) were used to calibrate the SEC. Elemental analysis was performed using a CE440 Elemental Analyser, Leeman Labs Inc. Inductively Coupled Plasma experiments were performed on a PS series ICP/AES apparatus, Leeman Labs Inc. FTIR was carried out with a Bruker Vector 22 spectrometer equipped with a Golden Gate ATR accessory. Cross polarization magic angle spinning NMR experiments were carried out using adamantane as standard on a CMX 600 Chemagnetics 150.9 MHz NMR spectrometer. The ^1H 90° pulse length was ~5 μs, giving a spin-lock and decoupling field strength of 37.5 kHz. The experiments were generally carried out with a 1 ms contact time. The MAS probe was 9.5 mm.

TYPICAL PROCEDURE FOR SYNTHESIS OF SUPPORTED CATALYST;

Phthalimidomethylated cross-linked polystyrene beads (PSX): To a stirred suspension of cross-linked chloromethylated beads (3 g, 12 mmol) in DMF (100 ml) was added potassium phthalimide (11.19 g, 60.4 mmol). The reaction mixture was subsequently heated at 110 °C for 7 h. After cooling, toluene (100 ml) was added and the reaction mixture was filtered and subsequently washed with water (100 ml), methanol (100 ml) and diethyl ether (100 ml). The product was dried under vacuum at ambient temperature for one day and subsequently at 60°C overnight in a vacuum oven. Yield = 4.15 g. IR: 1710, 1770 cm^{-1} (v C=O). EA: 80.64 %C, 5.85 %H, 3.49 %N

Aminomethylated cross-linked polystyrene beads: To a stirred suspension of the phthalimide functional beads (4.07 g, 16.3 mmol) in ethanol (150 ml) was added hydrazine monohydrate (4.6 ml, 0.147 mol). The reaction mixture was heated at 80 °C for 3 h prior to cooling to room temperature and being left overnight. The reaction mixture was filtered and the solid washed with water (100 ml), methanol (50 ml) and diethyl ether (50 ml). The solid was dried under vacuum at ambient temperature for one day and at 60 °C overnight in a vacuum oven. The product was recovered as a white solid, yield = 3.24 g. IR : 1650, 1600, 1490 cm^{-1}: EA: 76.61 %C, 6.56 %H, 8.48 %N.

Pyridiniminemethylated cross-linked polystyrene beads: To a suspension of amino functional support (1.94 g, 7.74 mmol NH$_2$) in toluene (50 ml) was added 2-pyridine carbaldehyde (1.661 g, 15.3 mmol). The mixture was heated under reflux (130 °C) in a soxhlet extractor where the thimble contained 3A molecular sieves. The support was removed by filtration and washed successively with THF (50 ml), methanol (50 ml) and diethyl ether (50 ml) to give, after drying under reduced pressure at ambient temperature and 60 °C overnight, to constant weight, an orange solid, yield = 2.18 g. IR: 1650, 1600, 1490 cm^{-1} (v C=N). EA: 81.06 %C, 6.5 %H, 8.05 %N.

Pyridiniminemethylated cross-linked silica: As above for polystyrene beads except 3-amino propyl functionalised silica gel. Support S1 isolated as a bright orange solid. S1 was obtained from a 6 hour reflux, S2 – S4 had overnight refluxing whilst S5 was prepared at ambient temperature.

TYPICAL POLYMERIZATION PROCEDURE

Crosslinked polystyrene support. In a typical reaction, CuBr (0.134 g, 9.34 x 10^{-4} mol) was placed in a predried Schlenk flask, which was evacuated and then flushed with nitrogen three times. The ligand was then added in the flask (except in the case of N-^{n}pentyl-2-pyridylmethanimine that was added by syringe after toluene). Deoxygenated toluene (20 ml, 66% v/v) and deoxygenated methyl methacrylate (10 mL, 9.36 x 10^{-2} mol) were added and the suspension stirred. The flask was heated in a thermostatted oil bath at 90 °C and after the temperature had equilibrated ethyl-2-bromoisobutyrate (0.137 mL, 9.34 x 10^{-4} mol, [MMA]$_0$:[In]$_0$=100:1) was added. Samples (1-2 ml) were taken periodically after initiator was added. Conversions were calculated by gravimetry heating sample to constant weight overnight at 90 °C under vacuum. The polymer was then diluted in THF and passed through basic aluminium oxide in order to remove the copper catalyst, which has gone into solution.

RESULTS AND DISCUSSION

SYNTHESIS AND CHARACTERISATION OF SUPPORTED CATALYSTS

Pyridinal imine functionalized crosslinked polystyrene beads (PSX) were synthesised as in Scheme 1, by reaction of pyridine 2-carbaldehyde with primary amino functional beads from reduction of phthalimide prepared from chloromethylated beads. Pyridinal imine functionalized silica (SX) is prepared by similar reaction of the aldehyde with commercially available amino propyl functional silica. Functionalization is most conveniently followed by ATR-FTIR, Figure 2, and ^{13}C MAS NMR.

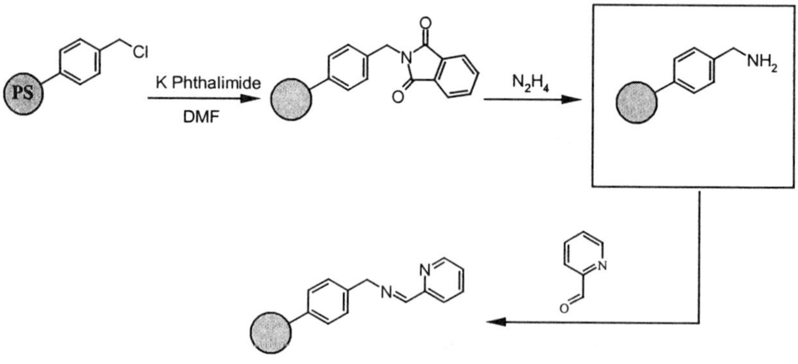

Scheme 1

POLYMERIZATION OF METHYL METHACRYLATE WITH SILICA SUPPORTED CATALYSTS

Addition of copper(I) bromide to pyridinal imine functionalized silica in a methyl methacrylate solution of toluene results in a dark brown/black suspension in a pale/colorless solution. Heating the solution for up to 6 hours at 90 °C (conventional conditions for homogeneous solution atom transfer using this type of ligand) with E2BI as initiator results in relatively high conversion to polymer, Table 1. ATP1 is a conventional homogeneous polymerization given for comparison purposes. SS1 and SS2 were carried out with the addition of primary amino functional silica, as control reaction without the addition of an initiator. Even when the catalyst is not efficiently

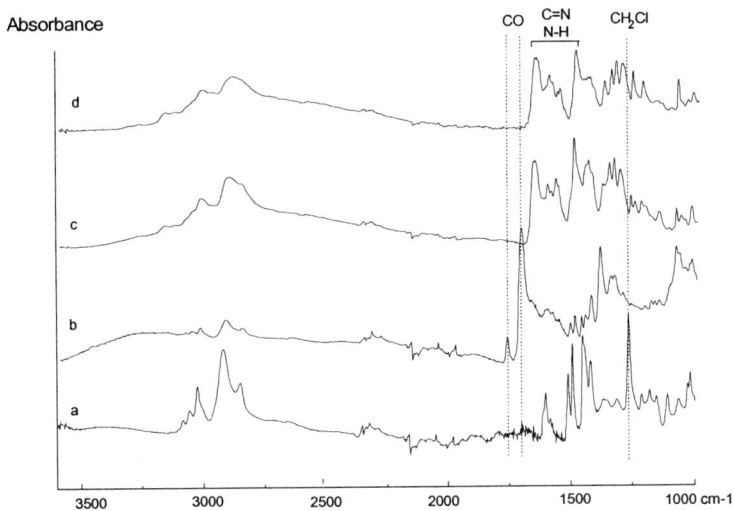

Figure 2. ATR-FTIR, showing the stepwise functionalization of cross-linked polystyrene beads to (a) -CH_2Cl (b) phthalimido (c) NH_2 and (d) pyridinal imine.

complexed to the support polymerization ensues. However, the Mn in these cases is very high, with PDi close to 2.0. It appears that under these conditions conventional free radical polymerization occurs; we are unsure of how the initiator is formed in this case. When pyridinal imine functionalized support is used polymerization proceeds at a higher rate, Figure 3a, reaching high conversion in 6 hours with PDi being significantly narrower. Figure 3b compares the rates of polymerization with supports S2, S3, S4 and S5. It is noticeable that silica supported ATP is reproducible in terms of kinetics, when the support has been synthesized at different times. For supports S2, S3 and S4, which have the same ligand loading, polymerization rates are essentially identical. Reproducibility within experiments with the same support is also observed. When the ligand functionalization is carried out at room temperature (support S5), the polymerization follows the same kinetics but gives higher molecular weights with PDi increasing to 2.1. It is noted that following the polymerization period when stirring is removed and the reaction mixture is cooled to ambient temperature the dark brown/black supported catalyst falls from solution *leaving a straw colored solution.*

One of the main aims for this work was to be able to isolate and re-use the catalyst. Therefore re-cycling experiments were carried out, Figure 3c. Polymerization was first carried out using 3 equivalents of silica supported ligand with respect to copper ([MMA]:[Cu]:[Si-lig S4]:[initiator] = [100]:[1]:[3]:[1]), subsequently the polymer

Table 1. Selected data for the atom transfer polymerization of MMA mediated by silica supported copper(I) catalyst.

	Support	[L]/[Cu]	Time (min)	Conv. (%)	Mn_{th} (g/mol)	$Mn_{(SEC)}$ (g/mol)	PDi
ATP1	solution	2	60	15	1 500	3 430	1.14
			360	80	8 010	9 050	1.11
SS1	SiNH$_2$	1	60	13	1 300		
			300	34	3 400	182 800	2.10
SS2	SiNH$_2$	2	60	19	1 900		
			300	52	5 200	146 300	1.94
SSATP1	S1	1	60	27	2 700	19 700	1.63
			360	67	6 700	18 500	1.80
SSATP2	S2	1	60	33	3 300	12 250	1.59
			360	75	7 510	15 950	1.56
SSATP3	S2	2	60	48	4 800	12 200	1.60
			360	98	9 810	14 900	1.68
SSATP5	S3	2	30	35	3 500	12 800	1.68
			250	86	8 610	15 500	1.71
SSATP6	S4	2	30	36	3 600	12 800	1.68
			260	91	9 110	16 350	1.78
SSATP7	S5	2	30	30	3 000	18 900	2.10
			300	91	9 110	16 500	2.10

$Mn_{(th)} = ([M_{MMA}]_0/[I]_0 \times MW_{MMA}) *$ conversion, where MW_{MMA} is the molecular weight of methyl methacrylate and $[M_{MMA}]_0/[I]_0$ is the initial concentration ratio of MMA to initiator.

solution was removed from the Schlenk tube via syringe following settling of the support. The support was washed 3 times with deoxygenated toluene and subsequently dried in vacuuo. During this procedure, the support was kept in contact with nitrogen in order to avoid any unnecessary deactivation by contact with air. The washed support was then reused for a second polymerization by introducing, in the following order, 20 ml of toluene, 10 ml of MMA and 0.137 ml of E2BI ([MMA]:[Cu]:[Silig]:[E2BI] = 100:1:3:1). In total three recycling polymerizations were experimented with the same support, Figure 3c and Table 2.

The rate of polymerization decreases upon each recycling until 2 recycling stages where the rate remains relatively constant. The decrease in rate is ascribed to a deactivation of the support by a build up of copper(II) through the termination of polymer chains by radical-radical termination reactions. No attempt to re-generate the support in-between experiments was carried out. An alternate explanation would be a leaching of copper from the support. However, no color is observed in the product solution and ICP measurements for residual copper on the silica support find no detectable trace of copper in the polymer solution following polymerization (detection limit of 0.0054 ppm*). This is typical for all copper(I) supported polymerizations.*

243

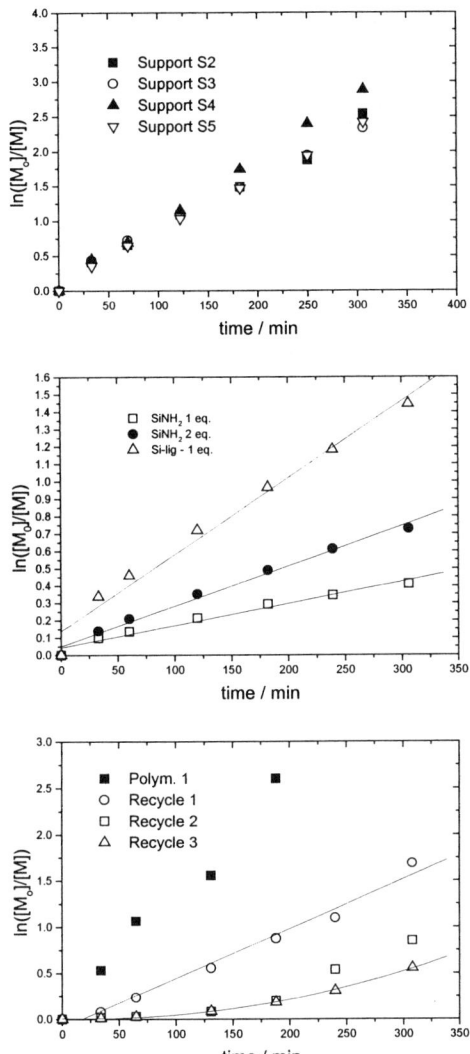

Figure 3. Silica supported atom transfer polymerization of MMA (a) rate of polymerization in the presence of aminated silica and complexed silica (b) rate of polymerization for supported pyridinal imine catalyst and (c) effect of re-cycling the supported catalyst

Table 2. Recycling experiments carried out with support S4 for the polymerization of MMA by silica supported atom transfer polymerization; [MMA]:[Cu]:[Si-lig]:[E2BI] = 100:1:3:1

Experiment	Time (min)	Conversion (%)	Mn_{th} (g/mol)	$Mn_{(SEC)}$ (g/mol)	PDI
First	30	41	4 100	11 600	1.76
	180	90	9 010	13 800	1.80
Recycling 1	130	43	4 300	13 900	1.75
	330	81	8 110	16 850	1.69
Recycling 2	130	8	800		
	360	57	5 700	17 100	1.69
Recycling 3	130	8	800		
	310	43	4 300	17 200	1.70

Experiments were carried out in order to determine at which loading of silica supported ligands the polymerization works most effectively. Moreover from Table 1 it is noticeable that when 2 equivalents of supported ligand are used the polymerization rate is increased. All reactions were carried out with the same conditions as before; [MMA]:[CuBr]:[Si-lig S2]:[E2BI] = 100:1:n:1, with n = 1, 2, 3 then 4 (where n is the proportion of ligand linked on silica compared to copper), Figure 4. As might be expected the rate of polymerization increases with an increasing amount of catalyst in the system. An effect on Mn on increasing the amount of support is observed. If the loading of supported ligand is increased, the molecular weight seems to be better controlled and the initiator efficiency is improved. An important aspect of living polymerizations is the ability to prepare block copolymers. Block copolymers were prepared by synthesis of poly(methyl methacrylate) macro-initiators by silica supported atom transfer polymerization. Reactions were stopped after approximately 60% conversion. Polymer solutions were separated from the support and subsequently used as initiators in homogeneous atom transfer polymerization for methyl and benzyl methacrylate. The targeted M_n of the second block was higher than 10 000 g mol^{-1}.

The macro-initiator polydispersity was considered to be 1.0 for these calculations, results are given in Table 3. SEC of the re-initiation products, *Figure 5*, shows effective re-initiation does indeed. However, a significant quantity of dead chains are observed as the tail on the SEC trace. The UV and RI traces from the SEC are very similar this being good evidence for the formation of block copolymers.

POLYSTYRENE SUPPORTED ATOM TRANSFER POLYMERIZATION

Atom transfer polymerization of methyl methacrylate mediated by cross-linked polystyrene supported catalysts (PS-SATP) was carried out under a range of

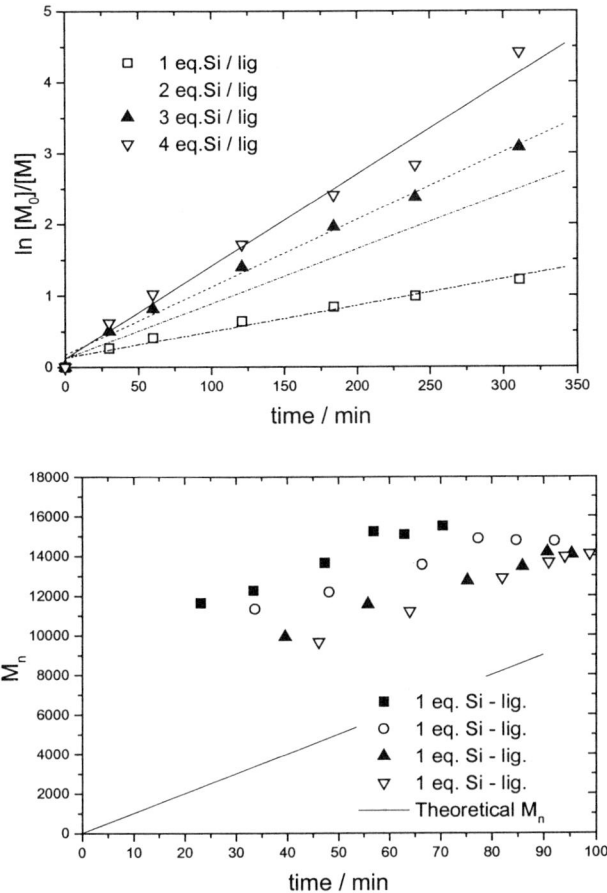

Figure 4, effect of varying the support to copper ratio on the polymerization of MMA (a) kinetic plots (b) dependence of Mn on conversion.

Table 3. Re-initiation experiments carried out by homogeneous ATP from SSATP-made P(MMA) macroinitiators

Experiment	Mass targeted	Time (min)	Conv % 2nd pol°	Mn_{th} (g/mol)	$Mn_{(SEC)}$ (g/mol)	PDi
R1	73 500	0	0		18 525	1.57
+ MMA		35	5.25	20 600	19 700	1.52
		310	34.2	36 500	30 800	1.62
R2	78 900	0	0		17 700	1.61
+ BzMA		35	12.7	25 500	22 800	1.73
		310	65.1	57 500	49 400	1.89

Table 4. Data summary for Polystyrene Supported Atom Transfer Polymerizations of MMA in toluene

Type	Support	[lig]/[Cu]	Time (min)	Conv. (%)	Mn_{th} (g/mol)	$Mn_{(SEC)}$ (g/mol)	PDI
ATP1	Solution	2	360	80.0	8 010	9 050	1.11
PS-SATP	PS2	1.25	300	77.5	7 760	15 590	1.66
PS-SATP	PS6 (A)	1.25	300	75.2	7 530	11 670	1.56
PS-SATP	PS6 (B)	1.25	300	73.5	7 360	11 370	1.42
PS-SATP	PS4	2	310	84.0	8 400	11 150	1.63
PS-SATP	PS7 (A)	1	300	53.3	5 300	12 050	1.45
PS-SATP	PS7 (B)	2	31	28.5	2 800	7 580	1.39

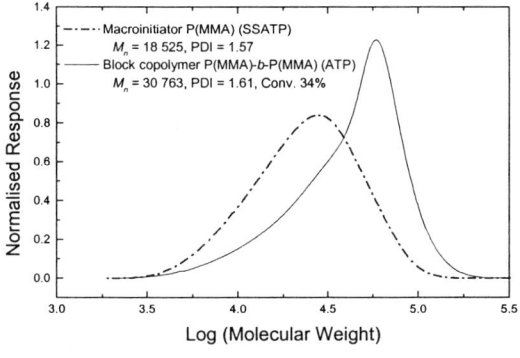

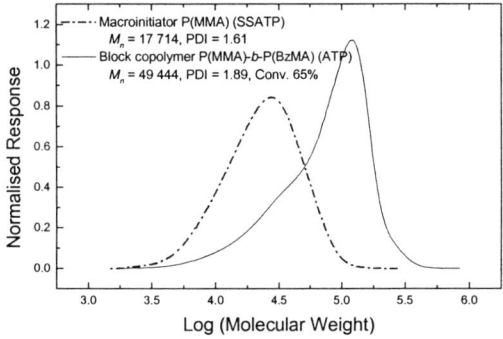

Figure 5. SEC traces for the re-initiation of macro-initiators from solid supported atom transfer polymerization (a) MMA as second block and (b) benzyl methacrylate as second block

conditions, Table 4. Data for the homogeneous polymerization with N-^{n}pentyl-2-pyridine methanimine ligand is given for comparison (ATP1). As with silica-supported mediated polymerization, polymerization occurs with poly(styrene) supported catalysts. In general poly(styrene) supports lead to better control of the polymerization. The Mn increases fairly linearly with conversion after the formation

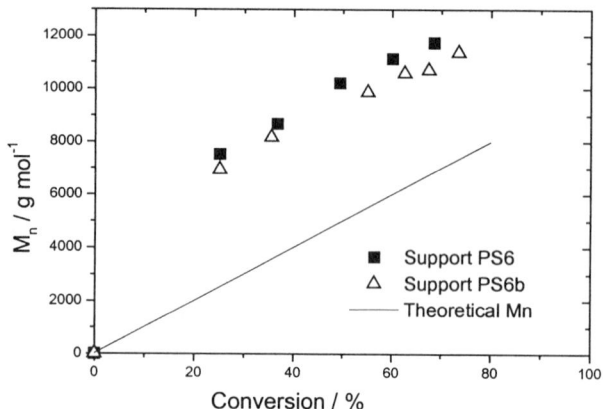

Figure 6. Molecular weights dependence versus conversion reproducibility using support PS6 for polystyrene supported atom transfer polymerization

of higher than expected Mn at the initial stages of the reaction, Figure 6. Observed M_n's are much lower than the ones obtained from silica supported mediated polymerization. Furthermore, PDi's are considerably lower (PDI ~ 1.4) than in the case of silica with *no* copper residues being detectable in the homogeneous medium

Polystyrene supported atom transfer polymerization is very reproducible, in terms of, kinetics, molecular weights and PDi's. Figure 7 shows kinetic plots for different using supports PS6 and PS2 at the same [support]:[Cu] ratio (1.25:1). The kinetic plots are almost identical, even though the supports have been synthesised at different times. A second polymerization with support PS6 was carried out (PS6B), in order to confirm the reproducibility of an experiment with a non-varying support. Again reproducible data was collected with very similar kinetic behaviour and molecular weight progression with PDi as low as 1.4.

Normal commercial crosslinked polystyrene beads are usually a mixture of the 2 and the 4 chloromethyl isomers in most of the experiments reported here the pure 4-chloromethyl styrene beads were utilized. In the case of PS4 a commercial mixed resin was used this is directly compared with PS7, Figure 8. At the beginning of the polymerizations, the kinetics are similar for approximately the first 2 hours after

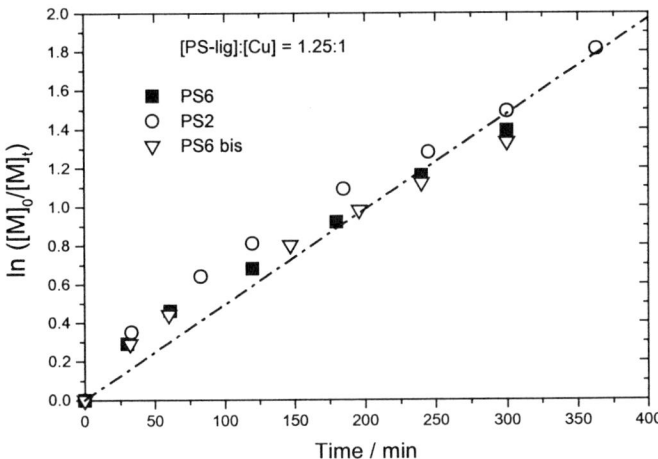

Figure 7. Kinetic plots for polystyrene supported atom transfer polymerization of MMA using supports PS2 and PS6.

which PS7 is slower. Molecular weight dependence on conversion remains similar with both resins but the PDi is higher for PS4 (PDI ~ 1.6). Consequently, using supports containing pure-para or mixture of meta and para chloromethylated styrene-co-styrene-co-divinyl benzene beads, after ligand functionalisation, has an effect on the subsequent atom transfer polymerization. This needs to be taken into consideration when using this approach.

RECYCLING OF SUPPORTED CATALYST

Figure 9 shows the first order kinetic plots for 2 successive recycles of the catalyst and table 5 final polymerization data. At each stage the catalyst is deactivated with a decrease in observed rate and a broadening of PDi. This is ascribed to a build up of copper(II) species on the support which will deactivate the polymerization. However, the catalyst is still active and the PDi remains quite acceptable. It is envisaged that an appropriate re-activation process could be applied to restore full catalyst activity.

250

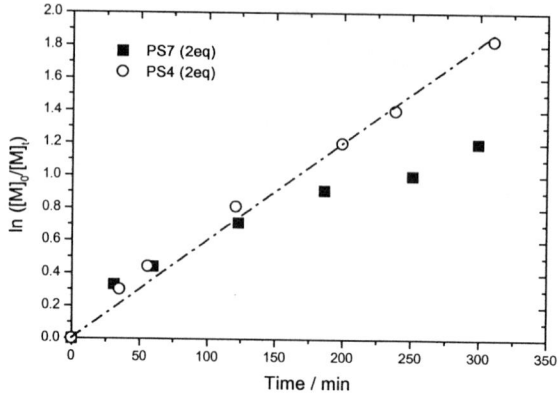

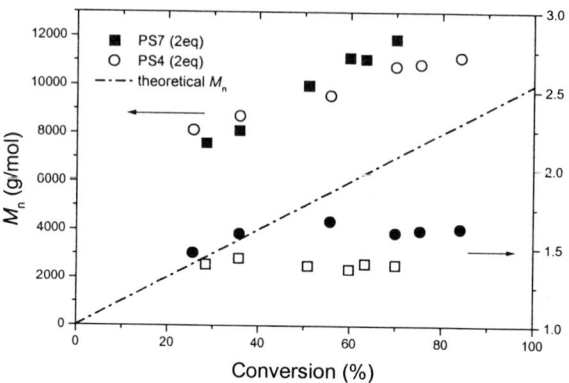

Figure 8. Mn and PDi (a) and first order kinetic plots (b) for poly(styrene) supported atom transfer polymerization of MMA using mixed and pure ligand functionalised polystyrene support PS4 and PS7.

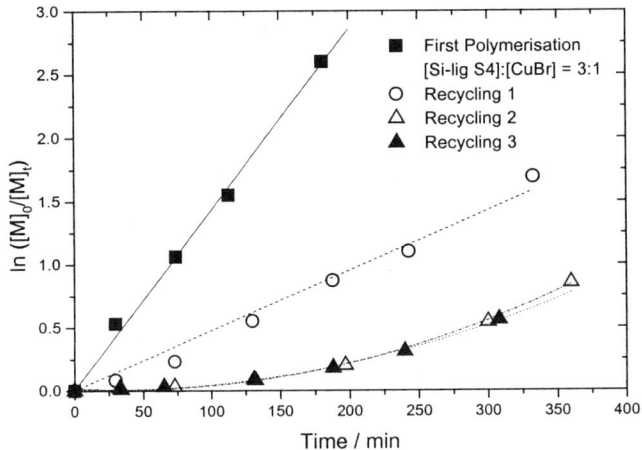

Figure 9. Recycling experiments carried out with support PS7 using the same conditions; [MMA]:[Cu]:[PS-lig]:[E2BI] = 100:1:2:1

Table 5. Recycling experiments carried out with support PS7 for the polymerization of MMA by polystyrene supported atom transfer polymerization; [MMA]:[Cu]:[PS-lig PS7]:[E2BI] = 100:1:2:1

Experiment	Time (min)	Conversion (%)	Mn_{th} (g/mol)	$Mn_{(SEC)}$ (g/mol)	PDi
First polym.	300	70.0	7 010	11 890	1.39
Recycling 1	329	52.1	5 210	16 560	1.61
Recycling 2	336	25.7	2575	13 310	1.65

CONCLUSIONS

Covalently bound pyridine imines to silica and crosslinked poly(styrene) may be easily synthesised by reaction of pyridine 2-carbaldehyde with primary amino functional resins. These supported ligands make ideal precursors for atom transfer polymerization catalysts by reaction with copper(I) bromide. Both types of support may be used to catalyze polymerization of methyl methacrylate under conventional atom transfer polymerization conditions. In general excellent control is achieved with an increase in Mn with conversion and approximately linear first order kinetic plots, indicative of living polymerizations. The PDi achieved is broader than under homogeneous polymerization conditions but less than 1.50 is achieved with organic supports. This approach leads to product with undetectable levels of metal and a supported catalyst, which may be re-used in a subsequent polymerization, with at present an associated loss in activity. Optimization of supported catalysts for atom transfer polymerization is an excellent method for producing catalyst free polymers.

ACKNOWLEDGMENTS

We thank D Pears and A Shooter (Avecia) for supply of polystyrene resins and EPSRC (D.K. GR/K90364, D.J.D. GR/L10314) and Elf-AtoChem (A.R.) for funding.

REFERENCES

1. Wang, J. S.; Matyjaszewski, K. *J. Am. Chem. Soc.* **1995**, *117*, 5614-5615.
2. Kato, M.; Kamigaito, M.; Sawamoto, M.; Higashimura, T. *Macromolecules* **1995**, *28*, 1721-1723.
3. Matyjaszewski, K.; Wang, J. L.; Grimaud, T.; Shipp, D. A. *Macromolecules* **1998**, *31*, 1527-1534.
4. Ueda, J.; Matsuyama, M.; Kamigaito, M.; Sawamoto, M. *Macromolecules* **1998**, *31*, 557-567.
5. Moad, G.; Solomon, D. H. *Aust. J. Chem.* **1990**, *43*, 215.
6. Georges, M. K.; Veregin, R. P. N.; Hamer, G. K. *Trends Polym. Sci.* **1994**, *2*, 66.
7. Veregin, R. P. N.; Odell, P. G.; Michalak, L. M.; Georges, M. K. *Macromolecules* **1996**, *29*, 2746.
8. Hawker, C. J. *Acc. Chem. Res.* **1997**, *30*, 373.
9. Uegaki, H.; Kotani, Y.; Kamigaito, M.; Sawamoto, M. *Macromolecules* **1998**, *31*, 6756-6761.
10. Uegaki, H.; Kotani, Y.; Kamigaito, M.; Sawamoto, M. *Macromolecules* **1997**, *30*, 2249.
11. Vander Sluis, M.; Barboiu, B.; Pesa, N.; Percec, V. *Macromolecules* **1998**, *31*, 9409-9412.
12. Lecomte, P.; Drapier, I.; DuBois, P.; Teyssie, P.; Jerome, R. *Macromolecules* **1997**, *30*, 7631.

13. Moineau, G.; Granel, C.; Dubois, P.; Jerome, R.; Teyssie, P. *Macromolecules* **1998**, *31*, 542-544.
14. Moineau, G.; Dubois, P.; Jerome, R.; Senninger, T.; Teyssie, P. *Macromolecules* **1998**, *31*, 545.
15. Haddleton, D. M.; Jasieczek, C. B.; Hannon, M. J.; Shooter, A. J. *Macromolecules* **1997**, *30*, 2190-2193.
16. Haddleton, D. M.; Heming, A. M.; Kukulj, D.; Duncalf, D. J.; Shooter, A. J. *Macromolecules* **1998**, *31*, 2016-2018.
17. Haddleton, D. M.; Crossman, M. C.; Dana, B. H.; Duncalf, D. J.; Heming, A. M.; Kukulj, D.; Shooter, A. J. *Macromolecules* **1999**, *32*, 2110-2119.
18. Haddleton, D. M.; Kukulj, D.; Duncalf, D. J.; Heming, A. H.; Shooter, A. J. *Macromolecules* **1998**, *31*, 5201-5205.
19. Haddleton, D. M.; Kukulj, D.; Radigue, A. P. *Chem. Comm.* **1999**, 99-100.
20. Haddleton, D. M.; Duncalf, D. J.; Kukulj, D.; Radigue, A. P. *Macromolecules* **1999**, *32*, 4769-4775.
21. Kickelbick, G.; Paik, H. J.; Matyjaszewski, K. *Macromolecules* **1999**, *32*, 2941-2947.
22. Keller, R. N.; Wycoff, H. D. *Inorg. Synth.* **1947**, *2*, 1.

Chapter 16

Mechanistic Aspects of Catalytic Chain Transfer Polymerization

Johan P. A. Heuts[1,2], Darren J. Forster, and Thomas P. Davis[2]

School of Chemical Engineering and Industrial Chemistry,
University of New South Wales, Sydney, New South Wales 2052, Australia

A range of experiments has been conducted with the aim of elucidating the mechanism of catalytic chain transfer in methacrylic and styrenic polymerizations. Evidence for a diffusion-controlled transfer step in methacrylate polymerization is presented. The measured chain transfer constant, C_S, for styrene is relatively low when compared with the methacrylates. This low C_S in styrene systems can be (partly) attributed to the formation of Co-C bonds reducing the concentration of active Co(II) catalyst. The precise identification of the elementary steps in catalytic chain transfer remains elusive. The experimental observation that the molecular weight distribution of the polymer formed in catalytic chain transfer remains constant with conversion is at present inexplicable.

Introduction

Free-radical methods can be used to polymerize a vast range of (vinyl) monomers, because free-radical polymerization is tolerant to a wide range of functional groups including acid, amine, hydroxy and amide and is compatible with a wide range of solvents including water. However, a weakness inherently associated with free-radical polymerization is the difficulty in controlling the selectivity of the free radical reactions, which creates problems in controlling polymer architecture[1].

[1]Current address: GE Plastics, Plasticslaan 1, 4600 AC Bergen op Zoom, The Netherlands.
[2]Corresponding authors.

Catalytic Chain Transfer Polymerization.

Over the past decade new methods have emerged for the production of free radical polymers with, e.g., controlled molecular weights, narrow polydispersity indices, controlled compositions (e.g., block copolymers), and particular end functionalities[2]. One of these controlled radical polymerization techniques is catalytic chain transfer polymerization, in which certain low-spin Co(II) complexes catalyze a hydrogen-transfer reaction from the growing polymeric radical to the monomer, thus causing chain transfer to monomer (Scheme 1) and providing a means of molecular weight control[3,4]. From Scheme 1 it is clear that chains terminated by this process are characterized by a vinyl endgroup and that those chains initiated by this process contain a hydrogen "initiator" fragment. It is thus possible to produce macromers consisting solely of monomer units by this technique. Typical low-spin Co(II) complexes that have been used for this purpose are Co(II) prophyrins and cobaloximes (See Scheme 2).

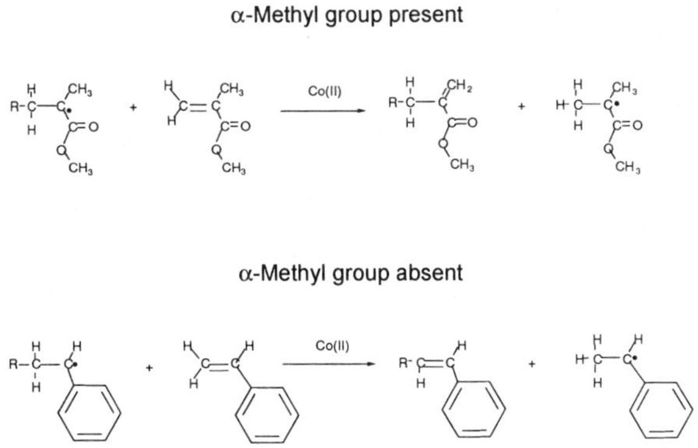

Scheme 1

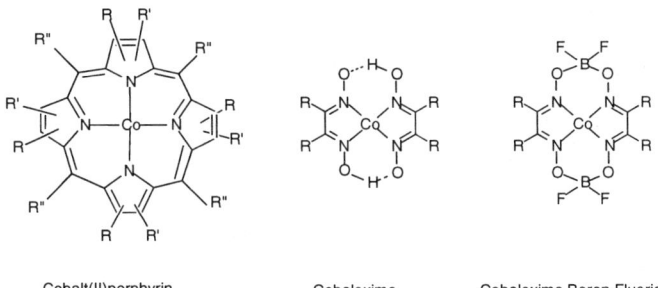

Cobalt(II)porphyrin   Cobaloxime   Cobaloxime Boron Fluoride

Scheme 2

Of the catalytic chain transfer agents investigated thus far, the cobaloximes have proven to be most active, with the BF_2-bridged derivatives being less prone to oxidation, and hence deactivation. Typical chain transfer constants (see below) for the cobalt porphyrins in methyl methacrylate polymerization at 60°C are 10^3 whereas those for the cobaloximes are typically an order of magnitude higher (mercaptans have chain transfer constants of ~ 0.1 – 10). These very high chain transfer constants combined with the catalytic nature of the process lead to a situation in which only ppm quantities of the catalytic chain transfer agent are required to produce degrees of polymerization as low as 10.

Applications.

As shown in Scheme 1 and stated earlier, catalytic chain transfer polymerization yields oligomeric species with a hydrogen initiator group and a vinyl endgroup, i.e. macromers. The reactive vinyl endgroups can potentially be used for a post-polymerization modification leading to the introduction of other functional groups, but they can also be used as reactive groups in a subsequent free-radical polymerization. In the latter case the behavior of the macromers will depend on the nature of the propagating radicals that add to the double bond of the macromer. If the propagating radical is of, for example, acrylic or styrenic nature, then the macromer will readily undergo a copolymerization, in which case an effective graft copolymerization is carried out (Scheme 3)[5]. However, if the propagating radical is of a methacrylic nature, then a β-scission takes place after the addition of the radical to the double bond, in which case the macromer effectively acts as a chain transfer agent and we have so-called addition-fragmentation polymerization (Scheme 4).[6]

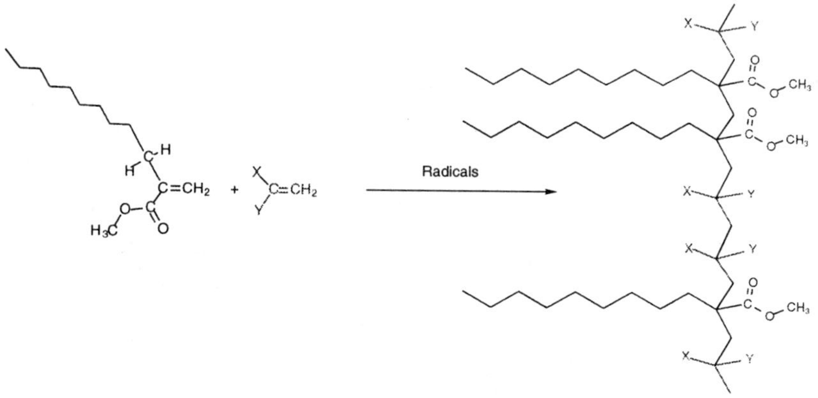

Scheme 3.

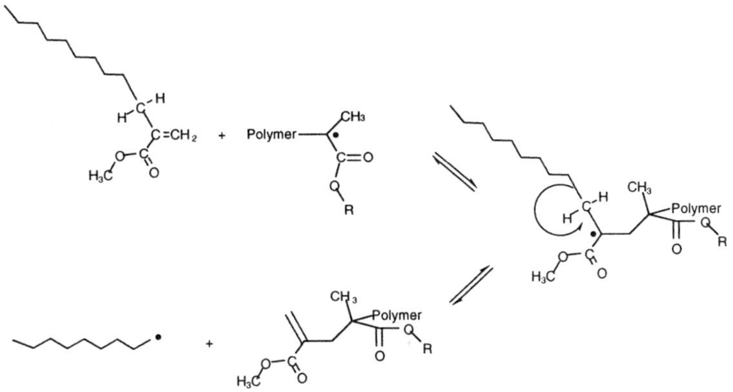

Scheme 4.

From the above it is clear that catalytic chain transfer polymerization provides a very effective technique for the production of oligomers with a vinyl end-functionality, which themselves can be used in a versatile way. Given the potential industrial importance of catalytic chain transfer polymerization, as evidenced by the numerous patents on this technique (see, for example, reference 3), it is remarkable that only very little evidence regarding the actual chain transfer mechanism is available. Despite many efforts to elucidate the mechanism by several research groups globally, the only uncircumstantial existing information[7] is that the β-hydrogens are abstracted from the growing radical and that they are somehow transferred to the monomer as shown in Scheme 1. In what follows we will make an attempt to describe our current understanding of catalytic chain transfer polymerization including some of the recent advances made in our group.

Experimental Procedures

Materials

The bis(methanol) complexes of the BF_2-bridged cobaloximes with R = CH_3 (i.e., bis[(difluoroboryl)dimethyl-glyoximato]cobalt(II) – COBF in the remainder of this chapter) and R = phenyl (i.e., bis[(difluoroboryl)diphenyyl-glyoximato]cobalt(II) – COPhBF) are prepared according to the method described by Bakac et al.[8], in the latter case replacing dimethyl glyoxime in the described procedure. Since purification and characterization of these catalysts is extremely difficult, only single batches of catalyst are used throughout a particular study. Monomers (typically: Aldrich, 99%), passed through a column of activated basic alumina (ACROS, 50-200 micron) prior to use, and solvents (typically of analytical grade) are purged with high purity nitrogen (BOC) for 1.5 hours prior to use. AIBN (DuPont) is recrystallized twice from methanol and used as initiatior.

Chain transfer constant measurements

Two stock solutions are prepared: (i) an initiator stock solution, and (ii) a catalyst stock solution. (i) The initiator solution is prepared by dissolution of approximately 400 mg of AIBN in 150 ml of monomer-solvent mixture. (ii) The catalyst stock solution was prepared by dissolution of approximately 3 mg of catalyst into 10 ml of solution (i) and a subsequent 10-fold dilution with solution (i). Five reaction mixtures are then prepared, each containing 4.0 ml of initiator solution (i) and 0.00, 0.10, 0.20, 0.30 and 0.40 ml of catalyst solution (ii), respectively. The reaction ampoules, specially modified for use with standard Schlenk equipment, are deoxygenated by two freeze-pump-thaw cycles and subsequently placed in a thermostated waterbath. At all times an oxygen-free atmosphere is maintained using standard syringe and vacuum techniques.

Polymerization reactions

A roundbottom flask is charged with the required amount of catalyst, AIBN and a magnetic stirrer bar, and subsequently sealed with a rubber septum. After (three times) evacuating and subsequent purging with nitrogen, the purged monomer and toluene are charged into the roundbottom flask using a canula, maintaining an oxygen-free system. At regular intervals samples are removed with a syringe, and the polymer is isolated by evaporation of the monomer. This polymer was subsequently used for conversion and molecular weight measurements.

Molecular weight analysis

Molecular weight distributions are determined by size exclusion chromatography using a GBC Instruments LC1120 HPLC pump, a Shimadzu SIL-10A Autoinjector, a column set consisting of a Polymer Laboratories 3.0 μm bead-size guard column (50 × 7.5 mm) followed by four linear PL columns (10^6, 10^5, 10^4 and 10^3) and a VISCOTEK Dual Detector Model 250 differential refractive index detector. Tetrahydrofuran (BDH, HPLC grade) is used as eluent at 1 ml/min. Calibration of the SEC equipment is performed with narrow polystyrene (Polymer Laboratories, molecular weight range: 580 - $3.0 \cdot 10^6$) and poly(methyl methacrylate) (Polymer Laboratories, molecular weight range: 200 - $1.6 \cdot 10^6$) standards.

Chain Transfer Constant Measurements

Chain Transfer Constants

The most important kinetic parameter describing the reactivity of a chain transfer agent is the chain transfer constant, C_S, which is the ratio of the chain transfer rate coefficient, $k_{tr,S}$, and the propagation rate coefficient, k_p:

$$C_S = \frac{k_{tr,S}}{k_p} \qquad (1)$$

This chain transfer constant is commonly obtained by using the Mayo[9] and/or chain length distribution (CLD)[10-12] procedures. Both procedures require knowledge of the overall kinetic scheme after which expressions are derived for the number average degree of polymerization (DP_n – Mayo procedure) and the high molecular weight slope of the number molecular weight distribution plotted in a semi-logarithmic way (Λ – CLD procedure). In both procedures, the overall chain transfer rates are required, and in order to obtain adequate expressions for the overall chain transfer rates, knowledge of the mechanism of the chain transfer reaction is required. For an "ordinary" chain transfer agent, e.g., a thiol, the following reaction occurs and reaction rate equation applies:

$$R_n^\bullet + S \xrightarrow{k_{tr,S}} P_n + S^\bullet \quad (2)$$

$$R_{tr,S} = k_{tr,S}[S]\sum_n [R_n^\bullet] \quad (3)$$

In these expressions $R_n^\bullet$ is a growing radical containing n monomer units, S the chain transfer agent, P_n a dead polymer chain containing n monomer units, $S^\bullet$ the chain transfer agent derived radical, and $R_{tr,S}$ the overall rate of chain transfer to S. The corresponding expressions for the determination of C_S in the Mayo and CLD procedures are then given by eqs 4 and 5, respectively, which simply express the respective parameters as the ratio of the rate of all chain stopping events (i.e., termination, chain transfer to monomer and chain transfer to chain transfer agent) and the rate of chain growth (i.e., propagation).

$$\frac{1}{DP_n} = (1+\lambda)\frac{\langle k_t \rangle [R^\bullet]}{k_p[M]} + C_M + C_S\frac{[S]}{[M]} \quad (4)$$

$$\Lambda = \lim_{M\to\infty}\frac{d\ln(P(M))}{dM} = -\left(\frac{\langle k_t \rangle [R^\bullet]}{k_p[M]} + C_M + C_S\frac{[S]}{[M]}\right)\frac{1}{m_0} \quad (5)$$

In these expressions, λ is the fraction of termination by disproportionation, $\langle k_t \rangle$ the average termination rate coefficient, $[R^\bullet]$ the overall radical concentration, [M] the monomer concentration, C_M the chain transfer constant to monomer, P(M) the number of polymer molecules with mass M (i.e., the chain length distribution), and m_0 is the monomer mass. Chain transfer constants are now obtained by measuring DP_n or Λ for a range of experiments in which the ratio of [S]/[M] is varied and by plotting DP_n^{-1} or $-\Lambda m_0$ against [S]/[M]. The resulting plots will be straight lines, provided that the termination term does not change significantly upon changing [S]/[M], of which the slopes are equal to C_S. In practice both these methods give identical and satisfactory results.

In the case of catalytic chain transfer chain transfer constants are determined in a similar way, also requiring an overall rate of chain transfer. As stated in the introduction, the actual mechanism of the catalytic chain transfer process has not yet been proven unambiguously, but the generally accepted working model is a two-step process which involves a hydrogen abstraction step by the Co(II) catalyst, and a subsequent hydrogen-transfer reaction between the resulting Co(III)–H complex and a monomer molecule:

$$R_n^{\bullet} + Co(II) \xrightarrow{k_{tr,Co}} P_n + Co(III)H \qquad (6)$$

$$Co(III)H + M \longrightarrow Co(II) + R_1^{\bullet} \qquad (7)$$

Assuming that the abstraction step of eq 6 is rate determining, then this situation is equivalent to the situation depicted in eqs 2 and 3, resulting in equivalent expressions for the Mayo (eq 4) and CLD (eq 5) procedures; in the case of catalytic chain transfer, the S in eqs 4 and 5 needs to be replaced by Co(II). A selection of chain transfer constants of catalytic chain transfer agents in a range of monomers, determined using the outlined procedures is given in Table 1. From this table it is clear that the rate coefficients for catalytic chain transfer reactions are very high, and that those obtained for cobaloximes in methyl methacrylate polymerization ($k_p \sim 10^3$ dm^3 $mol^{-1}s^{-1}$) are of the same order of magnitude as radical-radical termination rate coefficients ($\sim 10^7 - 10^8$ dm^3 mol^{-1} s^{-1}), which are known to be diffusion controlled.

Temperature Dependence

Valuable information regarding the mechanism of a particular reaction can be obtained from the Arrhenius parameters of the reaction rate coefficients. Several studies on methyl methacrylate polymerization reported in the literature[15,17,19] suggest that the chain transfer constants of cobaloximes are independent of temperature, which means that the Arrhenius frequency factor A_{tr} for the (bimolecular) chain transfer reaction is about 10^{10} dm^3 mol^{-1} s^{-1}, and the corresponding activation energy, E_{tr}, is about 23 kJ mol^{-1}. These Arrhenius parameters suggest that the rate determining step in the catalytic chain transfer reaction in methyl methacrylate polymerization is either a unimolecular reaction (in which case the units of A_{tr} are s^{-1}) or a diffusion-controlled one. Investigation of the former possibility is more difficult than that of the latter and therefore we investigated the latter possibility.

It was found previously by Myronichev et al.[20] that the chain transfer constant of a cobalt porphyrin continuously decreased with increasing ester chain length in a homologous series of methacrylates. This was ascribed to greater steric hindrances and particular ligation effects; these particular ligation effects should decrease with increasing temperature, but this was not further tested. An alternative explanation is that the observation can be ascribed to viscosity effects as the medium viscosity increases with increasing ester chain length in the monomer. In order to investigate these possibilities, we simultaneously investigated the temperature and viscosity effects in the chain transfer polymerizations of a range of methacrylates[17]. The results for the chain transfer constants of COBF in methyl (MMA), ethyl (EMA) and butyl methacrylate (BMA) at a range of temperatures is shown in Figure 1, and it is clear that the chain transfer constants of COBF in these three monomers do not display a significant temperature dependence, in accordance with earlier work reported slolely on MMA. It is also clear that the chain transfer constants decrease with increasing ester chain length in the monomer, which is consistent with the earlier study by Myronychev et al.[20]

Table 1. Chain Transfer constants for catalytic chain transfer reactions performed with a variety of monomers and catalysts

CCT agent	Monomer	T/°C	C_S	Ref.
cobalt porphyrins	styrene	60	$2 \cdot 10^2$	13
cobalt porphyrins	MMA	60	$2.4 \cdot 10^3$	13
cobalt porphyrins	BMA	60	$1.2 \cdot 10^3$	13
cobalt porphyrins	n-Nonyl MA	60	$1.05 \cdot 10^3$	13
cobalt porphyrins	n-Dodecyl MA	60	$8 \cdot 10^2$	13
COBF	styrene	60	$5 \cdot 10^2 - 1.5 \cdot 10^3$	14,15
COBF	MMA	60	$34 \cdot 10^3 - 40 \cdot 10^3$	14,15
COPhBF	styrene	60	$6.6 \cdot 10^2$	16
COPhBF	MMA	60	$25.3 \cdot 10^3$	16
COBF	BMA	60	$10 \cdot 10^3$	17
COBF	alphamethyl styrene	50	$8.9 \cdot 10^5$	18

In order to investigate whether the observed phenomenon can be explained by a diffusion-controlled reaction rate, it is interesting to consider the predicted kinetic behavior assuming a diffusion-controlled process. It is common to express the rate coefficient of a diffusion-controlled reaction in terms of the Smoluchowski equation, which states that the rate coefficient is proportional to the mutual diffusion coefficients of the two reactants, which can be approximated by the sum of the two self-diffusion coefficients:

$$k_{tr,Co} \propto D_{Co} + D_i \qquad (8)$$

In this expression D_{Co} is the diffusion coefficient of the cobalt catalyst and D_i the diffusion coefficient of an i-meric radical, which should be dependent on the chain length of the radical. Furthermore it is generally observed that diffusion coefficients are inversely related to the medium viscosity η in a manner similar to eq 9.

$$D \propto \frac{1}{\eta^\alpha} \qquad (9)$$

In this expression, the value of the exponent α lies generally between 0.5 and 1.0[17]. Combining eqs 8 and 9 (i.e., $k_{tr,Co} \propto \eta^{-\alpha}$) and realizing that C_S is defined s the ratio of $k_{tr,Co}$ and k_p, the following (approximate) proportionality is obtained for C_S in the case of a diffusion-controlled chain transfer reaction:

$$C_S \propto \frac{1}{k_p \eta^\alpha} \qquad (10)$$

which in turn implies that:

$$C_S k_p \eta^\alpha \approx \text{constant} \qquad (11)$$

This approximate equality was indeed found to be applicable to the catalytic chain transfer polymerizations of MMA, EMA and BA at 40, 50 and 60°C, as shown in Table 2 for α = 1 and COBF catalyst. The same observation was made for the larger COPhBF catalyst, which consistently displayed chain transfer constants aboput 50% less than those observed with COBF under otherwise identical conditions. This observation could possibly be explained by the larger cross-sectional area of COPhBF as compared to COBF which should lead to a slower diffusion, as was previously published by Haddleton et al.[19]

Two further tests of our diffusion-hypothesis were subsequently conducted. Firstly, we tested the validity of eq 11 for a very viscous methacrylate, i.e., 2-phenoxyethyl methacrylate (POEMA)[21].

Table 2. Comparison of Values for $C_S \cdot k_p \cdot \eta$ Determined at 40, 50, 60 and 70°C for Methyl, Ethyl and Butyl Methacrylate Polymerizations in the Presence of COBF

Temperature/°C	$C_S \cdot k_p \cdot \eta$		
	MMA	EMA	BMA
40	7.4×10^6	7.2×10^6	7.1×10^6
50	10×10^6	9.6×10^6	8.6×10^6
60	10×10^6	9.6×10^6	9.5×10^6
70	9.4×10^6	9.6×10^6	11×10^6

For this monomer, the propagation rate coefficient, chain transfer constant to COPhBF and viscosity were measured at 60°C and the product of these three parameters was found to be $5.8 \cdot 10^6$, which is in good agreement with that found for MMA (i.e., $5.7 \cdot 10^6$).

Secondly, we investigated the catalytic chain transfer polymerization of methyl methacrylate in the presence of COPhBF in supercritical CO_2, which has a very low viscosity[22]. Although we were not able to determine an accurate value for the chain transfer constant in this medium because of the very low-molecular weight polymer that was formed, the results suggested that the chain transfer rate coefficient ($k_{tr,Co}$) was at least a factor of ten higher.

Styrene Polymerizations.

The discussion regarding diffusion-controlled reaction rates in the catalytic chain transfer polymerization of the methacrylates does not seem to apply to the catalytic chain transfer polymerization of styrene. Firstly, reported chain transfer constants for COBF and COPhBF in styrene are about one to two orders of magnitude lower than those reported in MMA, leading to values of $k_{tr,Co}$ of the order of 10^5 dm^3 mol^{-1} s^{-1}, which are clearly too low for a diffusion-controlled reaction under ordinary free-radical polymerization conditions. Secondly, the only study[15] on the temperature dependence of the catalytic chain transfer polymerization of styrene that we are aware of that has appeared in the literature to date indicates very different trends when compared to the methacrylates. Although there is a large scatter in the data, a negative activation energy and a frequency factor of $\sim 10^3$ dm^3 mol^{-1} s^{-1} were found for $k_{tr,Co}$ in styrene. It is clear that a negative activation energy is not consistent with an ordinary chemically-controlled reaction rate, and that the low frequency factor is not consistent with a diffusion-controlled reaction rate. In what follows we will attempt to explain these (apparently) disparate and contradictory results.

Chain Transfer Mechanism

General Aspects.

Earlier we discussed the determination of chain transfer constants (C_S) in catalytic chain transfer polymerization and the possibility of a diffusion-controlled rate as an explanation for the observed chain transfer coefficients ($k_{tr,Co}$) in the methacrylates. Furthermore, we briefly discussed that the (relatively limited) data on the catalytic chain transfer polymerization of styrene suggest a very different behavior. The question now arises as to how this different behavior could possibly be explained. To answer this question, it is important to consider the actual reaction mechanism of the catalytic chain transfer process.

As already eluded to in the introduction, no conclusive evidence exists regarding the actual chain transfer mechanism other than that presented in Scheme 1, i.e., a β-hydrogen is abstracted from the growing radical and transferred to the monomer. A good understanding of the mechanism is obviously for practical reasons important, because it will enable the development of better catalyst systems. Alternatively, and more fundamentally, an incorrect understanding of the mechanism may lead to misinterpretation of experimental observations. At this point it is therefore important to state that the discussions in the previous section were based upon chain transfer constants which were determined by means of eqs 4 and 5, which in turn were derived *assuming* the mechanism given by eqs 6 and 7 is correct. Although this mechanism seems to be the most conceivable and generally accepted mechanism to date, it should be borne in mind that this mechanism has not yet been unambiguously proven and that therefore there is a chance that the chain transfer constants measured thus far have no other meaning than just being the result of a regression between C_S and [Co]/[M].

Possible Mechanisms

In this section, three conceivable, kinetically distinct alternatives of the catalytic chain transfer mechanism will be discussed, i.e., the two-step mechanism involving Co(III)H intermediates (mechanism I), the reaction of a radical coordinatively bound radical to the cobalt center with a monomer (mechanism II) or the reaction of a monomer-cobalt complex with a free radical (mechanism III).

Mechanism I
This mechanism [13], involving the two consecutive reactions as shown in eqs. 12 and 13, is the most widely accepted version of the catalytic chain transfer process

and it is this mechanism on which all the chain transfer constants published to date have been based upon (see eq.6).

$$R_n^\bullet + Co(II) \xrightarrow{k_{tr,Co}} P_n + Co(III)H \qquad (12)$$

$$Co(III)H + M \rightarrow Co(II) + R_1^\bullet \qquad (13)$$

This mechanism is able to explain many kinetic observations, and several studies indeed suggest the existence of a Co(III)H intermediate.

The question now arises as to how the hydrogen abstraction step takes place. Two possible scenarios are a direct radical pathway, similar to an "ordinary" disproportionation reaction between two radicals, or a coordination pathway, in which a β elimination of a hydrogen from a coordinated radical takes place (Scheme 5).

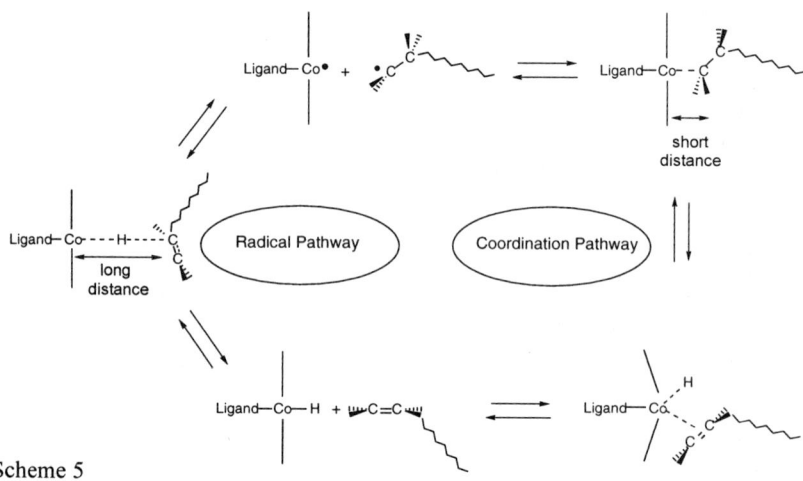

Scheme 5

In either case the overall reaction is given by eq 12, but the involved transition states in these two pathways are significantly different. It is conceivable that the radical pathway requires a less tight transition structure than the coordination pathway; this important for the design of new catalysts. We refer to the work of Gridnev et al. [23] for a clear and thorough additional discussion of this mechanism.

Mechanism II

This mechanism is somewhat related to the coordination pathway of mechanism I, but in this scenario the coordinated radical (eq 14) does not undergo a β elimination, but it reacts with the monomer directly (eq 15). The radical should in this case be considered as activated by the catalyst, analogous to many heterogeneous catalytic processes.

$$R_n + Co(II) \rightleftarrows [R_n \cdots Co] \qquad (14)$$

$$[R_n \cdots Co] + M \rightarrow \ldots \rightarrow P_n + R_1^\bullet + Co(II) \qquad (15)$$

Mechanism III

Similar to mechanism II, complexation of one of the substrates takes place, but in this case it is the monomer, which can act as an axial ligand of the cobalt center (eq 16). This complex then reacts with the radical, yielding the Co(II) catalyst, macromer and the monomeric radical (eq 17).

$$M + Co(II) \rightleftharpoons [M \cdots Co] \qquad (16)$$

$$[M \cdots Co] + R_n^\bullet \rightarrow \ldots \rightarrow P_n + R_1^\bullet + Co(II) \qquad (17)$$

If this mechanism were indeed operative, then it is conceivable that possible solvent effects can be explained by the formation of a relatively strong solvent-catalyst complexes analogous to eq 16, hence leading to a decrease in active catalyst concentration. This scenario was previously suggested by Haddleton and co-workers to explain the decrease in chain transfer constant of COBF in MMA when performing the same polymerization in a toluene solution[19]. However, we have not been able to observe any significant solvent effects of this nature except when using pyridine, which is known to be a very strong ligand.

Mechanisms II and III lead to a very different Mayo equations as compared to mechanism I. This is easily seen from the overall rate of catalytic chain transfer ($R_{tr,Co}$), which in the case of mechanism I only contains the concentrations of the radicals and the Co(II) catalyst, whereas in the cases of mechanisms II and III these also contain the concentration of monomer (eq 18).

$$R_{tr,Co} = k_{tr,Co} K_{complex} [Co(II)][R^\bullet][M] \qquad (18)$$

In this expression, $k_{tr,Co}$ is the overall rate coefficient of the rates of reactions shown in eq 15 or 17 and $K_{complex}$ is the equilibrium constant of the complex formation shown in either eq 14 or 16. Use of eq 18 then yield the following form of the Mayo equation:

$$\frac{1}{DP_n} = (1+\lambda)\frac{\langle k_t \rangle [R^\bullet]}{k_p[M]} + C_M + \frac{k_{tr,Co} K_{complex}}{k_p}[Co(II)] \qquad (19)$$

It is interesting to note that in contrast to eq 4, which is applicable for mechanism I, the catalytic chain transfer term does not contain the monomer concentration in the denominator. This situation is comparable to "ordinary" chain transfer to monomer, in which case there is neither a dependence on monomer concentration. Since catalytic chain transfer actually catalyzed chain transfer to monomer, it may even be more convenient for reasons that will become apparent in a later section, to write eq 19 as:

$$\frac{1}{DP_n} = (1+\lambda)\frac{\langle k_t \rangle [R^\bullet]}{k_p[M]} + C_M + C_{M,catalyzed} \qquad (20)$$

Comparison of the Mayo equations applicable to mechanism I (i.e., eq 6) and to mechanisms II and III (i.e., eq 21) one obvious difference clearly stands out.

Whereas the catalytic chain transfer term in eq 6 is dependent on conversion, this dependence is absent in the case of eq 20.

In summary it can be stated that different mechanisms will lead to significantly different polymerization behavior and in what follows a number of experimental observations will be discussed in the light of these different mechanisms.

Experimental Observations.

In the previous section we discussed three possible mechanisms and it is clear that a decision on which mechanism is the most conceivable can only be made after careful consideration of the available experimental data. As stated a few times before, mechanism I is the most promising candidate and much circumstantial evidence clearly supports this mechanism. We will therefore discuss some experimental data mainly in the light of this mechanism, point out any inconsistencies and investigate whether these can be explained in terms of the other mechanisms. Firstly, we will discuss the possible role of cobalt-carbon bond formation, as is commonly observed in the organo-cobalt chemistry and has also been shown to be present in catalytic chain transfer polymerization. Secondly we will discuss one very curious observation in catalytic chain transfer polymerization and that is the fact that constant molecular weight distributions are observed in time.

Cobalt-carbon bond formation.
It is known from the organometallic literature that many Co(II) complexes can form reversible Co–C bonds with organic radicals (eq 21), and this has also directly been shown by ESR studies to be significant in the catalytic chain transfer polymerization of styrene, but to be negligible in MMA[24,25].

$$R^\bullet + Co(II) \rightleftarrows R-Co(III) \qquad (21)$$

This is an interesting fact, as in both the coordination pathway of mechanism I and in mechanism II the formation of a Co–C bond is essential. However, if the catalytic chain transfer mechanism does not involve the formation of Co–C bonds, then this reaction may not be desirable. In the free-radical polymerization of acrylates in the presence of a cobalt(II) complex, no chain transfer is observed at all and the reversible Co–C bond formation even leads to a "living" radical polymerization (see Scheme 6).

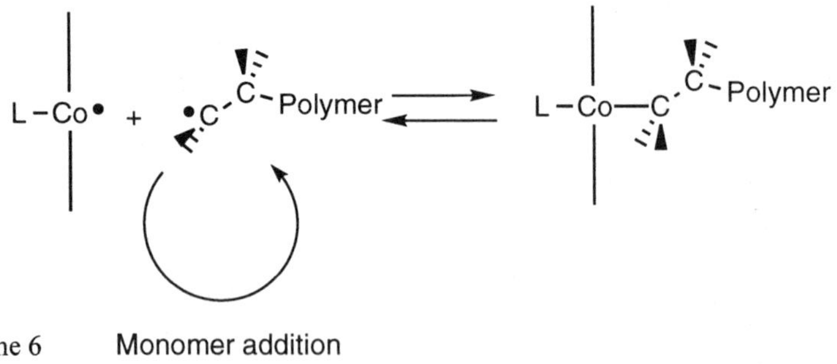

Scheme 6   Monomer addition

It may be concluded that the extend of Co–C bond formation depends on the type of radicals present in the system, and that the Co–radical complex is more stable when secondary (e.g., styrenic and acrylic) rather than tertiary (e.g. methacrylic) radicals are present. The question now remains as to what the effect is of the presence of these complexes. Firstly, if the formation of a R–Co(III) species plays a role in the catalytic chain transfer mechanism, then a large presence of these species indicates that the rate determining step in the mechanism is not the formation of these species, but the subsequent hydrogen abstraction. This would then be the case for styrene polymerization, as nearly all the Co(II) initially present is converted into R–Co(III) when polymerization takes place, whereas no significant amount of R–Co(III) is observed in MMA polymerization. Considered in this light, the much higher chain transfer constants observed in MMA (~40·10^3 for COBF) as compared to styrene (~1·10^3 for COBF) should then be explained by a much higher abstraction rate coefficient in the former case (which is likely to be diffusion-controlled, see above). However, the R–Co(III) needs to be formed first and this is clearly easier for acrylic and styrenic radicals as compared to methacrylic radicals. We cannot think of any conceivable reason other than steric requirements that could then explain why the hydrogen abstraction (or hydrogen transfer in mechanism III) would be much faster in the methacrylates. However, we think it quite inconceivable that this could affect the rate by several orders of magnitude.

An alternative explanation [24] for the lower chain transfer constants observed in styrene (and acrylate) polymerization is the fact that the equilibrium of eq 21 reduces the concentration of free active Co(II) catalyst. In this case, the lower chain transfer constant is then explained by realizing that it is determined by assuming that the actual concentration of catalyst is equal to its initial concentration, which in the case of, for example, styrene polymerization is clearly not the case. Since DP_n^{-1} is directly related to the product of the chain transfer constant and the concentration of chain transfer agent, the relationship between the observed and the true chain transfer constants is simply given by:

$$\left(C_S \times [Co(II)]\right)_{TRUE} = \left(C_S' \times [Co(II)]_0\right)_{OBSERVED} \quad (22)$$

where C_S' and $[Co(II)]_0$ are used to denote the observed chain transfer constant and initial Co(II) catalyst concentration, respectively. For the catalytic chain transfer polymerization of styrene, the actual [Co(II)] is much lower than $[Co(II)]_0$, which directly implies that $C_S \gg C_S'$. This result is a possible explanation for the observation that the catalytic chain transfer polymerization in MMA is diffusion-controlled and apparently chemically controlled in styrene; in light of the above, this difference may actually be much smaller and it is conceivable that catalytic chain transfer in styrene is also diffusion-controlled. The significant formation of R–Co(III) and its effect on the observed chain transfer constant is also likely to cause a very different temperature dependence of the chain transfer constant, as the equilibrium of eq 21 should shift towards an increasing [Co(II)] at higher temperatures. However, the only experimental study published on styrene thus far does not appear to be consistent with this argument; for reasons unknown at the time of that publication and which will be outlined below, the presented data may not be reliable enough to draw any mechanistic conclusions.

The formation of R–Co(III) was also shown to affect the initial kinetics of the polymerization, as significant inhibition times were observed dependent on the type of monomer and the amount of Co(II) complex. After the equilibrium of eq 21 has been established it is found that the overall polymerization rates can be modeled using conventional free-radical polymerization kinetics using a chain-length-dependent average termination rate coefficient. However, a different situation was found for the molecular weight evolution. The following expression was derived and confirmed experimentally for the conversion-dependence of the number average degree of polymerization:

$$DP_n = \frac{[M]_0 - [M]}{2f([I]_0 - [I]) + [R\text{-}Co(III)] + xC_s'[Co(II)]_0} \quad (23)$$

where $[M]_0$ is the initial monomer concentration, f the initiator efficiency, $[I]_0$ and $[I]$ the initial and actual initiator concentrations, respectively, and x the fractional conversion. This expression was derived assuming the validity of mechanism I, and taking into consideration that the R–Co(III) formation is reversible and that therefore the R–Co(III) species can act as "living" free-radical initiators. A graphical illustration of eq 23 is shown in Figure 1 and it is clear that the average degree of polymerization as predicted by the Mayo equation and based upon the experimentally determined chain transfer constant is only reached after about 5% conversion.

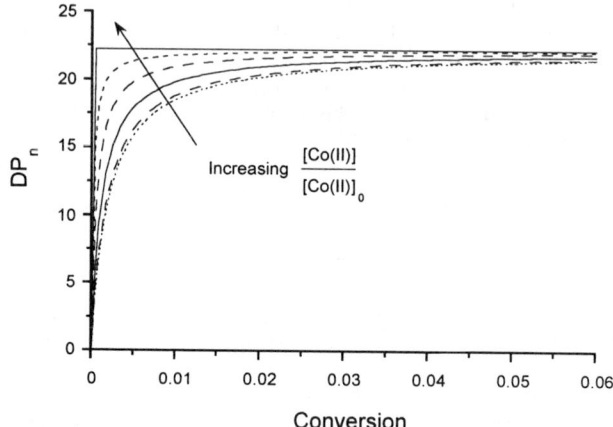

Figure 1. Conversion-dependence of the average degree of polymerization in catalytic chain transfer polymerization according to eq 23 for $[Co(II)]/[Co(II)]_0$ varying from 0 to 1.

If we further assume a negligible contribution to DP_n from the initiator fragments, the following expression is obtained for the inverse of DP_n:

$$\frac{1}{DP_n} \approx \frac{[R-Co(III)]}{x[M]_0} + C_S' \frac{[Co]_0}{[M]_0} \quad (24)$$

Defining the γ as $[R-Co]/[Co]_0$ then yields the following expression of the Mayo equation:

$$\frac{1}{DP_n} \approx \left(C_S' + \frac{\gamma}{x}\right)\frac{[Co]_0}{[M]_0} \quad (25)$$

This expression, which is visually represented in Figure 2, clearly suggests that the measured chain transfer constant may be highly dependent on the conversion at which it is measured. Considering the differences in the extent of Co-C bond formation in styrene and methyl methacrylate polymerizations, a relatively large conversion-dependence of the chain transfer constant is expected in styrene, whereas this should be negligible in methyl methacrylate. In the light of this finding, which is currently being tested experimentally, it may be concluded that an incredibly good control of conversion is required for catalytic chain transfer constant measurements in styrene, and that this may explain some of the unusual observations reported in the literature.

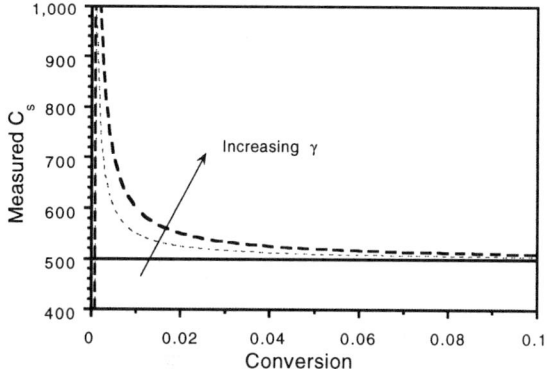

Figure 2. Conversion-dependence of measured apparent catalytic chain transfer constant according to eq 25. $C_S' = 500$: $\gamma = 0$ (–), $\gamma = 0.5$ (···), $\gamma = 1$ (- - -).

A final note is made with respect to the validity of eqs 23 - 25. They were derived assuming mechanismI is valid, but their results are applicable to either the coordination or the radical pathways. The results would be different for mechanisms II and III as these have different kinetic expressions for the chain transfer reaction.

Constant molecular weight distributions

Although the predictive and explanatory value of mechanism I is very high, it has one major shortcoming. According to eq 4 the molecular weight continuously decreases with increasing conversion, but experiments clearly indicate that the molecular weight distribution remains virtually unchanged during the course of polymerization as clearly shown in Figure 3. This behavior was observed for all monomers and monomer systems of which we studied the molecular weight evolution, i.e., styrene[24], MMA[24], 3-[tris(trimethylsilyloxy)silyl] propyl methacrylate[26], and the terpolymerization of styrene with MMA and 2-hydroxyethyl methacrylate[27]. A possible explanation for this observation may be irreversible catalyst poisoning or degradation, which we know are operative in the system. However, this seems very unlikely to us as it is very fortuitous that this poisoning then occurs at exactly the same rate as monomer conversion, and this in all the systems investigated.

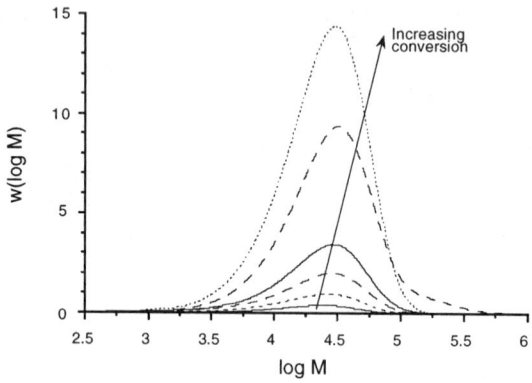

Figure 3. Evolution of cumulative molecular weight distribution of COPhBF-mediated catalytic chain transfer polymerization of methyl methacrylate at 60 °C.

An alternative explanation would be the direct participation of monomer in the hydrogen abstraction process as already discussed for mechanism II and III, which were shown to obey a Mayo equation in which the catalytic chain transfer contribution is only dependent on the catalyst concentration. Hence, if either of these two mechanisms were operative, then a virtually unchanging molecular weight distribution is expected in time for a fully chain transfer dominated polymerization, conform Figure 3.

The applicability of a Mayo equation with the functional form shown in eq 20 was tested[28] by performing chain transfer constant measurements in which the catalyst concentration was held constant and the monomer concentration changed by addition of an inert solvent (in contrast to the common procedure in which the catalyst concentration is changed and the monomer concentration is kept constant). The result of this test was a common Mayo plot with a positive slope (equal to C_S) as expected from eq 4, and not a horizontal line as expected from eq 20. This clearly suggests that mechanisms for which eq 20 is applicable (e.g., mechanisms II and III in this chapter) are unlikely candidates for the catalytic chain transfer mechanism, which brings us back to mechanism I as the most likely candidate, but with an experimentally observed evolution of the molecular weight distribution which is inexplicable at present.

Conclusions

The precise mechanism of catalytic chain transfer is still unclear despite a number of careful mechanistic studies. Further work on evaluating the evolution of the molecular weight distribution with conversion (both in high and low conversion regions) is required in conjunction with *in situ* monitoring of the cobalt(II) concentration by ESR. The transfer reaction with methacrylates appears to be diffusion-controlled and this may well be a general result for all vinyl monomers, though this needs to be confirmed. The precise role of cobalt-carbon bond formation in catalytic chain transfer is still not entirely clear and further studies on polymerization and small radicals are required to complete understanding of this mechanistic feature.

Acknowledgments

Discussions relating to this work with Alexei Gridnev, Dave Haddleton, Dax Kukulj, Graeme Moad and Bunichiro Yamada are all gratefully acknowledged, as is financial support by ICI Plc and the Australian Research Council.

References

(1) Moad, G.; Solomon, D. H. *The Chemistry of Free Radical Polymerization*; Pergamon: Oxford, 1995.
(2) Matyjaszewski, K. *Controlled Radical Polymerization*; Matyjaszewski, K., Ed.; American Chemical Society: Washington, DC, 1998; Vol. 685.
(3) Davis, T. P.; Haddleton, D. M.; Richards, S. N. *J. Macromol. Sci., Rev. Macromol. Chem. Phys.* **1994**, *C34*, 243-324.
(4) Davis, T. P.; Kukulj, D.; Haddleton, D. M.; Maloney, D. R. *Trends Polym. Sci.* **1995**, *3*, 365-373.
(5) Muratore, L. M.; Steinhoff, K.; Davis, T. P. *J. Mater. Chem.* **1999**, *9*, 1687-1691.
(6) Moad, C. L.; Moad, G.; Rizzardo, E.; Thang, S. H. *Macromolecules* **1996**, *29*, 7717-7726.
(7) Smirnov, B.R.; Morozova, I.S.; Marchenko, A.P.; Markevich, M.A.; Pushchaeva, L.M.; Enikolopyan, N.S. *Dokl. Akad. Nauk SSSR* **1980**, *253*, 891-895.
(8) Bakac, A.; Brynildson, M.E.; Espenson, J.H. *Inorg. Chem.* **1986**, *25*, 4108.
(9) Mayo, F. R. *J. Am. Chem. Soc.* **1943**, *65*, 2324.
(10) Heuts, J. P. A.; Davis, T. P.; Russell, G. T. *Macromolecules* **1999**, *32*, 6019-6030.
(11) Christie, D. I.; Gilbert, R. G. *Macromol. Chem. Phys.* **1996**, *197*, 403-412.

(12) Christie, D. I.; Gilbert, R. G. *Macromol. Chem. Phys.* **1997**, *198*, 663.
(13) Enikolopyan, N. S.; Smirnov, B. R.; Ponomarev, G. V.; Belgovskii, I. M. *J. Polym. Sci., Polym. Chem. Edn.* **1981**, *19*, 879.
(14) Suddaby, K. G.; Maloney, D. R.; Haddleton, D. M. *Macromolecules* **1997**, *30*, 702-713.
(15) Kukulj, D.; Davis, T. P. *Macromol. Chem. Phys.* **1998**, *199*, 1697-1708.
(16) Heuts, J. P. A.; Kukulj, D.; Forster, D. J.; Davis, T. P. *Macromolecules* **1998**, *31*, 2894-2905.
(17) Heuts, J. P. A.; Forster, D. J.; Davis, T. P. *Macromolecules* **1999**, *32*, 3907-3912.
(18) Kukulj, D.; Heuts, J. P. A.; Davis, T. P. *Macromolecules* **1998**, *31*, 6034-6041.
(19) Haddleton, D. M.; Maloney, D. R.; Suddaby, K. G.; Muir, A. V. G.; Richards, S. N. *Macromol. Symp.* **1996**, *111*, 37-46.
(20) Mironychev, V. Y.; Mogilevich, M. M.; Smirnov, B. R.; Shapiro, Y. Y.; Golikov, I. V. *Polym. Sci. USSR* **1986**, *28*, 2103.
(21) Forster, D. J.; Heuts, J. P. A.; Davis, T. P. *Polymer* **1999**, *in press.*
(22) Forster, D. J.; Heuts, J. P. A.; Lucien, F.; Davis, T. P. *Macromolecules* **1999**, *32*, 5514-5518.
(23) Gridnev, A. A.; Ittel, S. D.; Wayland, B. B.; Fryd, M. *Organometallics* **1996**, *15*, 5116 - Supplementary material.
(24) Heuts, J. P. A.; Forster, D. J.; Davis, T. P.; Yamada, B.; Yamazoe, H.; Azukizawa, M. *Macromolecules* **1999**, *32*, 2511-2519.
(25) Gridnev, A. A.; Belgovskii, I. M.; Enikolopyan, N. S. *Dokl. Akad. Nauk SSSR (Engl. Transl.)* **1986**, *289*, 1408.
(26) Muratore, L. M.; Heuts, J. P. A.; Davis, T. P. *Macromol. Chem. Phys.* **1999**, *in press.*
(27) Heuts, J. P. A.; Muratore, L. M.; Davis, T. P. *Macromol. Chem. Phys.* **1999**, *in press.*
(28) Heuts, J. P. A.; Forster, D. J.; Davis, T. P. *Macromol. Rapid Commun.* **1999**, *20*, 299-302.

INDEXES

Author Index

Boffa, Lisa Saunders, 1
Britcher, L. G., 127
Camurati, Isabella, 174
Davis, Thomas P., 254
Dong, Yuping, 146
Duncalf, David J., 236
Fait, Anna, 174
Forster, Darren J., 254
Gupta, Shashi, 24
Haddleton, David M., 236
Heuts, Johan P. A., 254
Hong, Jun-Bae, 94
Huang, Diyun, 24
Jun, Chul-Ho, 94
Kamigaito, Masami, 196
Kanda, Chika, 77
Katano, Hirotomo, 77
Kimura, Yoshiharu, 77
Kotani, Yuzo, 196
Kukulj, Dax, 236
Lee, Dae-Yon, 94
Lee, Hyuk, 94
Lee, Priscilla P. S., 146
Levy, Amy T., 224
Londergan, Timothy M., 24
Lu, Yingying, 165
Lu, Zejian, 165

Mabry, Joseph M., 24
Matisons, J. G., 127
Matyjaszewski, Krzysztof, 207
Miyamoto, Masatoshi, 77
Patten, Timothy E., 224
Paulasaari, Jyri K., 24
Penelle, J., 59
Peng, Han, 146
Piemontesi, Fabrizio, 174
Radigue, Arnaud, 236
Resconi, Luigi, 174
Salhi, Fouad, 146
Sargent, Jonathan R., 24
Sawamoto, Mitsuo, 196
Sun, Qunhui, 146
Tang, Ben Zhong, 146
Tartarini, Stefano, 174
Tsai, Ming-der, 108
Tsiang, Raymond Chien-chao, 108
Tsuda, Tetsuo, 38
Uegaki, Hiroko, 196
Weber, William P., 24
Wu, Qing, 165
Xia, Jianhui, 207
Xu, Kaitian, 146
Yang, Wen-shen, 108
Zhang, Xuan, 207

Subject Index

A

Acetylenes. *See* Substituted acetylene polymerizations
Acid-catalyzed siloxane equilibration polymerization synthesis and analysis of monomers and polymers, 32–35
synthesis of reactive '-monomers for, 27–28
See also Carbonylbis(triphenylphosphine) ruthenium catalysis
Acrylates. *See* Living radical polymerization of acrylates
Activators. *See* Nucleophilic polymerization of methyl methacrylate activated by Brønsted acid
Activity
  activity dependence on ligand on catalyst for hydrosilation, 130
  catalytic activity in hydrosilation, 139–143
  cobaloximes in chain transfer polymerization, 256
  comparison between complexes for hydrosilation, 141
  complex in styrene atom transfer radical polymerization, 229–231
  effect of added ligand on activity of styrene polymerization, 232
  improvements in catalyst activity for solid-supported atom transfer polymerization, 237–238
  influence of monomer concentration for polypropylene, 182–184
  influence of propylene concentration on activity of *rac*-Me$_2$C(3-*t*-Bu-1-Ind)$_2$ZrCl$_2$/MAO (1/MAO) and *rac*-C$_2$H$_4$(Ind)$_2$ZrCl$_2$/MAO (2/MAO), 182–184
  polymerization activity of 1/MAO, 183*f*
Addition polymerization transition metal catalysis, 145
See also Norbornene
Alkyne cycloaddition polymerization, transition metal-catalyzed
  cobalt-catalyzed diyne/nitrile cycloaddition forming poly(pyridine)s, 42, 45
  cobalt-catalyzed diyne/norbornadiene/CO cycloaddition forming poly(enone)s, 54, 55
  copolymerization of 1,4-bis(phenylethynyl)benzene with equimolar amounts of aryl isocyanates, 42, 43
  copolymerization of 1,4-diethynylbenzene and 1,11-dodecadiyne with *N*-octylmaleimide, 48, 51
  cycloaddition to 2,3,4,5-tetraphenylthiophene, 44, 45
  Diels–Alder polymerizations, 39
  diyne cycloaddition copolymerization, 39–47

275

diyne cyclopolymerization
  terpolymerization, 54
diyne/maleimide double-
  cycloaddition copolymerization,
  49
diyne/transition metal complex
  cycloaddition forming transition
  metal-containing polymers, 47,
  49
formations of 2-pyrone and 2-
  pyridone, 40
intramolecular diyne cycloaddition,
  39, 40
ladder poly(2-pyridone) formation,
  46t
method of polymer synthesis, 38–
  39
monoyne and diyne double-
  cycloaddition copolymerization,
  47–48, 52
nickel-catalyzed cyclic diyne/CO_2
  and isocyanate cycloaddition
  forming ladder poly(2-pyrone)s
  and poly(2-pyridone)s, 44, 46
nickel-catalyzed diyne/CO_2
  cycloaddition forming poly(2-
  pyrone)s, 41
nickel-catalyzed
  diyne/dimaleimide double-
  cycloaddition forming
  poly(imide) with pendant
  carbocycle, 52, 53
nickel-catalyzed diyne/isocyanate
  cycloaddition forming poly(2-
  pyridone)2, 41–42, 43
nickel-catalyzed diyne/maleimide
  double-cycloaddition forming
  poly(bicyclo[2.2.2]oct-7-ene)s,
  48, 52
nickel-catalyzed
  monoyne/dimaleimide double-
  cycloaddition forming
  poly(imide)s, 52, 53t
nickel-catalyzed
  monoyne/maleimide double-
  cycloaddition forming
  bicyclo[2.2.2]oct-7-enes, 47–48
one-step construction of polymer
  repeat unit, 54, 56
palladium-catalyzed
  copolymerization of 1,4-bis(4'-
  hexylphenylethynyl)benzene, 44,
  45
palladium-catalyzed
  diyne/elemental sulfur
  cycloaddition forming
  poly(thiophene)s, 42, 44
poly(2-pyridone)s by reaction of
  poly(2-pyrone) with amines, 42,
  45
preparation of bicyclo[2.2.2]oct-7-
  enes with *exo,exo-*
  stereochemistry, 48, 50
regiocontrol, 56
side diyne cyclotrimerization in
  diyne cycloaddition
  copolymerization, 40
3,11-tetradecadiyne/CO_2
  cycloaddition forming poly(2-
  pyrone), 41, 43
three types, 54
Aluminum porphyrin initiators
  chain terminating reaction, 12
  living behavior with Lewis acid
    additives, 11–12
  polymerization of methyl
    methacrylate, 12
  polymerizations of both primary
    and branched acrylate
    monomers, 13
  (tetraphenylporphinato)aluminum
    methyl [(TPP)AlMe], 11–13
Anionic polymerization
  1,2- rather than 1,4-attack of
    initiator on monomer, 4
  t-butylmagnesium bromide for

highly isotactic PMMA, 5
chain transfer, 5
common-ion salts for imparting control, 5–6
complexation of counterion rather than anionic endgroup, 6
concerted mechanism, 3
hindered lithium alkyl or enolate complex as initiator, 6
intramolecular backbiting, 4
ligated, 5
lithium 2-(2-methoxyethoxy)ethoxide (LiOEEM), 6
lithium *t*-butoxide, 6
mechanisms, 2–3
methacrylates and acrylates, 2–6
precoordination mechanism, 3
side reactions, 4–5
solvents, 2
strategies for living polymerization, 5–6
terminating side reactions, 4
Aqueous polymerization conditions. *See* Substituted acetylene polymerizations
Architectures
 atom transfer radical polymerization (ATRP), 14–15
 development of new, 23
 group transfer polymerization (GTP), 8–9
 lanthanide(III) initiating systems, 11
 special for acrylics, 2
 working on strategies for well-controlled, 60
Atom transfer radical polymerization (ATRP)
 advantages over anionic, 13
 alkyl halide as initiator, 13
 basic concept, 207
 block and random copolymers, 14–15
 conversion and polydispersity, 13–14
 copper-mediated ATRP of methyl methacrylate, 14
 dynamic equilibrium between active and dormant species, 208
 halide initiators for end-functionalized PMMA, 15
 lacking direct structure-activity relationships for catalyst systems, 224–225
 proposed mechanism for transition metal-mediated ATRP, 207–208
 range of monomers, conditions, and metal complexes, 224
 reverse, 211
 telechelic oligomers, 15
 well-defined polymer structures, 13
 See also Copper(I) atom transfer radical polymerization (ATRP) catalysts; Copper-mediated atom transfer radical polymerization (ATRP); Solid-supported catalysts for atom transfer polymerization

B

Backbiting, intramolecular, anionic polymerization side reaction, 4–5
Bicyclo[2.2.2]oct-7-enes, nickel-catalyzed monoyne/maleimide double-cycloaddition, 47–48
Bidentate ligands
 bipyridine derivatives, 211–212
 diamines, 212–214
 diimines, 214
 pyridine-imines, 214–215
 See also Ligands
Bipyridines
 bidentate nitrogen ligands, 210–211

derivatives used in copper-mediated atom transfer radical polymerization (ATRP), 211–212
derivatives with CuBr for atom transfer radical polymerization (ATRP) of styrene, 227–229
ligands in controlled radical polymerizations of styrene and (meth)acrylates, 208–209
See also Copper(I) atom transfer radical polymerization (ATRP) catalysts; Ligands
1,9-Bis(trimethylsilyl)-1,8-nonadiyne, polymerization, 159
Block copolymers. See Hydrogenation of polystyrene-b-polybutadiene-b-polystyrene block copolymers
Brønsted acid
 appropriate activator for polymerization of methyl methacrylate (MMA), 87–88
 See also Nucleophilic polymerization of methyl methacrylate activated by Brønsted acid
n-Butyl acrylate (BA)
 living radical polymerization, 201
 See also Living radical polymerization of acrylates
n-Butyllithium. See Hydrogenation of polystyrene-b-polybutadiene-b-polystyrene block copolymers
t-Butylmagnesium bromide, highly isotactic poly(methyl methacrylate) (PMMA), 5

C

Carbonylbis(triphenylphosphine)ruthenium catalysis

acid catalyzed siloxane equilibration polymerization of 1,8-bis(3,3,5,5,5-pentamethyl-4-oxa-3,5-disilahexanyl)thioxanthen-9-one (III), 27f
advantages, 30
bond distances and angles for complex, 26t
carbosilane or carbosilane/siloxane units alternating with aromatic ketone groups, 25
complex of divinyldimethylsilane and (Ph$_3$P)$_2$RuCO (1:1), 25–26
complication of ^{1}H NMR spectra for copolymers, 30
differential scanning calorimetry (DSC) method, 31
elemental analysis method, 31
experimental, 30–35
gel permeation chromatography (GPC) analytical methods, 30–31
^{1}H and ^{13}C NMR methods, 30
hydrolytic stability of poly(silyl ethers), 29
hydrosilylation polymerization of 4-dimethylsiloxyacetophenone (XI), 29
hydrosilylation polymerization of ketones and ,-bis(SiH)siloxanes, 29–30
^{31}P NMR of 1:1 complex of divinyldimethylsilane and (Ph$_3$P)$_2$RuCO, 27f
poly(silyl ethers) via hydrosilylation reaction, 30f
preparation and analysis of complex of divinyldimethylsilane and (Ph$_3$P)$_2$RuCO, 31–32
preparation and analysis of monomers and polymers, 32–35

preparation and analysis of (Ph$_3$P)$_2$RuCO, 31
reagents, 31
synthesis of *alt*-copoly[3,3,5,5-tetramethyl-4-oxa-3,5-disila-1,7-heptanylene/2',2"-diacetyl-5',5"(1,3-benzene)-3',3"-dithiophenylene] (X), 28*f*
synthesis of *alt*-copoly[3,3,5,5-tetramethyl-4-oxa-3,5-disila-1,7-heptanylene/2',2"-diacetyl-5',5"(1,4-benzene)-3',3"-dithiophenylene (VII), 28*f*
synthesis of reactive '-monomers for acid catalyzed silane equilibration polymerization, 27–28
thermogravimetric analysis (TGA) method, 31
X-ray structure analysis method, 32
X-ray structure and dimensions of complex, 26*f*

Catalytic chain transfer polymerization
activity of cobaloximes, 256
addition-fragmentation polymerization, 256–257
alternative explanation for lower chain transfer constants in styrene and acrylate polymerization, 267
applicability of Mayo equation, 270–271
applications, 256–257
chain transfer constant measurements, 258
chain transfer constants, 258–260
chain transfer constants for variety of monomers and catalysts, 261*t*
chain transfer constants of cobaloxime boron fluoride (COBF) in methyl, ethyl, and butyl methacrylates over range of temperatures, 260, 262*t*
cobalt-carbon bond formation, 266–269
cobalt(II)porphyrin, cobaloxime, and COBF structures, 255
constant molecular weight distributions, 269–271
conversion-dependence of average degree of polymerization for [Co(II)]/[Co(II)]$_0$ varying from 0 to 1, 268*f*
conversion-dependence of measured apparent catalytic chain transfer constant, 269*f*
conversion-dependence of number-average degree of polymerization, 268
effective graft polymerization, 256
equation for diffusion-controlled chain transfer reaction, 261
evolution of cumulative molecular weight distribution of COPhBF-mediated polymerization of methyl methacrylate, 270*f*
experimental materials, 257
experimental observations, 266–271
experimental procedures, 257–258
expressions for chain transfer constant determination by Mayo and chain length distribution (CLD) procedures, 259
expression suggesting measured chain transfer constant dependent on conversion, 269
extent of Co–C bond formation dependence on radical type present, 267
formation of R–Co(III) affecting initial polymerization kinetics, 268
general aspects of chain transfer

mechanism, 263
generally accepted working model for mechanism, 259–260
"living" radical polymerization for reaction lacking chain transfer, 266
low-spin Co(II) complex catalyzing hydrogen-transfer reaction from growing polymer radical to monomer, 255
mechanism related to coordination pathway of one possible mechanism, 264
mechanism requiring complexation of monomer to catalyst, 265–266
molecular weight analysis, 258
polymerization reactions, 258
possible mechanisms, 263–266
production of oligomers with vinyl end-functionality, 257
radical pathway and coordination pathways of possible mechanism, 263–264
rate coefficient of diffusion-controlled reaction in terms of Smoluchowski equation, 261
relationship between true and observed chain transfer constants, 267
styrene polymerizations, 262–263
temperature dependence for chain transfer constant, 260–262
typical low-spin Co(II) complexes, 255
Chain length distribution (CLD), determining chain transfer constant, 258–260
Chain transfer constants. *See* Catalytic chain transfer polymerization
Chain transfer reactions
influence of monomer concentration for polypropylene, 186–188
See also Catalytic chain transfer polymerization; Propylene polymerization
Cobalt catalysts
hydroformylation of olefins, 96
See also Alkyne cycloaddition polymerization, transition metal-catalyzed; Catalytic chain transfer polymerization
Cobaltocene. *See* Hydrogenation of polystyrene-*b*-polybutadiene-*b*-polystyrene block copolymers
Controlled radical polymerization
organometallic polymer synthesis, 195
suppressing side reactions for "living" behavior, 195
See also Living radical polymerization of acrylates
Copolymers
block and random by atom transfer radical polymerization (ATRP), 14–15
group transfer polymerization (GTP), 8–9
hydrophilic/hydrophobic using lanthanide(III) initiators, 11
Copper catalysts
extensive study in atom transfer radical polymerization (ATRP), 208–209
mediating ATRP, 13–14
See also Solid-supported catalysts for atom transfer polymerization
Copper(I) atom transfer radical polymerization (ATRP) catalysts
coordination chemistry of copper(I) halides with 2,2'-bipyridine and 1,10-phenanthroline ligands, 225–226

dependence of apparent rate of styrene polymerization on ligand-to-metal ratio, 233
4,4'-di-(neophyldimethylsilylmethyl)-2,2'-bipyridyl (dNEObipy) and CuBr (complex 1) preparation, 227–228
dynamic behavior of copper(I) halide complexes with 2,2'-bipyridines and 1,10-phenanthroline, 226
effect of added ligand on activity of styrene polymerization, 232
experimental materials and characterizations, 227
^{1}H NMR spectra of complex-1, 228–229
methyl methacrylate ATRP, 233–234
plot of apparent rate constants as function of ratio of moles of dNEObipy to moles copper(I) centers, 233f
polymerization activity of complex-1 in standard styrene ATRP, 229–231
procedure for solution polymerization of styrene using complex-1, 227
semilogarithmic kinetic plot and molecular weight and molecular weight distribution plot for ATRP of styrene with complex-1, 230f
structural chemistry, 226
structural types for copper(I) halide complexes with 2,2'-bipyridines and 1,10-phenanthroline, 225f
structure of complex-1, 228f
styrene polymerizations using CuBr/1 dNEObipy and CuBr/2 dNEObipy as copper(I) catalysts, 231

Copper-mediated atom transfer radical polymerization (ATRP)
bidentate ligands with mixed nitrogen types, 214–215
bidentate nitrogen ligands, 210–215
bipyridine (bpy) derivatives, 211–212
classifying nitrogen donors in multidentate ligands, 210
conformational change of ligand during atom transfer, 221
controlled ATRP of acrylates using tripodal tetramine, 219
Cu-mediated ATRP of styrene in bulk using different bpy ligands, 212t
Cu-mediated ATRP using pyridine-imine ligands, 216t
diamine ligands, 212–214
diimine ligands, 214
effect of ligand on polymerization rates, 220
effect of linker with bidentate amines, 214t
electron-donating groups on bpy, 212
future outlook, 221
importance of spacer between coordinating nitrogens, 220–221
influence of ligand choice on effectiveness of catalyst, 209–210
linear, cyclic, and branched (tripodal) tetradentate amine ligands, 219f
1,10-phenanthroline and derivatives in styrene, 212
proposed copper(I) and copper(II) species using bpy as ligand, 210
proposed mechanism for transition metal-mediated ATRP, 207–208
role of ligand in ATRP, 209
stability of metal complex, 221

steric effect with bidentate amines, 213t
structures of pyridine-imines, 215f
structures of some bpy derivatives, 211f
synthesis of N,N-bis(2-pyridylmethyl)octylamine (BPMOA), 218f
tetradentate ligands with mixed nitrogen types, 220
tetradentate nitrogen ligands, 219–220
tetramine ligands, 219
triamine ligands, 216–217
tridentate ligands with mixed nitrogen types, 217–218
tridentate nitrogen ligands, 216–218
tripyridine (tpy) ligands, 216
use of cyclic amines, 217t
using N-(2-pyridylmethyl)-(2-pyridyl)methanimine (PMPMI) or N-(2-N-(dimethyl)-ethyl)-(2-pyridyl)methanimine (DMEPMI) as ligand, 218t
using redox potential of metal complex in catalyst design, 220
See also Atom transfer radical polymerization (ATRP)
Counterion complexation, anionic polymerization, 6
Cyanoacetylene monomers
molecular structures, 148
See also Substituted acetylene polymerizations
Cyanocyclopropanecarboxylates, polymerization, 68
Cycloaddition polymerization. See Alkyne cycloaddition polymerization, transition metal-catalyzed
Cycloalkane polymerization. See Ring-opening polymerization of small cycloalkanes
Cyclobutanes
ring-opening polymerization of activated, 69–70
See also Ring-opening polymerization of small cycloalkanes
Cyclopropane-1,1-dicarboxylates
dependence on solubility of final polymers, 67
first-order kinetic plot, 64f
influence of counterion on polymerization of diisopropyl cyclopropane-1,1-dicarboxylate, 66t
number-average molecular weight and polydispersity index with conversion, 64f
polymerization at elevated temperatures, 65, 66f
ring-opening polymerization, 63–67
Cyclopropanedinitrile, polymerization, 68
Cyclopropanes
anionic polymerization, 61–62
reactivity of monomers, 73
See also Ring-opening polymerization of small cycloalkanes

D

1,9-Decadiyne, polymerization, 158–159
Diamine ligands, copper complexes in atom transfer radical polymerization (ATRP), 212–214
Diblock copolymers, special architecture, 2

Diels–Alder polymerizations, cycloaddition, 39
Diethyl sulfide, initiating system for polymerization of methyl methacrylate (MMA), 88–90
Diffuse reflectance Fourier transform infrared (DRIFT) spectroscopy
analysis of supported complexes, 133–135
instrumental method, 132
See also Hydrosilation
Diffusion-controlled reaction. See Catalytic chain transfer polymerization
Diimine ligands, copper complexes in atom transfer radical polymerization (ATRP), 214
Divinyldimethylsilane. See Carbonylbis(triphenylphosphine)ruthenium catalysis
Diyne cycloaddition copolymerization. See Alkyne cycloaddition polymerization, transition metal-catalyzed
Diyne cycloaddition terpolymerization
cobalt-catalyzed diyne/norbornadiene/CO cycloaddition forming poly(enone)s, 54, 55
See also Alkyne cycloaddition polymerization, transition metal-catalyzed
Diynes
^{13}C NMR spectrum of chloroform-d solution of poly(3,9-dodecadiyne), 160f
cyclotrimerization reactions, 147
evidence of polyene structures, 160
intermolecular cyclotrimerization to hyperbranched poly(phenylenealkenes), 157
new catalysts for polymerizations, 156–160
polymerization behavior of 1,9-decadiyne, 158–159
polymerization of 1,8-nonadiyne and 1,9-decadiyne, 158t
polymerization of 1,9-bis(trimethylsilyl)-1,8-nonadiyne, 159t
polymerization of 3,9-dodecadiyne, 159, 160t
polymerization results of terminal, 157–158
3,9-Dodecadiyne, polymerization, 159–160
Double-cycloaddition copolymerizations
nickel-catalyzed diyne/dimaleimide forming poly(imide) with pendant carbocycle, 52, 53
nickel-catalyzed diyne/maleimide forming poly(bicyclo[2.2.2]oct-7-ene)s, 48, 52
nickel-catalyzed monoyne/dimaleimide forming poly(imide)s, 52, 53t
nickel-catalyzed monoyne/maleimide forming bicyclo[2.2.2]oct-7-enes, 47–48
See also Alkyne cycloaddition polymerization, transition metal-catalyzed

E

E-glass fibers. See Hydrosilation
Elastomers. See Hydrogenation of polystyrene-b-polybutadiene-b-polystyrene block copolymers
Epimerization. See Propylene polymerization

F

Free radical polymerization
difficulty controlling selectivity, 254
little control in polymethacrylates and polyacrylates, 1
side reactions, 1–2

G

Glass fibers
preparing supported catalysts, 130–131
See also Hydrosilation
Group transfer polymerization (GTP)
activating initiator, 7
architecturally interesting polymers, 8–9
associative silyl GTP mechanism, 7, 8
dissociative silyl GTP mechanism, 7, 8
mechanism, 7
silyl GTP, 7–9
zirconocene GTP, 9

H

Heterogeneous catalysts, hydrosilation, 129–130
Homogeneous catalysts, comparison to supported catalysts, 128*t*
Hybrid-type polymerization (HTP)
alternative to replace conventional radical polymerization, 80–81
availability of Brønsted acid as activator for HTP of methyl methacrylate (MMA), 80
concept of HTP, 79
mechanism, 80
See also Nucleophilic polymerization of methyl methacrylate activated by Brønsted acid
Hydroacylation
chelation-assisted, 97–99
imine-mediated, 98
Hydroacylation of polybutadiene
^{13}C NMR spectra of phenyl terminated polybutadiene (PTPB), 105*f*
general procedure for hydroacylation of PTPB with aldehyde, 103
^1H NMR spectra of PTPB, 101*f*, 105*f*
hydroacylation of PTPB with various primary alcohols, 104*t*
hydroiminoacylation, 99–102
hydroiminoacylation of PTPB with ferrocene-carboxaldimine, 100
imine-mediated, 102–103
IR spectra of PTPB, 105*f*
IR spectra of PTPB and hydroacylated PTPB, 101*f*
procedure for hydroiminoacylation of PTPB with ferrocene-carboxaldimine, 102
procedure of hydroacylation of PTPB with benzyl alcohol, 104, 106
simultaneous hydrogenation and hydroacylation, 103–106
See also Polymer modification
Hydrocarboxylation, polybutadiene modification, 96
Hydroformylation, polybutadiene modification, 95–96
Hydrogenation, simultaneous with hydroacylation of polybutadiene, 103–106
Hydrogenation of polystyrene-*b*-polybutadiene-*b*-polystyrene block copolymers

back calculating degree of hydrogenation, 111
catalytic mechanism of metallocene dichloride in presence of methyl aluminoxane (MAO), 117
comparing metallocene catalysts, 109
comparison among various metallocenes, 111
comparison among various reducing agents, 111, 116
effect of hydrogen pressure on extent of hydrogenation, 116
effect of ratio of n-butyllithium to metallocene, 116
effect of reaction temperature, 116
experimental materials, 109–110
extents of hydrogenation of polybutadiene segment at various temperatures, 118f, 119f
extents of hydrogenation of polybutadiene segment using cobaltocene with various reducing agents, 114f, 115f
extents of hydrogenation of polybutadiene segment using various metallocenes with n-butyllithium, 112f, 113f
^{1}H NMR spectra of cobaltocene plus n-butyllithium and lithium cyclopentadienide, 122f, 123f
hydrogenation efficiencies at various hydrogen pressure, 124t
hydrogenation efficiencies at various n-butyllithium to cobaltocene ratios, 124t
hydrogenation procedures, 110
mechanistic studies of metallocene/reducing agent system, 117
sample analyses methods, 110–111
structures of cobaltocene, nickelocene, and titanocene, 109
UV–VIS spectra of n-butyllithium, cobaltocene, and their mixture, 120f, 121f
Hydrogen concentration, influence for propylene polymerization, 189–190
Hydroiminoacylation
polybutadiene, 99–102
polymer modification, 97–98
Hydrophilic/hydrophobic materials, lanthanide(III) initiators, 11
Hydrosilation
ACI E-glass fibers (complex 1) modified with mercaptopropyltrimethoxysilane, 134f
activity comparison between complex 1 and Ahlstrom glass fiber (complex 2), 141
advantages of inorganic supports over polymer supported complexes, 129
analysis of supported complexes, 133–139
atomic concentrations for untreated E-glass fibers and complexes 1 and 2, 136t
catalyst activity dependence on ligand, 130
catalyst deactivation, 142
catalytic activity, 139–143
comparison of activity between complexes 1 and 2 supported mercaptoplatinum complexes, 142f
comparison of homogeneous and supported catalysts, 128t
diagram of possible way mercaptosilane adsorbing to glass, 137f
diffuse reflectance Fourier transform infrared (DRIFT)

spectroscopy method, 132
DRIFT spectra of Ahlstrom E-
  glass fibers and ACI E-glass
  fibers, 134f
DRIFT spectroscopy, 133–135
DRIFT spectrum of complex 2,
  135f
experimental, 131–132
extent of surface treatment and
  metal adsorption for complexes
  1 and 2, 138–139
^{1}H NMR of hydrosilation of
  hexene with SiH containing
  siloxane using supported
  complexes as catalysts, 141f
heterogeneous catalysts, 129–130
hydrosilation of hexene using glass
  supported platinum complexes
  as catalysts, 131–132
immobilization of
  hexachloroplatinic acid on ion
  exchange resins, 130f
immobilized sulfur-platinum
  complexes on E-glass fibers as
  support, 132–133
instrumental methods, 132
nuclear magnetic resonance
  (NMR) method, 132
poly(methylhydrogen)siloxane
  with hexene, 139–140
preparation of E-glass fiber
  supported complexes, 131
preparation of sulfur containing
  silica compounds with attached
  platinum complexes, 137–138
reaction, 129f
reaction for evaluating supported
  complexes, 140f
side reactions, 140–143
sulfoxide and sulfide complexes of
  platinum, 143
supported catalysts, 129–131

supported mercaptosiloxane
  platinum complexes, 138f
untreated E-glass fibers, 131
using glass fibers, 130–131
XPS data for supported complexes,
  136t
XPS results, 135–139
X-ray photoelectron spectroscopy
  (XPS) method, 132
Hydrosilylation, polybutadiene
  modification, 96
Hydrosilylation polymerization
  ketones and -bis(SiH)siloxanes,
  29–30
  See also
    Carbonylbis(triphenylphosphine)
    ruthenium catalysis

I

Initiators
  alkyl halides for atom transfer
    radical polymerization (ATRP),
    13
  aluminum porphyrin, 11–13
  lanthanide(III), 9–11
  thiophenolate salts for ring-
    opening cycloalkanes, 62–63
Intramolecular backbiting, anionic
  polymerization side reaction, 4–5
Ion exchange resins
  immobilization of
    hexachloroplatinic acid, 130f
  supports for platinum catalysts,
    129
Isomerization reactions. See
  Propylene polymerization
Isotacticity
  influence of polymerization
    temperature for polypropylene,
    177–178

influence of propylene
concentration, 184–186
See also Propylene polymerization

K

Kharasch addition, 207

L

Ladder polymers
  group transfer polymerization
    (GTP), 9
  nickel-catalyzed cyclic diyne/CO_2
    and isocyanate cycloaddition
    copolymerizations forming
    poly(2-pyrone)s and poly(2-
    pyridone)s, 44, 46
Lanthanide(III) initiators
  all-acrylic ABA triblock
    preparation, 10–11
  hydrophilic/hydrophobic materials,
    11
  lanthanidocenes, 10
  living polymerization, 9–10
  polymerization of (meth)acrylates,
    10
  samarocene, 10–11
  tacticity, 9–10
Ligands
  bidentate ligands with mixed
    nitrogen types, 214–215
  bidentate nitrogen, 210–215
  bipyridine (bpy) derivatives, 211–
    212
  coordination chemistry of 2,2'-
    bipyridine and 1,10-
    phenanthroline with copper(I)
    halides, 225–226
  diamines, 212–214
  diimines, 214
  dynamic behavior of 2,2'-
    bipyridine and 1,10-
    phenanthroline with copper(I)
    halides, 226
  effect on ATRP, 220–221
  influence on effectiveness of
    catalyst, 209–210
  pyridine-imines, 214–215
  role in atom transfer radical
    polymerization (ATRP), 209
  tetradentate ligands with mixed
    nitrogen types, 220
  tetramines, 219
  triamines, 216–217
  tridentate ligands with mixed
    nitrogen types, 217–218
  tripyridines, 216
  See also Copper-mediated atom
    transfer radical polymerization
    (ATRP)
Ligated anionic polymerization
  (LAP), suppressing termination
  in anionic polymerization, 5
Lithium *t*-butoxide, -bridging
  additive, 6
Lithium 2-(2-
  methoxyethoxy)ethoxide
  (LiOEEM), preparing
  monodisperse poly(*n*-butyl
  acrylate), 6
Living polymerization
  forming stable organic free
    radicals, 236–237
  lanthanidocenes, 10
  strategies, 5–6
Living radical polymerization of
  acrylates
  n-butyl acrylate (BA), 201
  controlling molecular weights and
    molecular weight distribution
    (MWD), 197
  effects of additives on
    polymerization of MA, 203*f*

experimental materials, 204
factors affecting dormant-active species equilibrium, 197–198
^{1}H NMR spectrum of poly(MA), 204f
measurement methods, 205
metal complexes generating radical species from organic precursors, 197
metal complexes mediating controlled formation of radicals, 197
methyl acrylate (MA), 198–200
methyl methacrylate (MMA), 201, 202f
M_n, M_w/M_n, and MWD of poly(MA) using different iodides at varying monomer conversion, 199, 200f
polymerization of MA with R–X/ReO$_2$I(PPh$_3$)$_2$/Al(Oi-Pr)$_3$ in toluene at 80°C, 199f
polymerization procedures, 204–205
polymer terminal structure, 202–203
quenching experiments, 202
reaction schematic, 198
ReO$_2$I(PPh$_3$)$_2$ with alkyl iodide or bromide as initiator, 198

M

Mayo equation
 determining chain transfer constant, 258–260
 See also Catalytic chain transfer polymerization
Mechanisms
 aluminum porphyrin initiators, 11–12
 anionic polymerization, 2–3
 associative silyl group transfer polymerization (GTP), 7, 8
 atom transfer radical polymerization, 13
 development of new, 23
 dissociative silyl GTP, 7, 8
 hybrid-type polymerization (HTP), 79–81
 hydroiminoacylation, 97–98
 imine-mediated hydroacylation, 98
 lanthanide(III) initiators, 10
 polymerization via activated monomer mechanism (PAMM), 78–79
 possible for catalytic chain transfer polymerization, 263–266
 proposed for transition metal-mediated atom transfer radical polymerization, 207–208
 studies of metallocene/reducing agent system for hydrogenation of block copolymers, 117
Mercaptosiloxane platinum complexes. See Hydrosilation
Metallocene catalysts
 polymerizations, 108–109
 See also Hydrogenation of polystyrene-b-polybutadiene-b-polystyrene block copolymers
Metathesis polymerization, regularly unsaturated hydrocarbon polymers, 23
Methacrylates. See Organometallic polymerization of (meth)acrylates
Methyl acrylate (MA)
 effects of additives, 203f
 effects of potential terminators, 202
 ^{1}H NMR spectrum, 204f
 living radical polymerization, 198–200

M_n, M_w/M_n, and MWD with iodide at varying monomer conversion, 200f
molecular weights and distributions, 199f
polymer terminal structure, 202–203
quenching experiments, 202
See also Living radical polymerization of acrylates
Methylaluminoxane (MAO)
effect on norbornene polymerization, 168–169
See also Norbornene
Methyl methacrylate (MMA)
living radical polymerization, 201, 202f
needing catalytic studies for atom transfer radical polymerization (ATRP), 233–234
See also Catalytic chain transfer polymerization; Living radical polymerization of acrylates; Nucleophilic polymerization of methyl methacrylate activated by Brønsted acid; Solid-supported catalysts for atom transfer polymerization
Molecular weight
influence of monomer concentration for polypropylene, 186–188
influence of polymerization temperature for polypropylene, 180, 181f
Molybdenum catalysts. *See* Substituted acetylene polymerizations

N

Nickel catalysts
polymerizing norbornene, 166

See also Alkyne cycloaddition polymerization, transition metal-catalyzed
Nickelocene. *See* Hydrogenation of polystyrene-*b*-polybutadiene-*b*-polystyrene block copolymers
Nitrogen-containing ligands. *See* Ligands
1,9-Nonadiyne, polymerization, 157–158
Norbornadiene. *See* Substituted acetylene polymerizations
Norbornene
analysis and characterization, 167–168
^{13}C NMR spectrum of polynorbornene sample from CpTi(OBz)$_3$/MAO catalyst, 172f
catalyst type determining course of reaction, 165
cationic palladium(II) complex, 166
effect of cocatalyst methylaluminoxane (MAO), 168–169
effect of CpTi(OBz)$_3$ concentration on polymerization with MAO cocatalyst, 170t
effect of monomer concentration on polymerization with CpTi(OBz)$_3$/MAO, 170t
effect of temperature on polymerization with CpTi(OBz)$_3$/MAO, 170t
experimental materials, 167
IR spectrum of polynorbornene sample from CpTi(OBz)$_3$/MAO catalyst, 171f
polymer characterization, 171–172
polymerization by different titanocenes activated with MAO, 168t
polymerization catalyzed by CpTi(OBz)$_3$ activated with

various MAOs at different Al/Ti ratios, 169t
polymerization conditions, 169
polymerization procedure, 167
polymerizations with different titanocene compounds, 168
two polymerization modes, 165
Nucleophilic polymerization of methyl methacrylate activated by Brønsted acid
alternative to replace conventional radical polymerization, 80–81
appropriate Brønsted acid as activator, 87–88
bulk polymerization of MMA with methyl trimethylsilyl dimethylketene acetal (MTDA) and trifluoromethanesulfonic acid (TfOH), 90t
diethyl sulfide initiating systems, 88, 90
effect of Brønsted acid on MMA polymerization initiated with PPh$_3$, 87t
effect of concentrations of initiator and activator on polymerization, 86t
effect of feed ratios of initiator and activator to monomer, 86
effect of polymerization temperature, 84–85
expected mechanism, 81
experimental, 81–82
factors for selection of activator/initiator combination, 82
^{1}H NMR spectrum of polymer with M$_n$=10,300, 85f
hybrid-type polymerization of MMA, 84
molecular weight by gel permeation chromatography (GPC), 83
polymerization of MMA with diethyl sulfide (Et$_2$S) as initiator, 89t
solvent effect, 82–83
time dependence of MMA polymerization with Et$_2$S and TfOH, 90t
typical polymerization procedure, 82

O

Olefin/alkyne addition polymerization, transition metal catalysis, 145
Olefinic unsaturations
^{1}H NMR analysis, 175–176
See also Propylene polymerization
Organic counterions, metal free, controlling anionic polymerization, 6
Organic polymerization processes, development of new mechanisms or architecture, 23
Organometallic polymerization of (meth)acrylates
aluminum porphyrin initiators, 11–13
anionic polymerization, 2–6
associative silyl group transfer polymerization (GTP) mechanism, 8
atom transfer radical polymerization (ATRP), 13–15
block and random copolymers, 14–15
t-butylmagnesium bromide for highly isotactic PMMA, 5
chain transfer in anionic, 5
common-ion salts for imparting control, 5–6
complexation of counterion rather

than anionic endgroup, 6
concerted mechanism for methacrylate polymerization, 3
copper-mediated ATRP of methyl methacrylate, 14
dissociative silyl GTP mechanism, 8
group transfer polymerization (GTP), 7–9
lanthanide(III) initiators, 9–11
lanthanidocenes, 10
ligated anionic polymerization (LAP), 5
methyl methacrylate polymerization with photoactivated immortal aluminum porphyrin initiator, 12
precoordination mechanism for anionic methacrylate polymerization, 3
samarocene, 10–11
side reactions of anionic polymerization, 4–5
silyl GTP, 7–9
strategies for living polymerization, 5–6
zirconocene GTP, 9

P

Palladium catalysts
polymerizing norbornene, 166
See also Alkyne cycloaddition polymerization, transition metal-catalyzed
Polar substituted acetylenes. See Substituted acetylene polymerizations
Poly(2-pyridone)s
1,4-bis(phenylethynyl)benzene with equimolar amounts of aryl isocyanates, 42, 43
ladders by nickel-catalyzed cyclic diyne/CO_2 and isocyanate cycloaddition copolymerizations, 44, 46
nickel-catalyzed diyne/isocyanate cycloaddition copolymerization, 41–42
reaction of poly(2-pyrone) with amines, 42, 45
Poly(2-pyrone)s
ladders by nickel-catalyzed cyclic diyne/CO_2 and isocyanate cycloaddition copolymerizations, 44, 46
nickel-catalyzed diyne/CO_2 cycloaddition copolymerization, 41, 43
Polyacetylenes
air- and moisture-sensitivity of conventional catalysts, 147
conjugated polymers, 146–147
effective catalysts for preparation, 147
See also Substituted acetylene polymerizations
Polyacrylates. See Polymethacrylates and polyacrylates
Poly(bicyclo[2.2.2]oct-7-ene)s, nickel-catalyzed diyne/maleimide double-cycloaddition copolymerization, 48, 52
Polybutadiene, reactive sites in course of reaction, 95
Polybutadiene modification by transition metal catalysts
chelation-assisted hydroacylation, 97–99
chelation-assisted hydroacylation with primary alcohol instead of aldehyde as substrate, 98–99
examples of metal-mediated reactions, 94

functionalization of polybutadiene, 95–96
hydroacylation of polybutadiene, 99–106
hydrocarboxylation, 96
hydroformylation, 95–96
hydrogenation using homogeneous or heterogeneous catalysis, 95
hydroiminoacylation of polybutadiene, 99–102
hydroiminoacylation using aldimine, 97–98
imine-mediated hydroacylation of polybutadiene, 102–103
imine-mediated intermolecular hydroacylation, 98
mechanism of hydroiminoacylation, 97f
mechanism of imine-mediated hydroacylation, 98f
reactive sites for chelation-assisted hydroacylation, 99
reactive sites in course of polymerization, 95
rhodium and cobalt-based catalysts for hydroformylation of olefins, 96
silane-modified polymers by hydrosilylation, 96
simultaneous hydrogenation and hydroacylation of polybutadiene, 103–106
See also Hydroacylation of polybutadiene
Poly(enone)s, cobalt-catalyzed diyne/norbornadiene/CO cycloaddition, 54, 55
Poly(imide)s
nickel-catalyzed diyne/dimaleimide double-cycloaddition copolymerization forming, with pendant carbocycle, 52, 53
nickel-catalyzed monoyne/dimaleimide double-cycloaddition copolymerization, 52, 53t
Polymer modification
transition metals, 93
See also Polybutadiene modification by transition metal catalysts
Polymer synthesis
relation to organic synthesis, 38
See also Carbonylbis(triphenylphosphine) ruthenium catalysis
Polymerization
hybrid-type polymerization (HTP), 79–81
polymerization via activated monomer mechanism (PAMM), 78–79
propagation between reactive propagating end and monomer, 77, 78
Polymerization via activated monomer mechanism (PAMM)
concept, 78
2-isopropenyl-2-oxazoline as example, 79
requirements, 78
Polymers, working on strategies for well-controlled architectures, 60
Polymethacrylates and polyacrylates
anionic polymerization, 2–6
applications, 1
little control with radical polymerization, 1–2
telechelic, 2
Poly(methyl methacrylate) (PMMA)
properties, 1
See also Methyl methacrylate (MMA)

Polynorbornene. *See* Norbornene
Polypropylene (PP). *See* Propylene polymerization
Poly(pyridine)s, cobalt-catalyzed diyne/nitrile cycloaddition copolymerization, 42, 45
Poly(silyl ethers)
 complication of ^{1}H NMR spectra, 30
 equilibration of dialkoxy-silanes and -diols, 29
 (Ph$_3$P)$_2$RuCO catalyzed hydrosilylation reaction, 29–30
 See also Carbonylbis(triphenylphosphine) ruthenium catalysis
Polystyrene. *See* Copper(I) atom transfer radical polymerization (ATRP) catalysts
Polystyrene-*b*-polybutadiene-*b*-polystyrene block copolymers. *See* Hydrogenation of polystyrene-*b*-polybutadiene-*b*-polystyrene block copolymers
Polystyrene-supported catalysts. *See* Solid-supported catalysts for atom transfer polymerization
Poly(thiophene)s, palladium-catalyzed diyne/elemental sulfur cycloaddition copolymerization, 42, 44, 45
Poly(trimethylene-1,1-dicarboxylate) salts, synthesis, 68–69
Poly(trimethylene)s
 choice of initiators and reaction conditions, 62–63
 influence of architecture on local conformation, 70–73
 local stiffness influencing thermal properties, 73
 ring-opening of cyclopropane-1,1-dicarboxylates, 63–67

 space filling models for segments, 71*f*
 structural and physical properties, 70–73
 synthetic strategy, 61–62
 theoretical conformational analysis of *n*-butane and 2,2,5,5-tetramethylhexane, 71, 72*f*
 thermal behavior, 72*f*
 See also Ring-opening polymerization of small cycloalkanes
Post-polymerization modification of polymers, methodologies, 93
Pressure, effect on hydrogenation of polystyrene-*b*-polybutadiene-*b*-polystyrene block copolymers, 116, 118*f*, 119*f*
Propylene polymerization
 catalyst *rac*-C$_2$H$_4$(Ind)$_2$ZrCl$_2$/MAO (2/MAO), 175
 catalyst system *rac*-Me$_2$C(3-*t*-Bu-1-Ind)$_2$ZrCl$_2$/MAO (1/MAO), 174–175, 190
 chain transfer reactions operating with 1/MAO, 176
 dependence of polypropylene (PP) molecular weight on monomer concentration, 188
 epimerization via double bond reorientation or reversible formation of cation, 186
 equation describing dependence of isotacticity on monomer concentration, 184
 equation for rate of polymerization, 182
 equation showing change of P$_n$ with monomer concentration, 188
 experimental procedures, 191
 formation of allyl intermediate accounting for formation of

internal unsaturations, 188
four different olefinic groups, 175
^{1}H NMR analysis of olefinic unsaturations, 175–176
influence of molecular hydrogen on performance of 1/MAO, 189–190
influence of monomer concentration, 180 182–188
influence of monomer concentration with 1/MAO, 182t
influence of polymerization temperature on isotacticity, 177–178
influence of polymerization temperature on molecular weight, 180
influence of polymerization temperature with 1/MAO, 177t
influence of propylene concentration on activity of 1/MAO and 2/MAO, 182–184
influence of propylene concentration on chain transfer reactions and molecular weight, 186–188
influence of propylene concentration on isotacticity, 184–186
ln[b_{obs}/(1-b_{obs})] versus 1/T_p for i-PP samples from 1/MAO and 2/MAO, 179f
ln P_n versus 1/T_p for i-PP from 1/MAO and 2/MAO, 181f
1/MAO lacking detectable secondary propylene insertions, 189
mechanisms explaining epimerization, 186
molecular weight data from ^{1}H NMR, 178t
olefinic region of proton spectrum of typical sample prepared in liquid propylene, 176f
P_n versus monomer concentration for 1/MAO and 2/MAO, 187f
polymerization activity of 1/MAO, 183f
polypropylene (PP) isotacticity versus monomer concentration for 1/MAO, 185f
primary-transfer reactions showing preferred chirality, 176
proposed reaction scheme for formation of internal vinylidene, 188
studying influence of polymerization parameters on catalyst performance, 175
unsaturated group composition from ^{1}H NMR, 178t
wide range of microstructures, 174
Pyridine-imine, mixed nitrogen ligands, 214–215

R

Radical polymerization. *See* Controlled radical polymerization; Living radical polymerization of acrylates
Recycling catalysts. *See* Solid-supported catalysts for atom transfer polymerization
Reducing agents. *See* Hydrogenation of polystyrene-*b*-polybutadiene-*b*-polystyrene block copolymers
Reverse atom transfer radical polymerization, 211
Rhenium(V) complexes. *See* Living radical polymerization of acrylates
Rhodium catalysts
complex for polymerizing phenylacetylene in water, 152–155

hydroformylation of olefins, 96
See also Substituted acetylene polymerizations
Ring-opening metathesis polymerization (ROMP). *See* Norbornene
Ring-opening polymerization of small cycloalkanes
activated cyclobutanes, 69–70
characterization confirming absence of side reactions, 64–65
choice of initiators and reaction conditions, 62–63
conformational preferences in new substitution architecture, 73–74
cyclopropane-1,1-dicarboxylates, 63–67
dependence on solubility of final polymers, 67
effect of architecture on local conformation of main chain, 71
effect of side substituents, 71
first-order kinetic plot and evolution of number-average molecular weight and polydispersity index with conversion, 64f
influence of architecture on local conformation, 70–73
influence of counterion on polymerization of diisopropylcyclopropane-1,1-dicarboxylate, 66t
insensitivity to traces of water, 62
local stiffness influencing ultimate thermal properties, 73
MALDI–ToF mass spectrum of poly(diisopropylcyclopropane-1,1-dicarboxylate), 65f
obtaining highly functionalized backbones, 60
polymerization of cyclopropanedinitrile and cyanocyclopropanecarboxylates, 68
polymerization of cyclopropanes at elevated temperatures, 65, 66f
polytrimethylene synthetic strategy, 61–62
reactivity of cyclopropane monomers, 73
ring-opening 1,1-dicyanocyclobutane by sodium thiophenolate, 70
role of counterion in rate and selectivity of polymerization, 66
space filling models for segments of poly(vinylidene cyanide) and poly(cyclopropane-1,1-dicarbonitrile), 71f
structural and physical properties of substituted poly(trimethylenes), 70–73
synthesis of poly(trimethylene-1,1-dicarboxylate) salts, 68–69
theoretical conformational analysis of *n*-butane and 2,2,5,5-tetramethylhexane, 72f
thermal behavior of polymers, 72f
typical polymerization experiment, 63
Ruthenium catalysis. *See* Carbonylbis(triphenylphosphine)ruthenium catalysis

S

Side reactions, anionic polymerization, 4–5
Silica-supported catalysts. *See* Solid-supported catalysts for atom transfer polymerization
Siloxane equilibration

polymerization, acid-catalyzed synthesis of reactive α,α'-monomers for, 27–28
See also Carbonylbis(triphenylphosphine) ruthenium catalysis
Silyl group transfer polymerization (GTP)
  activating initiator, 7
  associative mechanism, 7, 8
  dissociative mechanism, 7, 8
  low polydispersities, 7–8
  mechanism, 7
Smoluchowski equation, rate coefficient of diffusion-controlled reaction, 262
Solid-supported catalysts for atom transfer polymerization
  aminomethylated cross-linked polystyrene (PSX) beads, 239
  condensation of pyridine 2-carbaldehyde with aminated support to give supported ligand, 237f
  crosslinked polystyrene support, 239
  effect of recycling supported catalyst, 243f
  effect of varying support to copper ratio on MMA polymerization, 245f
  effect on M_n with increasing amount of support, 244
  first order kinetic plots of two successive recycles of catalyst, 251f
  general methods and analyses, 238
  improvements in catalyst activity, 237–238
  infrared showing stepwise functionalization of various PSX beads, 241f
  isolation and re-use of catalysts, 241–242
  kinetic plots for PS6 and PS2 supports at same support to copper ratio, 249f
  methyl methacrylate (MMA) polymerization with silica supported catalysts, 240–244
  MMA polymerization data by silica supported copper(I) catalyst, 242t
  molecular weight data and kinetic plots for polystyrene supported MMA polymerization using mixed and pure ligand supports PS4 and PS7, 250f
  molecular weight dependence versus conversion reproducibility using support PS6 for polystyrene supported ATP, 248f
  phthalimidomethylated PSX beads, 238–239
  polymerization rate of MMA for supported pyridinal imine catalyst, 243f
  polymerization rate of MMA in presence of aminated silica and complexed silica, 243f
  polystyrene supported atom transfer polymerization of MMA, 244, 246–249
  pyridiniminemethylated cross-linked silica, 239
  pyridiniminemethylated PSX beads, 239
  rate of polymerization with each recycling, 242, 244
  recycling experiments with PS7 support for ATP of MMA, 251t
  recycling experiments with S4 silica support for MMA

polymerization, 244t
recycling of supported catalyst, 249
re-initiation experiments using PMMA macroinitiators, 246t
schematic of pyridinal imine functionalized crosslinked PSX, 240
Schiff base ligand for homogeneous atom transfer polymerization, 237f
size exclusion chromatography (SEC) traces for re-initiation of macro-initiators with MMA and benzyl methacrylate as second blocks, 247f
synthesis and characterization of supported catalysts, 240
typical polymerization procedure, 239
typical procedure for synthesis, 238–239
Solvent effect, nucleophilic polymerization of methyl methacrylate by Brønsted acid, 82–83
Solvents, anionic polymerization, 2
Step-growth copolymerization. *See* Carbonylbis(triphenylphosphine)ruthenium catalysis
Structure-property relationships, scientific accomplishments, 60
Styrene polymerization. *See* Copper(I) atom transfer radical polymerization (ATRP) catalysts
Styrene polymerizations
  lower chain transfer constants with catalytic chain transfer polymerizations, 262–263
  See also Catalytic chain transfer polymerization
Substituted acetylene polymerizations
  ^{13}C NMR spectrum of chloroform-d solution of poly(3,9-dodecadiyne), 160f
  catalysts without cocatalysts and photoirradiation, 155–156
  checking Mo and W carbonyl complexes for phenylacetylene (PA) polymerization, 155t
  5-{[(4'-cyano-4-biphenylyl)oxy]carbonyl}-1-pentyne (2b) polymerization, 151t
  dioxane as solvent to eliminate solvent polymerization, 151–152
  evidence of polyene structures, 160
  experimental materials, 161–162
  GPC curves of products from W and Rh catalysts, 149–150
  GPC traces of polymerization products of 2b catalyzed by WCl_6 in THF and WCl_6–Ph_4Sn in dioxane, 151f
  ^1H and ^{13}C NMR spectra of PPAs by water-soluble complex Rh(cod)(tos)(H_2O) in water and in neat monomer, 154f
  instrumentation, 162
  intermolecular cyclotrimerization to hyperbranched poly(phenylenealkenes), 157
  new catalysts for polymerizations, 156–160
  polymerization behavior of 1,9-decadiyne, 158–159
  polymerization catalyzed by WCl_6 in THF, 150–151
  polymerization in aqueous media, 152–155
  polymerization of 1,9-bis(trimethylsilyl)-1,8-nonadiyne, 159t

polymerization of 1-chloro-1-
  octyne catalyzed by M(CO)$_x$L$_y$,
  156t
polymerization of 3,9-
  dodecadiyne, 159, 160t
polymerization of 1,8-nonadiyne
  and 1,9-decadiyne, 158t
polymerization of PA by
  organorhodium catalysts, 153t
polymerization of PA catalyzed by
  water-soluble organorhodium
  catalysts Rh(diene)(tos)(H$_2$O),
  153t
polymerization procedure of 1-
  chloro-1-octyne catalyzed by
  Mo(CO)$_4$(nbd), 163
polymerization procedure of 1,8-
  nonadiyne, 163
polymerization procedure of PA in
  aqueous media, 162–163
polymerization procedure of polar
  acetylene 1a, 162
polymerization procedures, 162–
  163
polymerization results of terminal,
  157–158
polymerizing {4-[({n-[(4'-cyano-4-
  biphenylyl)oxy]alkyl}oxy)-
  carbonyl]phenyl} acetylene
  using classical metathesis
  catalysts, 148–149
robust catalysts for
  polymerizations of polar
  substituted acetylenes, 148–152
structures of two synthesized
  cyanoacetylene monomers, 148
synthesis of 1,9-bis(trimethylsilyl)-
  1,8-nonadiyne, 161
synthesis of 1-chloro-1-octyne, 161
water-soluble Rh complexes in PA
  polymerization, 152–153
water-soluble Rh complexes
  polymerizing other acetylene
  derivatives, 154

Supported catalysts
  advantages of homogeneous and
    heterogeneous catalysts, 127
  two general classes, 127–128
  See also Hydrosilation; Solid-
    supported catalysts for atom
    transfer polymerization

T

Telechelic polymers
  atom transfer radical
    polymerization (ATRP), 14–15
  special architecture, 2
Temperature
  dependence on chain transfer
    constants for catalytic chain
    transfer polymerization, 260–
    262
  effect on hydrogenation of
    polystyrene-*b*-polybutadiene-*b*-
    polystyrene block copolymers,
    116, 118f, 119f
  effect on nucleophilic
    polymerization of methyl
    methacrylate by Brønsted acid,
    84–85
  influence on isotacticity of
    polypropylene, 177–178
  influence on molecular weight of
    polypropylene, 180
Terminating side reactions, anionic
  polymerization, 4–5
Tetradentate ligands
  linear, cyclic, and branched
    (tripodal), 219f
  mixed nitrogen types, 220
  tetramines, 219
  See also Ligands
Thermoplastic elastomers. *See*
  Hydrogenation of polystyrene-*b*-
    polybutadiene-*b*-polystyrene
    block copolymers

Titanocene
  compounds for polymerizing norbornene, 168
  effect of cocatalyst methylaluminoxane (MAO), 168–169
  effect of titanocene concentration on norbornene polymerization, 170t
  See also Hydrogenation of polystyrene-*b*-polybutadiene-*b*-polystyrene block copolymers; Norbornene
Transfer reactions. *See* Propylene polymerization
Transition metal catalysts
  addition polymerization, 145
  post-polymerization modification, 93
  radical reactions from uncontrollable to controllable, 196–197
  See also Living radical polymerization of acrylates; Polybutadiene modification by transition metal catalysts; Substituted acetylene polymerizations
Transition metal-containing polymers, diyne/transition metal complex cycloaddition copolymerization, 47, 49
Transition metal-mediated polymerizations
  advantages in polymerization control, 23
  See also Alkyne cycloaddition polymerization, transition metal-catalyzed
Tridentate ligands
  copper-mediated atom transfer radical polymerization (ATRP), 216–218
  mixed nitrogen types, 217–218
  triamines, 216–217
  tripyridines, 216
  use of cyclic amines in ATRP, 217t
  See also Ligands
Tungsten catalysts. *See* Substituted acetylene polymerizations

X

X-ray photoelectron spectroscopy (XPS)
  analysis of supported complexes, 135–139
  instrumental method, 132
  See also Hydrosilation

Z

Ziegler–Natta catalysts, addition or ring-opening metathesis polymerizations, 165
Zirconocene catalysts
  group transfer polymerization (GTP), 9
  influence of propylene concentration on molecular weight, stereoregularity, and end-group structure, 180
  producing polypropylenes with range of microstructures, 174
  See also Propylene polymerization
Zirconocene–methylaluminoxane (MAO), polymerization of strained cyclic olefin, 166

Highlights from ACS Books

Desk Reference of Functional Polymers: Syntheses and Applications
Reza Arshady, Editor
832 pages, clothbound, ISBN 0-8412-3469-8

Chemical Engineering for Chemists
Richard G. Griskey
352 pages, clothbound, ISBN 0-8412-2215-0

Controlled Drug Delivery: Challenges and Strategies
Kinam Park, Editor
720 pages, clothbound, ISBN 0-8412-3470-1

Chemistry Today and Tomorrow: The Central, Useful, and Creative Science
Ronald Breslow
144 pages, paperbound, ISBN 0-8412-3460-4

A Practical Guide to Combinatorial Chemistry
Anthony W. Czarnik and Sheila H. DeWitt
462 pages, clothbound, ISBN 0-8412-3485-X

Chiral Separations: Applications and Technology
Satinder Ahuja, Editor
368 pages, clothbound, ISBN 0-8412-3407-8

Molecular Diversity and Combinatorial Chemistry: Libraries and Drug Discovery
Irwin M. Chaiken and Kim D. Janda, Editors
336 pages, clothbound, ISBN 0-8412-3450-7

A Lifetime of Synergy with Theory and Experiment
Andrew Streitwieser, Jr.
320 pages, clothbound, ISBN 0-8412-1836-6

For further information contact:
Order Department
Oxford University Press
2001 Evans Road
Cary, NC 27513
Phone: 1-800-445-9714 or 919-677-0977
Fax: 919-677-1303

Bestsellers from ACS Books

The ACS Style Guide: A Manual for Authors and Editors (2nd Edition)
Edited by Janet S. Dodd
470 pp; clothbound ISBN 0-8412-3461-2; paperback ISBN 0-8412-3462-0

Writing the Laboratory Notebook
By Howard M. Kanare
145 pp; clothbound ISBN 0-8412-0906-5; paperback ISBN 0-8412-0933-2

Career Transitions for Chemists
By Dorothy P. Rodmann, Donald D. Bly, Frederick H. Owens, and Anne-Claire Anderson
240 pp; clothbound ISBN 0-8412-3052-8; paperback ISBN 0-8412-3038-2

Chemical Activities (student and teacher editions)
By Christie L. Borgford and Lee R. Summerlin
330 pp; spiralbound ISBN 0-8412-1417-4; teacher edition, ISBN 0-8412-1416-6

Chemical Demonstrations: A Sourcebook for Teachers, Volumes 1 and 2, Second Edition
Volume 1 by Lee R. Summerlin and James L. Ealy, Jr.
198 pp; spiralbound ISBN 0-8412-1481-6
Volume 2 by Lee R. Summerlin, Christie L. Borgford, and Julie B. Ealy
234 pp; spiralbound ISBN 0-8412-1535-9

The Internet: A Guide for Chemists
Edited by Steven M. Bachrach
360 pp; clothbound ISBN 0-8412-3223-7; paperback ISBN 0-8412-3224-5

Laboratory Waste Management: A Guidebook
ACS Task Force on Laboratory Waste Management
250 pp; clothbound ISBN 0-8412-2735-7; paperback ISBN 0-8412-2849-3

Reagent Chemicals, Ninth Edition
768 pp; clothbound ISBN 0-8412-3671-2

Good Laboratory Practice Standards: Applications for Field and Laboratory Studies
Edited by Willa Y. Garner, Maureen S. Barge, and James P. Ussary
571 pp; clothbound ISBN 0-8412-2192-8

For further information contact:
Order Department
Oxford University Press
2001 Evans Road
Cary, NC 27513
Phone: 1-800-445-9714 or 919-677-0977